FAIRNESS 2013: FAIR NExt generation of ScientistS 2013

Journal of Physics: Conference Series Volume 503

Berlin, Germany
16-21 September 2013

ISBN: 978-1-63439-283-9
ISSN: 1742-6588

Printed from e-media with permission by:

Curran Associates, Inc.
57 Morehouse Lane
Red Hook, NY 12571

CURRAN ASSOCIATES INC. proceedings.com

Some format issues inherent in the e-media version may also appear in this print version.

Printed by Curran Associates, Inc. (2013)

Institute of Physics
Dirac House, Temple Back
Bristol BS1 6BE UK

Phone: 44 1 17 929 7481
Fax: 44 1 17 920 0979

techtracking@iop.org

Additional copies of this publication are available from:

Curran Associates, Inc.
57 Morehouse Lane
Red Hook, NY 12571 USA
Phone: 845-758-0400
Fax: 845-758-2634
Email: curran@proceedings.com
Web: www.proceedings.com

TABLE OF CONTENTS

PAPERS

Journal of Physics: Conference Series **503** (2014) 011001 doi:10.1088/1742-6596/503/1/011001

Preface

FAIRNESS 2013 was the second edition in a series of workshops designed to bring together excellent international young scientists with research interests focused on physics at FAIR (Facility for Antiproton and Ion Research) and was held on 16–21 September 2013 in Berlin, Germany. The topics of the workshop cover a wide range of aspects in both theoretical developments and current experimental status, concentrated around the four scientific pillars of FAIR. FAIR is a new accelerator complex with brand new experimental facilities, that is currently being built next to the existing GSI Helmholtzzentrum for Schwerionenforschung close to Darmstadt, Germany.

The spirit of the conference is to bring together young scientists, e.g. advanced PhD students and postdocs and young researchers without permanent position to present their work, to foster active informal discussions and build up of networks. Every participant in the meeting with the exception of the organizers gives an oral presentation, and all sessions are followed by an hour long discussion period. During the talks, questions are anonymously collected in box to stimulate discussions. Since the physics program of FAIR is very broad, this is reflected in the wide range of topics covered at the Conference:

- Physics of hot and dense nuclear matter, QCD phase transitions and critical point
- Nuclear structure, astrophysics and reactions
- Hadron spectroscopy, Hadrons in matter and Hypernuclei
- Special emphasis is put on the experiments CBM, HADES, PANDA, NuSTAR, as well as NICA and the RHIC low beam energy scan
- New developments in atomic and plasma physics

For all of these different areas one invited speaker was selected to give a longer introductory presentation. The write-ups of the talks presented at FAIRNESS 2013 are the content of this issue of *Journal of Physics: Conference Series* and have been refereed according to the IOP standard for peer review. This issue constitutes therefore a collection of the forefront of research that is dedicated to the physics of FAIR.

February 2014.

Organizers of FAIRNESS 2013: Marco Destefanis, Tetyana Galatyuk, Fernando Montes, Diana Nicmorus, Hannah Petersen, Claudia Ratti, Laura Tolos, and Sascha Vogel.

Journal of Physics: Conference Series **503** (2014) 011001 doi:10.1088/1742-6596/503/1/011001

Support for holding the conference was provided by:

Peer review statement

All papers published in this volume of *Journal of Physics: Conference Series* have been peer reviewed through processes administered by the proceedings Editors. Reviews were conducted by expert referees to the professional and scientific standards expected of a proceedings journal published by IOP Publishing.

Journal of Physics: Conference Series **503** (2014) 012001 doi:10.1088/1742-6596/503/1/012001

A theory overview on the Compressed Baryonic Matter Experiment at FAIR

Marlene Nahrgang

Department of Physics, Duke University, Durham, North Carolina 27708-0305, USA
Frankfurt Institute for Advanced Studies (FIAS), Ruth-Moufang-Str. 1, 60438 Frankfurt am Main, Germany

E-mail: `marlene.nahrgang@phy.duke.edu`

Abstract. The Compressed Baryonic Matter (CBM) experiment at FAIR offers for the first time in heavy-ion physics the opportunity to investigate extremely baryon-dense strongly interacting matter with large data samples as a basis for high precision measurements. This will allow us to put theories at test, answer questions about the structure of the phase diagram of QCD and the transport properties of the medium. In this overview I will highlight some recent advances on several key questions, which will be addressed by the CBM experiment.

1. Introduction

The latest generation of heavy-ion collision experiments are conducted at the LHC, CERN, and RHIC, BNL, in a center-of-mass beam energy range of $\sqrt{s_{\rm NN}} = 0.2 - 5.1$ TeV [1, 2, 3, 4, 5]. In the years 2010 and 2011 during the beam energy scan at RHIC [6] data at lower energies ($\sqrt{s_{\rm NN}} = 11.5 - 200$ GeV) were taken, although energies below the injection energy remain challenging for the RHIC accelerator.

The new Facility for Antiproton and Ion Research (FAIR) will further extend our understanding of strongly interating matter by reaching highest net-baryon densities both during the expansion [7] and at hadronic freeze-out [8]. It will thus become possible to explore new regions in the QCD phase diagram. In addition the FAIR accelerator will provide highest beam intensities, which will largely increase the data sample and will allow us to investigate observables which require high accuracy and to measure rare probes with the CBM detector [9].

In this overview, I will first discuss theoretical challenges concerning the investigation of the phase structure of QCD in section 2 and then discuss open charm physics at FAIR energies in section 3 as an example for the various opportunities of studying rare probes at CBM.

2. The phase structure of dense baryonic matter

Our current understanding of the QCD phase diagram with the parameters temperature T and baryochemical potential μ_B is based on combining different approaches. Due to the nonperturbative character of QCD at the scale $\Lambda_{\rm QCD}$, perturbative QCD calculations are reliable only at very high energies. The most promising first-principle approach to QCD are lattice QCD calculations. At vanishing μ_B the liberation of color degrees of freedom above a transition temperture T_c is seen in thermodynamic quantities [10, 11, 12]. For various quantities, e. g. the energy density, the chiral susceptibility, quark number susceptibilities or the Polyakov-loop, it

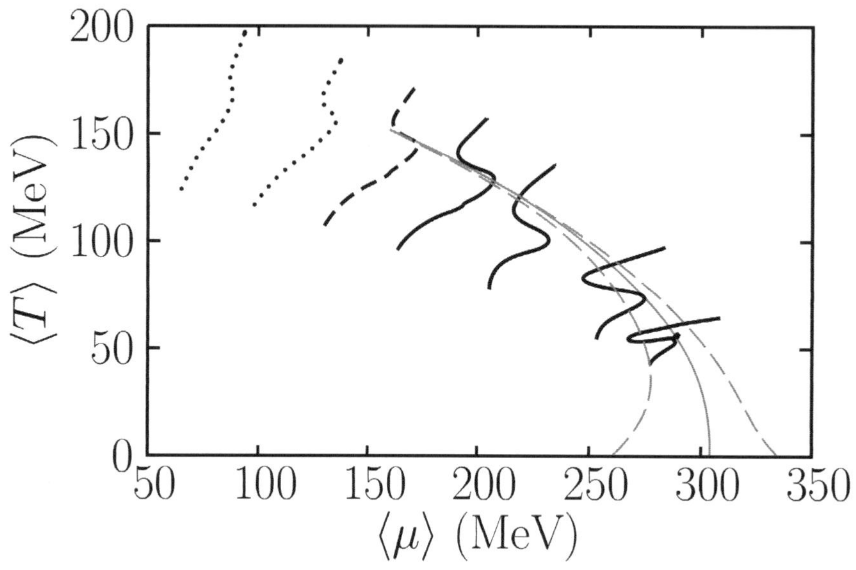

Figure 1. Averaged trajectories calculated from nonequilibrium chiral fluid dynamics. The order parameters of the phase transition are propagated explicitly and stochastically. In the region of the first-order phase transition, here shown in gray including the spinodal region limited by the gray dashed lines, one observes a back-bending of the trajectories. The system spends a significant amount of time in the spinodal region. The underlying model is the PQM model. Figure from [39].

has been found that the transition is an analytic crossover with T_c between 145 − 165 MeV. Due to the fermionic sign problem lattice QCD calculations at finite μ_B are hardly feasible numerically and although a couple of approaches exist [13, 14, 15, 16, 17] there are no conclusive results about the phase structure of QCD at finite μ_B yet. Besides lattice QCD there are other analytical approaches to calculate the phase structure of QCD, e. g. within a Dyson-Schwinger approach [18], which suggest the existence of a critical point at larger μ_B and a first-order phase transition line beyond it. Here, the quantitative results still depend on various approximations, which need to be made.

In low-energy effective models of QCD, e. g. Quark-Meson (QM) models or Nambu-Jona-Lasino (NJL) models, one tries to capture essential features like chiral symmetry and to mimic confinement via the inclusion of an effective Polyakov-loop field (PQM or PNJL models). The model-dependent results strongly differ according to the particular parameters, ingredients and approaches to solve the thermodynamic quantities, if in mean-field or beyond. Toward higher net-baryon densities most of these models find a critical point and a first-order phase transition line [19, 20, 21, 22, 23, 24].

Experimentally, there are a lot of indications for the formation of a color-deconfined plasma, the quark-gluon plasma (QGP) at high collision energies [1, 2, 3, 4, 5]. Going to lower energies the question of the existence and the location of the conjectured critical point, as well as the onset of deconfinement, is of highest priority for the CBM experiment. There are two strategies for a discovery of the critical point, either via directly seeing unambiguous signals of a critical point, or by evidence for a first-order phase transition starting at the location of the critical point and growing stronger toward lower beam energies.

For the search of the critical point a nonmonotonic behavior of fluctuations in multiplicities of conserved quantities as a function of beam energy is the most advocated signal [25, 26]. It is based on the divergence of the correlation length ξ of fluctuations in the order parameter in a thermodynamical system. The moments of these fluctuations are proportional to powers of ξ: the variance $\langle(\delta N)^2\rangle \propto \xi^2$, the skewness $\langle(\delta N)^3\rangle \propto \xi^{4.5}$ and the kurtosis $\langle(\delta N)^4\rangle - 3\langle(\delta N)^2\rangle^2 \propto \xi^7$ [27].

A system created in a heavy-ion collision is, however, inhomogenous, finite in space and time and highly dynamical. These conditions will weaken the growth of the correlation length and thus the critical phenomena are reduced. Most limiting might be the effect of critical slowing down: at a critical point not only the correlation diverges but so does the relaxation time. This means, that any system evolving through a critical point in a finite amount of time, will necessarily be driven out of equilibrium even if it was equilibrated at higher temperatures. In a phenomenological approach [28] including the dynamical critical exponents [29] it was demonstrated that the correlation length at best grows to $\sim$ 2 fm. For a first order phase transition, however, the situation is opposite: a strong nonequilibrium situation is necessary to see fluctuations coming from the instability of slow modes in the spinodal region [30, 31, 32, 33]. Including this spinodal region means that one has to avoid equilibrium equation of states, which consist of a high-temperature part and low-temperature part joined together by a Maxwell construction. If instead one uses an equation of state that allows unstable phases it was demonstrated, that initial density irregularities are enhanced during the phase transition [34].

In order to understand the phase transition dynamics in the CBM energy regime, it is extremely important to develop realistic dynamical simulations of the critical point and the first-order phase transition. Models of nonequilibrium chiral fluid dynamics [35, 36, 37, 38, 39] have recently addressed this challenge by coupling the stochastic dynamics of the order parameter of the phase transition to a fluid dynamical expansion of the bulk matter. It showed an increase of irregularities in the net-baryon density developing dynamically at the phase transition. These results have been obtained within a PQM model and studies need to be improved by applying a more realistic equation of state, which reproduces the baryon-rich environment and the phenomenology of saturated nuclear matter, e. g. the one presented in [40]. At high net-baryonic densities it is still under debate which slow mode is the real critical mode, the sigma mode, the density or the thermal mode [41]. The propagation of fluid dynamical fluctuations has recently been studied in a simplified Bjorken expansion and found to induce correlations extending over a unity in rapidity [42]. Applying a thermal conductivity from mode coupling theory and a parametrized equation of state with a critical point [43] it was shown in [44] that along a Bjorken-trajectory near the critical point these correlations are enhanced.

Further theoretical studies need to put special emphasis on realistically modeling the many sources of non-critical fluctuations in heavy-ion collisions, such as initial fluctuations and fluctuations due to final hadronic interactions in order to predict reliable signals of the critical point and the first-order phase transition. Recent results from the STAR collaboration on the higher moments of net-proton number fluctuations do show some deviations from a simulated Skellam distribution at lower beam energies [45]. A clear statement about a non-monotonic behavior is, however, not possible. It needs to be noted that the experimentally feasible measurement of net-protons does not directly compare to theoretical predictions on fluctuations in the conserved net-baryon number. Due to isospin randomization it is possible to generally relate fluctuations in net-protons to the fluctuations in net-baryon number, deviations from thermal models will however be hidden in the net-protons distributions [46, 47]. If efforts could be made to measure also neutrons it would be highly desirable because the comparison to theoretical models would be a lot cleaner.

3. Open charm production and propagation

At highest LHC and top-RHIC energies the production and propagation of open charm has been studied extensively, both theoretically and experimentally. Due to the higher masses charm-anticharm pairs are predominantly produced in the initial hard nucleon-nucleon scatterings. They are expected to not thermalize completely with the bulk medium, what makes them an interesting probe to study the properties of the QGP and the interaction mechanisms. While charm quarks are relatively abundant at LHC energies the production cross sections fall off rapidly toward lower beam energies. In the energy range of the CBM experiments charm quark production is at threshold. This makes charm quarks at CBM a very rare probe and the high beam intensities will pay off here. To study the medium modifications of charm quark spectra a reliable proton reference is needed. Initial spectra have been parametrized within the HSD transport approach [48] and can also be obtained from PYTHIA simulations [49]. Once reference proton-proton collisions are run at CBM these theoretical approaches need to be either validated or re-investigated.

Recently, first calculations of the charm quark observables, the nuclear modification factor R_{AA} and the elliptic flow v_2, within a Langevin+UrQMD hybrid model approach have been presented [50]. It was found that the initial spectra still had a large influence on the final observables, which is understandable due to lower temperatures and thus less thermalization at the CBM energies. Meaningful spectra only extend to $p_T \lesssim 1.5$ GeV such that the initially very soft spectra are pushed toward slightly higher p_T yielding a pronounced rise of the R_{AA}. In this study the transport coefficients were obtained from pQCD calculations including resonance scattering [51, 52]. Here, it was found that due to the baryon-rich environment the scattering cross sections are especially enhanced for anticharm quarks as compared to charm quarks. This leads to a significant difference between the final R_{AA} of D and $\bar{D}$ mesons and could thus enable us to obtain additional insight about the energy loss mechanism of heavy quarks, which as such cannot be investigated at LHC and RHIC. It should also motivate other groups to extend their calculations to lower energies. Interactions between D mesons and the hadronic medium are also expected to affect observables like the elliptic flow already at LHC energies. Understanding their contribution becomes more important at lower energies. Recently, the drag and diffusion coefficient of D mesons have been evaluated from an effective Lagrangian consistent with chiral and heavy-quark spin symmetries [53]. It was found that due to additional interactions with the baryons the relaxation times at CBM energies are reduced compared to those at LHC. It remains an interesting task to include those hadronic interactions in a dynamical setup.

4. Summary

The CBM experiment offers a broad range of theory questions to be addressed by reaching highest net-baryon densities with highest beam intensities. In this small overview I highlighted the opportunities to learn about the phase structure of QCD and the energy loss mechanism of charm quarks in a partonic medium and D mesons in a hadronic environment. This is of course not exhaustive and there are a lot more of interesting theory questions, like electromagnetic probes, their relation to chiral symmetry restauration and their hadronic or partonic origin, hidden charm production and propagation, signatures of color superconductivity and quarkyonic matter, the equation of state of neutron stars and nonequilibrium effects in transport approaches.

Acknowledgments

I acknowledge the continuous support by the Hessian LOEWE initiative Helmholtz International Center for FAIR and would like to thank the organizers of FAIRNESS 2013 for the inspiring conference and the many fruitful discussions.

Journal of Physics: Conference Series **503** (2014) 012001 doi:10.1088/1742-6596/503/1/012001

References

[1] J. Adams *et al.* [STAR Collaboration], Nucl. Phys. A **757** (2005) 102.
[2] B. B. Back *et al.* [PHOBOS Collaboration] Nucl. Phys. A **757** (2005) 28.
[3] I. Arsene *et al.* [BRAHMS Collaboration], Nucl. Phys. A **757** (2005) 1.
[4] K. Adcox *et al.* [PHENIX Collaboration], Nucl. Phys. A **757** (2005) 184.
[5] B. Abelev *et al.* [ALICE Collaboration], Phys. Rev. Lett. **109** (2012) 252301.
[6] H. Caines [STAR Collaboration], arXiv:0906.0305 [nucl-ex].
[7] I. C. Arsene, L. V. Bravina, W. Cassing, Y. .B. Ivanov, A. Larionov, J. Randrup, V. N. Russkikh and V. D. Toneev *et al.*, Phys. Rev. C **75** (2007) 034902 [nucl-th/0609042].
[8] J. Randrup and J. Cleymans, Phys. Rev. C **74** (2006) 047901.
[9] B. Friman, C. Hohne, J. Knoll, S. Leupold, J. Randrup, R. Rapp and P. Senger, Lect. Notes Phys. **814** (2011) 1.
[10] Y. Aoki, G. Endrodi, Z. Fodor, S. D. Katz and K. K. Szabo, Nature **443** (2006) 675.
[11] S. Borsanyi *et al.* [Wuppertal-Budapest Collaboration], JHEP **1009** (2010) 073.
[12] A. Bazavov, T. Bhattacharya, M. Cheng, C. DeTar, H. T. Ding, S. Gottlieb, R. Gupta and P. Hegde *et al.*, Phys. Rev. D **85** (2012) 054503.
[13] Z. Fodor and S. D. Katz, JHEP **0203** (2002) 014 [hep-lat/0106002].
[14] R. V. Gavai and S. Gupta, Phys. Rev. D **78** (2008) 114503 [arXiv:0806.2233 [hep-lat]].
[15] P. de Forcrand, PoS LAT **2009** (2009) 010 [arXiv:1005.0539 [hep-lat]].
[16] O. Kaczmarek, F. Karsch, E. Laermann, C. Miao, S. Mukherjee, P. Petreczky, C. Schmidt and W. Soeldner *et al.*, Phys. Rev. D **83** (2011) 014504 [arXiv:1011.3130 [hep-lat]].
[17] G. Endrodi, Z. Fodor, S. D. Katz and K. K. Szabo, JHEP **1104** (2011) 001 [arXiv:1102.1356 [hep-lat]].
[18] J. Luecker, C. S. Fischer, L. Fister and J. M. Pawlowski, PoS CPOD **2013** (2013) 057 [arXiv:1308.4509 [hep-ph]].
[19] O. Scavenius, A. Mocsy, I. N. Mishustin and D. H. Rischke, Phys. Rev. C **64** (2001) 045202.
[20] K. Fukushima, Phys. Lett. B **591** (2004) 277.
[21] B. -J. Schaefer and J. Wambach, Nucl. Phys. **A757** (2005) 479
[22] C. Ratti, M. A. Thaler, W. Weise, Phys. Rev. D **73** (2006) 014019.
[23] B. -J. Schaefer, J. M. Pawlowski, J. Wambach, Phys. Rev. D **76** (2007) 074023.
[24] V. Skokov, B. Friman and K. Redlich, Phys. Rev. C **83** (2011) 054904
[25] M. A. Stephanov, K. Rajagopal and E. V. Shuryak, Phys. Rev. Lett. **81** (1998) 4816
[26] M. A. Stephanov, K. Rajagopal and E. V. Shuryak, Phys. Rev. D **60** (1999) 114028.
[27] M. A. Stephanov, Phys. Rev. Lett. **102** (2009) 032301.
[28] B. Berdnikov and K. Rajagopal, Phys. Rev. D **61** (2000) 105017.
[29] P. C. Hohenberg and B. I. Halperin, Rev. Mod. Phys. **49** (1977) 435.
[30] L. P. Csernai and I. N. Mishustin, Phys. Rev. Lett. **74** (1995) 5005.
[31] I. N. Mishustin, Phys. Rev. Lett. **82** (1999) 4779.
[32] O. Scavenius, A. Dumitru, E. S. Fraga, J. T. Lenaghan and A. D. Jackson, Phys. Rev. D **63** (2001) 116003
[33] J. Randrup, Phys. Rev. C **82** (2010) 034902.
[34] J. Steinheimer and J. Randrup, Phys. Rev. Lett. **109** (2012) 212301
[35] M. Nahrgang, S. Leupold, C. Herold and M. Bleicher, Phys. Rev. C **84** (2011) 024912.
[36] M. Nahrgang, S. Leupold and M. Bleicher, Phys. Lett. B **711** (2012) 109.
[37] M. Nahrgang, C. Herold, S. Leupold, I. Mishustin and M. Bleicher, J. Phys. G **40** (2013) 055108.
[38] C. Herold, M. Nahrgang, I. Mishustin and M. Bleicher, Phys. Rev. C **87** (2013) 014907.
[39] C. Herold, M. Nahrgang, I. Mishustin and M. Bleicher, arXiv:1304.5372 [nucl-th].
[40] V. A. Dexheimer and S. Schramm, Phys. Rev. C **81** (2010) 045201 [arXiv:0901.1748 [astro-ph.SR]].
[41] Y. Minami and T. Kunihiro, Prog. Theor. Phys. **122** (2010) 881 [arXiv:0904.2270 [hep-th]].
[42] J. I. Kapusta, B. Muller and M. Stephanov, Phys. Rev. C **85** (2012) 054906 [arXiv:1112.6405 [nucl-th]].
[43] J. I. Kapusta, Phys. Rev. C **81** (2010) 055201 [arXiv:1001.3650 [nucl-th]].
[44] J. I. Kapusta and J. M. Torres-Rincon, Phys. Rev. C **86** (2012) 054911 [arXiv:1209.0675 [nucl-th]].
[45] L. Adamczyk *et al.* [STAR Collaboration], arXiv:1309.5681 [nucl-ex].
[46] M. Kitazawa and M. Asakawa, Phys. Rev. C **85** (2012) 021901 [arXiv:1107.2755 [nucl-th]].
[47] M. Kitazawa and M. Asakawa, Phys. Rev. C **86** (2012) 024904 [Erratum-ibid. C **86** (2012) 069902] [arXiv:1205.3292 [nucl-th]].
[48] W. Cassing, E. L. Bratkovskaya and A. Sibirtsev, Nucl. Phys. A **691** (2001) 753 [nucl-th/0010071].
[49] T. Sjostrand, S. Mrenna and P. Z. Skands, JHEP **0605** (2006) 026 [hep-ph/0603175].
[50] T. Lang, H. van Hees, J. Steinheimer and M. Bleicher, arXiv:1305.1797 [hep-ph].
[51] H. van Hees and R. Rapp, Phys. Rev. C **71** (2005) 034907 [nucl-th/0412015].
[52] H. van Hees, V. Greco and R. Rapp, Phys. Rev. C **73** (2006) 034913 [nucl-th/0508055].

[53] L. Tolos and J. M. Torres-Rincon, Phys. Rev. D **88** (2013) 074019 [arXiv:1306.5426 [hep-ph]].

Investigation of surface homogeneity of mirrors for the CBM-RICH detector and low-mass di-electron feasibility studies

E Lebedeva, C Hoehne
Justuts-Liebig University Giessen, Germany
E-mail: e.lebedeva@gsi.de,Claudia.Hoehne@exp2.physik.uni-giessen.de

Abstract. The Compressed Baryonic Matter (CBM) experiment at the future FAIR facility will investigate high net-baryon density matter at moderate temperatures in A+A collisions from 4-45 *A*GeV. One of the key observables of the CBM physics program is electromagnetic radiation as a probe of strongly interacting matter in heavy-ion collisions, carrying undistorted information on its conditions to the detector. This includes detailed investigations of low-mass vector mesons in their di-electron channel. A clean and efficient identification of electrons is required for such measurements. In CBM the electron identification will be performed by a Ring Imaging Cherenkov detector and several layers of Transition Radiation Detectors. The RICH detector will be operated with CO_2 radiator gas, MAPMTs as photodetector and spherical glass mirrors as focusing elements. A high quality of the mirrors in terms of reflectivity and surface homogeneity is required. In the first part of the contribution results on measurements of the mirror surface homogeneity are presented. Results on the feasibility studies of low-mass di-electron measurements with realistic detector response are discussed in the second part of the contribution.

1. Introduction

The Compressed Baryonic Matter (CBM) experiment [1] is designed for precision measurements of multidimensional observables including particles with very low production cross sections using high-intensity heavy-ion beams provided by the future FAIR accelerator. One of the essential requirements of CBM is good electron identification, which will be performed by a Ring Imaging CHerenkov (RICH) detector and Transition Radiation Detectors (TRD). A Time-of-flight detector (TOF) is used to improve electron identification for low momenta electrons ($p < 1.5\ GeV/c$).

2. D_0 measurements

A gaseous RICH detector is being planned in a standard projective geometry with focusing mirror elements and photon detector planes [2, 3]. One of the important criteria for the selection of appropriate mirrors is their optical surface quality. It defines the imaging quality of projected Cherenkov rings, and directly effects the ring finding and fitting performance. The optical surface quality can be estimated by the so called D_0 measurement: D_0 then is the diameter of a circle containing 95% of total light intensity reflected by a mirror surface when illuminated with a point source. In the ideal case the reflected light image on the camera is point-like. In reality it is a non homogeneous spot due to mirror surface deformations caused by the manufacturing processes (such as polishing, cutting), the mirror holder and deformations due to gravity.

A sketch of the D_0 measurement setup is shown in Figure 1. It consists of a mirror (a) and a CCD camera with close-by laser point-like source (b+c) which illuminates the whole mirror surface. CCD camera and point source (diameter $\approx 200\ \mu$m) are located exactly at the distance equals to the mirror curvature radius R. The projected image of the point source is centered on the CCD sensor. Camera and laser are installed together on a stepper motor (50 mm range) allowing movement along the optical axis of the mirror. Andor iKon CCD camera with 1024×1024 pixel (13.3×13.3 mm) and a laser point source with wavelength 650 nm were used.

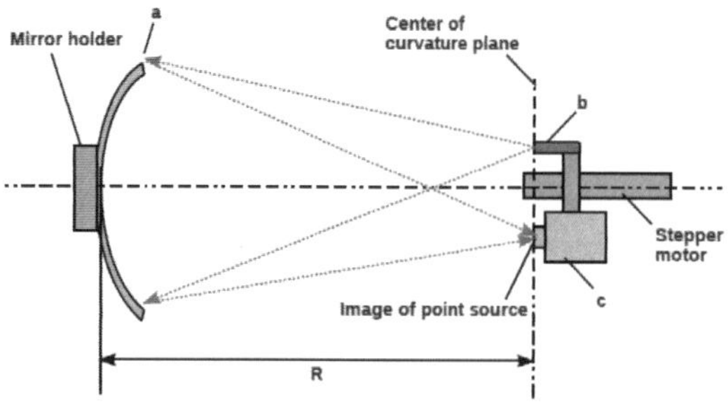

Figure 1. Experimental setup for D_0 measurements.

Four mirrors from JLO Olomouc for the RICH prototype were tested. In addition D_0 measurements were done for a mounted mirror in order to investigate the effect of gluing the mirror holder to the glass tile. Two different measurements of D_0 were performed: 1) at the nominal radius $R_{nominal}$, 2) at the distance R' where the best results of D_0 were obtained. Figure 2 shows spots of reflected light in the CCD chip. The results for all mirrors are summarized in Table 1.

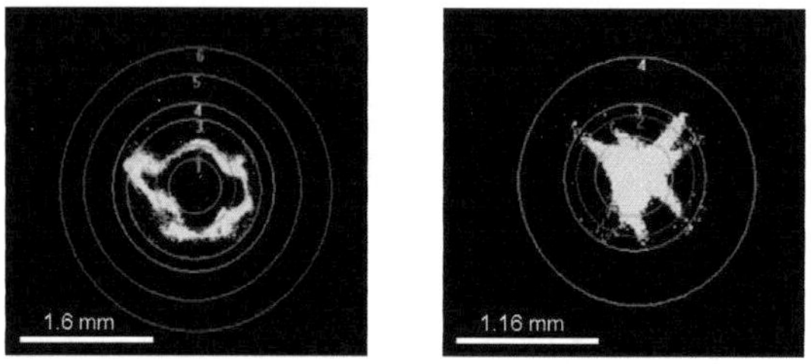

Figure 2. CCD camera image of reflected light. Left: for $R_{cam.} =$ 3 m. Right: for smallest spot at $R_{cam.} = R'$.

All tested mirrors have effective radii R' larger than the nominal one ($R_{nominal} = 3$ m). The difference $\Delta R = R' - R_{nominal}$ ranges between 5 mm and 12 mm. The results on D_0 (below 3 mm) are well below the anticipated photon detector pixel granularity of 5-6 mm. Also no differences were observed comparing the results of the mounted mirror with non-mounted mirrors. Presumably the mirror thickness of 6 mm ensures enough stability.

3. Low mass vector meson reconstruction

Feasibility studies of a measurement of low mass di-electrons in central Au-Au collisions at 8 AGeV (SIS100) and 25 AGeV (SIS300) beam energies are discussed in this section. All simulations were performed using the CbmRoot framework [4]. The following setup with most up-to-date realistic detector descriptions was used: STS (Silicon Tracking System), RICH, TRD and TOF detectors. For simulations at 8 AGeV beam energy the TRD detector was excluded. Hadrons (including π^0 and η) and photons were generated using UrQMD [5]. GEANT3 [6] performs the particle transport through detectors. The phase space distributions of $e^{\pm}$ from purely leptonic and semi-leptonic decays of light vector mesons (ρ, ω, ϕ and ω-Dalitz) were obtained using the PLUTO event generator [7]. To enhance statistics for these rare decays

Table 1. Summary results of D_0 and radius of curvature measurements for the 4 prototype mirror tiles (SP01 - SP04) and a mirror tile with glued holders.

Mirror	D_0 [mm]	ΔR [mm]	D_0 ($R = 3$ m) [mm]
SP01	1.15	+5	1.59
SP02	1.42	+12	2.58
SP03	0.88	+5	1.81
SP04	1.3	+13	2.89
Mounted	0.98	+3	1.4

one decay of a vector meson was embedded into each UrQMD event. Then in the analysis the contributions from these artificially enhanced contributions were normalized according to the multiplicities as predicted by the HSD model [8] and the respective branching ratios.

In the analysis, electron candidates are selected requiring them to be a primary vertex track reconstructed in all detectors (STS, RICH, TRD and TOF) and identified as electron in all ID detectors [9]. Dominant sources of the background are random combinations of $e^{\pm}$ from π^0-Dalitz decays and γ-conversion. About 700 $\pi^{\pm}$ are produced in central Au-Au events at 25 AGeV beam energy and can potentially contribute to the background if being misidentified. The combination of all ID detectors provides a pion suppression factor of 10^4 for momenta below 6 GeV/c, thus misidentified pions are a minor background source only. This factor is around 5000 for the measurement at 8 AGeV beam energy because the TRD detector is excluded from the setup. In order to efficiently reject misidentified pions an additional upper momentum cut of 5.5 GeV/c is applied in this case. To avoid a significant contribution of γ-conversion in the first place a single 25μm gold target was used. Further suppression of background contributions from γ-conversion and π^0-Dalitz decays was done by track cuts in three steps. The strategy for background subtraction was described in details in previous works [10, 11, 12].

γ-conversion cut: Conversion pairs have very small invariant masses and are thus essentially located below 25 MeV/c^2. All reconstructed pairs which have an invariant mass smaller than 25 MeV/c^2 are assumed to originate from γ-conversion. Tracks forming such pairs are fully removed from the further analysis.

Track topology cuts: The next two cuts are track topology cuts trying to reject electron candidate tracks stemming from an γ-conversion or π^0-Dalitz decay where the second partner is lost due to detector acceptance or the final particle identification efficiencies. Figure 3 shows the different track types used in the cut. A *track segment* is a track reconstructed only in the STS detector. These are low momentum tracks, after reconstruction only the momentum information is available. The second track type is a *global track* which is reconstructed in all detectors but is not identified as electron. The third track type is an electron track candidate (*electron track*). First, for each electron track candidate the track segment with the smallest opening angle ($\theta_{e^{\pm},segment}$) is searched for. A two dimensional cut is placed which is based on the opening angle between the identified electron track to this closest neighbour track segment ($\theta_{e^{\pm},segment}$) and the momentum product of these two tracks ($\sqrt{p_{e^{\pm}} \cdot p_{segment}}$). This cut is illustrated in Figure 4. The same holds for electrons from π^0-Dalitz decays. The second track topology cut works similarly, only this time all global tracks are combined with electron candidates.

Transverse momentum cut: Last, a transverse momentum cut may be applied. Typically a lower cut of 0.2 GeV/c is used.

The contributions of the different background sources to the combinatorial background after applying all cuts are shown in Figure 5. In central Au-Au collisions at 25 AGeV beam energy in 84% of the cases at least one of the $e^{\pm}$ tracks comes from π^0-Dalitz decays and in 50.1% of the cases at least one of the tracks comes from γ-conversion.

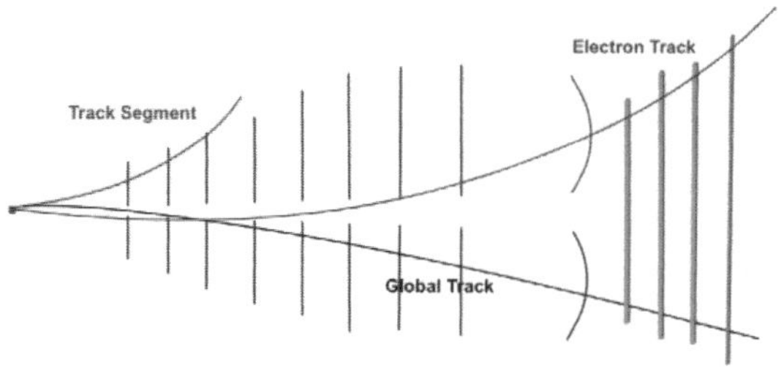

Figure 3. Schematic view of the different track types used in the track topology cuts.

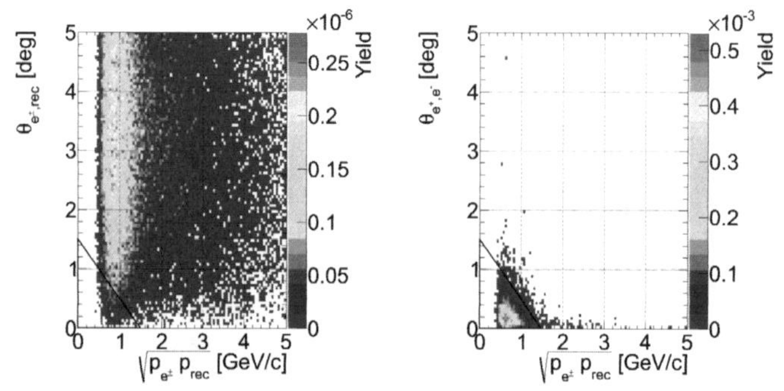

Figure 4. Track topology cut combining track segments with electron track candidates. Left: combinations of an electron track from a ρ-decay and any track segment. Right: combinations of an electron from γ-conversion if being combined with its true partner track segment. The black triangle shows the cut area.

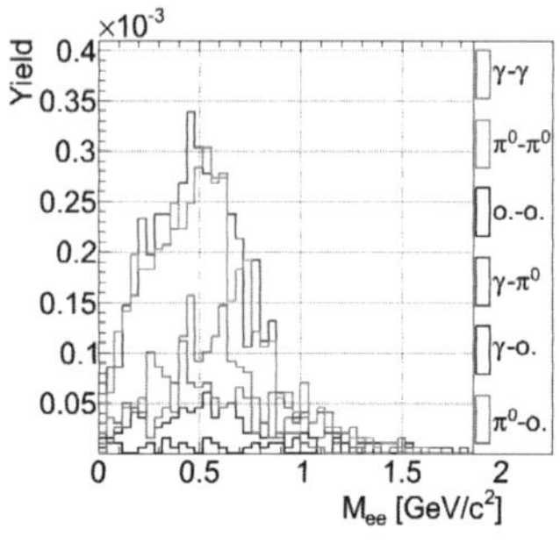

Figure 5. Contribution of the different background sources to the combinatorial background after applying all cuts (central Au-Au collisions, 25 AGeV). Abbreviations in the figure: γ – $e^{\pm}$ from γ-conversion, π^0 – $e^{\pm}$ from π^0-Dalitz decay, o. – other sources.

The final invariant mass spectrum of electron pairs is shown in the left panel of Figure 6 and Figure 7. The right panel shows the transverse momentum versus rapidity distribution of measured ρ^0 mesons after applying all cuts. For collisions at 25 AGeV beam energy the region around mid-rapidity is well covered while at 8 AGeV the acceptance is shifted slightly into the forward region. For both cases a very good coverage of low transverse momenta is achieved.

Summary results of low mass vector meson reconstruction after full event reconstruction, electron identification and applying all background rejection cuts are presented in Table 2.

The cocktail-to-background ratio around 100 for masses around 0.5 GeV/c^2 is very well compatible with existing di-lepton experiments and will allow for a precise investigation of low-mass di-electron pairs.

Acknowledgments

This work was supported by the Hessian LOEWE initiative through the Helmholtz International Center for FAIR (HIC for FAIR), Helmholtzzentrum GSI, and the German BMBF-Verbundforschung (05P12RGFCG).

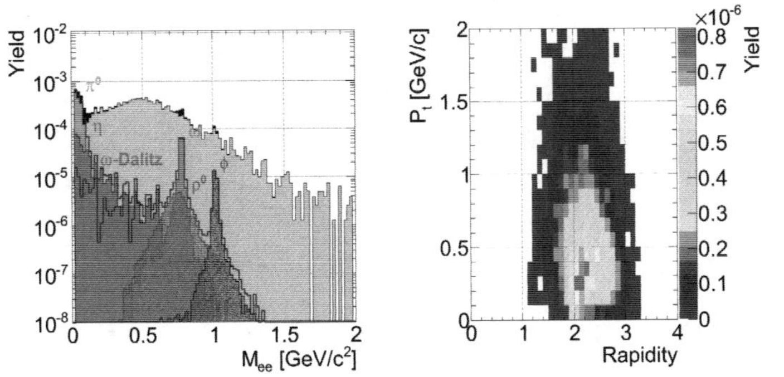

Figure 6. Left: invariant mass spectra after applying all cuts. Right: transverse momentum versus rapidity distribution for the ρ^0 meson after applying all cuts. Mid-rapidity lies at 2. The simulation was performed for 200k central Au-Au collisions at 25 *A*GeV.

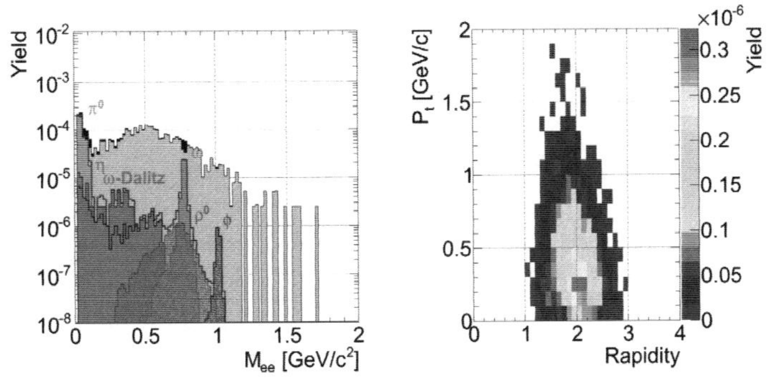

Figure 7. Left: invariant mass spectra after applying all cuts. Right: transverse momentum versus rapidity distribution for ρ^0 meson after applying all cuts. Mid-rapidity lies at 1.4. The simulation was performed for 100k central Au-Au collisions at 8 *A*GeV.

Table 2. Summary results of low mass vector meson reconstruction and different invariant mass regions for central Au-Au collisions at 25 *A*GeV beam energy with and without transverse momentum p_t cut. Efficiencies are normalized to 4π emission of the respective particle.

p_T cut	$p_T > 0.2\ GeV/c$		No p_T cut	
Mesons, mass ranges	S/B	efficiency [%]	S/B	efficiency [%]
ρ^0	—	4.39	—	5.27
ω	0.31	5.53	0.25	6.15
ϕ	0.11	7.08	7.24	0.10
$M_{ee} < 0.2\ GeV/c^2$	1.44	—	0.55	—
$0.2 < M_{ee} < 0.6\ GeV/c^2$	0.019	—	0.010	—
$0.6 < M_{ee} < 1.2\ GeV/c^2$	0.053	—	0.042	—

References

[1] Compressed Baryonic Matter Experiment. `http://www.gsi.de/documents/DOC-2005-Feb-447-1.pdf`
[2] Höhne C et al 2008 *Nucl. Inst. and Meth. A* **595** 187
[3] Höhne C et al 2011 *Nucl. Inst. and Meth. A* **639** 294
[4] M Al-Turany et al 2012 *J. Phys.: Conf. Series* **396** 022001
[5] Bass S et al 1998 *Prog. Part. Nucl. Phys.* **41** 255–370
[6] GEANT - Detector Description and Simulation Tool. CERN Program Library Long Writeup W5013.
[7] Froehlich I et al 2007 arXiv:0708.2382v2 [nucl-ex]
[8] Cassing W, Bratkovskaya E 1999 *Phys. Reports* **308** 65–233
[9] Lebedev S et al 2012 *J. Phys.: Conf. Series* **396** 022029
[10] Galatyuk T 2009 *Ph.D. thesis* at the Uni. Frankfurt `http://www.gsi.de/documents/DOC-2009-Oct-252-1.pdf`
[11] Höhne C 2008 *J. Phys. G: Nucl. Part. Phys.* **35** 104160
[12] Galatyuk T et al. 2006 *CBM note* `http://www.gsi.de/documents/DOC-2006-Feb-110-1.pdf`

Nonequilibrium dynamics and transport near the chiral phase transition of a quark-meson model

A Meistrenko[1], C Wesp[1], H van Hees[1,2] and C Greiner[1]

[1] Institut für Theoretische Physik, Goethe-Universität Frankfurt, Max-von-Laue-Straße 1, D-60438 Frankfurt, Germany

[2] Frankfurt Institute for Advanced Studies, Ruth-Moufang-Straße 1, D-60438 Frankfurt, Germany

E-mail: `meistrenko@th.physik.uni-frankfurt.de`

Abstract. Based on the 2PI quantum effective action of the linear sigma model with constituent quarks, we develop a transport approach to study systems out of equilibrium. In particular, we focus on the chiral phase transition as well as the critical point, where nonequilibrium effects near the phase transition give rise to critical behavior such as the fluctuation of the baryon number density. Predictions for long-range correlations and fluctuations of observables in our model could be used to study fundamental properties of the QCD phase transition. In the previous version of our transport model the chiral fields are implemented as mean fields, whereas quarks are treated as on-shell particles in the Vlasov equation with a dynamical force term. The current update includes also the distribution functions of sigma mesons and pions in a self-consistent way. On this basis a dissipation kernel between the mean fields and particle modes can be implemented.

1. Introduction

We concentrate our study on nonequilibrium effects and the chiral phase transition within the linear sigma model with constituent quarks, which is an effective theory of QCD in the low-energy limit. Our motivation is to investigate critical phenomena at the phase transition and the critical point, where time-dependent long-range correlations can arise. Such effects can be studied in a (3+1)-dimensional numerical approach. We present our model in the following section 2 and discuss an improved set of transport equations in section 3. Finally, we summarize the progress in section 4 and give an outlook to further improvements.

2. Classical transport equations within a linear sigma framework

The linear sigma model with constituent quarks is a well known $O(4)$-model [1], which is suited for studying the chiral phase transition. Because of the spontaneously and explicitly broken chiral symmetry the mesonic part of this theory consists of a massive scalar sigma field and three isoscalar pion fields, which form the chiral field $\Phi = (\sigma, \vec{\pi})$. The sigma field represents the order parameter for the chiral phase transition and mimics the properties of the quark condensate in QCD, since both transform equally under chiral transformation. Without explicit symmetry breaking a $\mathrm{SU_L}(2) \times \mathrm{SU_R}(2)$ symmetry transformation would let the Lagrangian invariant. It has the following form:

$$\begin{aligned}\mathcal{L} = &\sum_i \bar{\psi}_i \left[i\partial\!\!\!/ - g\left(\sigma + i\gamma_5 \vec{\pi}\cdot\vec{\tau}\right)\right]\psi_i + \frac{1}{2}\left(\partial_\mu\sigma\partial^\mu\sigma + \partial_\mu\vec{\pi}\partial^\mu\vec{\pi}\right) \\ &- \frac{\lambda}{4}\left(\sigma^2 + \vec{\pi}^2 - \nu^2\right)^2 + f_\pi m_\pi^2 \sigma + U_0 ,\end{aligned} \tag{1}$$

where the field shift term and the zero potential constant are given by $\nu^2 = f_\pi^2 - m_\pi^2/\lambda$, $U_0 = m_\pi^4/(4\lambda) - f_\pi^2 m_\pi^2$. The sum runs over included quark flavor ψ_i and the parameters of the model are adjusted to match the vacuum values of the pion decay constant $f_\pi = 93\,\mathrm{MeV}$, the pion mass $m_\pi = 138\,\mathrm{MeV}$ as well as an estimated sigma mass $m_\sigma \approx 604\,\mathrm{MeV}$.

The inhomogeneous Klein-Gordon equations of motion for the mean fields follow directly from the functional derivative of the classical action with respect to the chiral field components and treating quarks at one-loop level

$$\begin{aligned}\partial_\mu\partial^\mu\sigma + \lambda\left(\sigma^2 + \vec{\pi}^2 - \nu^2\right)\sigma - f_\pi m_\pi^2 + g\left\langle\bar{\psi}\psi\right\rangle = 0 ,\\ \partial_\mu\partial^\mu\vec{\pi} + \lambda\left(\sigma^2 + \vec{\pi}^2 - \nu^2\right)\vec{\pi} + g\left\langle\bar{\psi}i\gamma_5\vec{\tau}\psi\right\rangle = 0 .\end{aligned} \tag{2}$$

The scalar and pseudoscalar densities are given by

$$\begin{aligned}\left\langle\bar{\psi}\psi\right\rangle(t,\vec{x}) &= g d_q \sigma(t,\vec{x}) \int d^3\vec{p}\, \frac{f_q(t,\vec{x},\vec{p}) + f_{\bar{q}}(t,\vec{x},\vec{p})}{E(t,\vec{x},\vec{p})} ,\\ \left\langle\bar{\psi}i\gamma_5\vec{\tau}\psi\right\rangle(t,\vec{x}) &= g d_q \vec{\pi}(t,\vec{x}) \int d^3\vec{p}\, \frac{f_q(t,\vec{x},\vec{p}) + f_{\bar{q}}(t,\vec{x},\vec{p})}{E(t,\vec{x},\vec{p})} ,\end{aligned} \tag{3}$$

where f_q and $f_{\bar{q}}$ denote the phase-space distribution functions of quarks and antiquarks with their degeneracy factor d_q. Numerically the quarks are treated as test particles, which propagate according to the Vlasov equation

$$\left[\partial_t + \frac{p}{E(t,\vec{x},\vec{p})}\cdot\nabla_{\vec{x}} - \nabla_{\vec{x}}E(t,\vec{x},\vec{p})\,\nabla_{\vec{p}}\right] f_{q|\bar{q}}(t,\vec{x},\vec{p}) = 0 . \tag{4}$$

Thereby the force term is dynamically generated through the effective mass of quarks, depending on the mean-field values

$$\begin{aligned}E(t,\vec{x},\vec{p}) &= \sqrt{\vec{p}^2(t) + M^2(t,\vec{x})} ,\\ M^2(t,\vec{x}) &= g^2\left[\sigma^2(t,\vec{x}) + \vec{\pi}^2(t,\vec{x})\right] .\end{aligned} \tag{5}$$

The resulting equilibrium properties of such a classical system are shown in Fig. 1 for the order parameter and in Fig. 2 for the effective sigma mass as a function of the temperature. Depending on the Yukawa coupling constant g different order of a phase transition can be generated. More results, also with a constant binary cross section for quarks, can be found in [2].

3. Transport equations from the 2PI effective action

Since we are interested in nonequilibrium dynamics at the phase transition, a consistent real-time description of the interaction between the mean-field (soft) and the particle (hard) modes of the chiral field as well as quarks is needed. The model should also include a dissipation term, which would drive the system to thermal equilibrium. In the following we use the 2PI quantum effective action to derive a set of coupled equations of motion. Thereby 2PI stands for two-particle irreducible, which means that a 2PI diagram does not become disconnected by cutting 2 inner lines (propagators). This effective action for the mean fields and propagators preserves global symmetries of the original theory, guarantees thermodynamic consistency [3] and can

Journal of Physics: Conference Series **503** (2014) 012003 doi:10.1088/1742-6596/503/1/012003

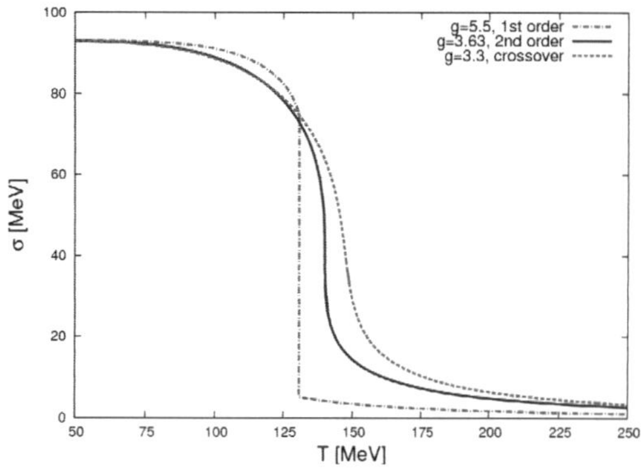

Figure 1. Order parameter sigma as a function of the temperature T for 3 different values of the Yukawa coupling g.

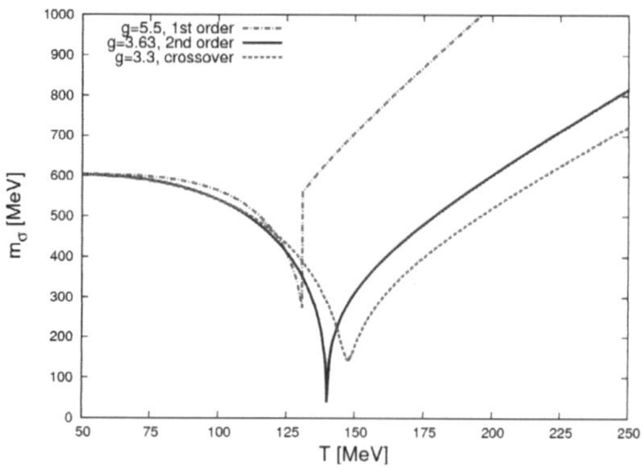

Figure 2. Same as Fig. 1 but for the sigma mass spectrum.

be renormalized with vacuum counter terms [4, 5]. At the same time such a self-consistent approximation is suitable to derive off-shell transport equations, which also respect conservation laws and the correct equilibrium limit. Furthermore, a systematic inclusion of collisional memory effects is possible [6, 7, 8, 9].

A more phenomenological study, motivated by the Langevin equation, is discussed in [2], where a new statistical approach for the scattering effects between quarks as quasi particles and mean fields is introduced. Within the linear sigma model with constituent quarks (1) the 2PI effective action reads

$$\begin{aligned}\Gamma[\sigma,\vec{\pi},G,D] =& S_{cl}[\sigma,\vec{\pi}] + \frac{i}{2}\cdot \mathrm{Tr}\log G^{-1} + \frac{i}{2}\cdot \mathrm{Tr}\, G_0^{-1}G \\ & - i\cdot \mathrm{Tr}\log D^{-1} - i\cdot \mathrm{Tr}\, D_0^{-1}D + \Gamma_2[\phi,\vec{\pi},G,D]\,,\end{aligned} \tag{6}$$

where Γ_2 includes only 2PI diagrams and S_{cl} stands for the classical action, defined by the Lagrangian (1). G_0, D_0 denote the free and G, D the fully dressed propagators for bosons and fermions, which fulfills the Schwinger-Dyson equation and are formally given by

$$\begin{aligned} G^{-1}(x,y) &= G_0^{-1}(x,y) - \Pi(x,y)\,, \qquad & \Pi(x,y) &= 2i\frac{\delta\Gamma_2}{\delta G(x,y)}\,, \\ D^{-1}(x,y) &= D_0^{-1}(x,y) - \Sigma(x,y)\,, \qquad & \Sigma(x,y) &= -i\frac{\delta\Gamma_2}{\delta D(y,x)}\,. \end{aligned} \tag{7}$$

Here, Π and Σ denote the bosonic and fermionic self-energies. These are derived from the 2PI part of the effective action, where we include only 1- and 2-point diagrams as shown in Fig. 3, since a first order gradient expansion of Kadanoff-Baym and Schwinger-Dyson equations in Wigner space then reduces to a Markov-like collisional dynamics without memory effects [6, 7, 8, 9]. Unfortunately, finite truncations of the 2PI effective action violate Ward-Takahashi-identities of global and local symmetries in the first neglected order of the expansion parameter, that is a direct consequence of the resummation in the 2-point function. It follows, that also the Goldstone theorem is violated [10, 11], and the pions acquire a non-vanishing (temperature dependent) mass in the broken phase of the linear sigma model, even without explicitly broken chiral symmetry. Possible modifications for a symmetry-improved 2PI effective action are discussed in [12].

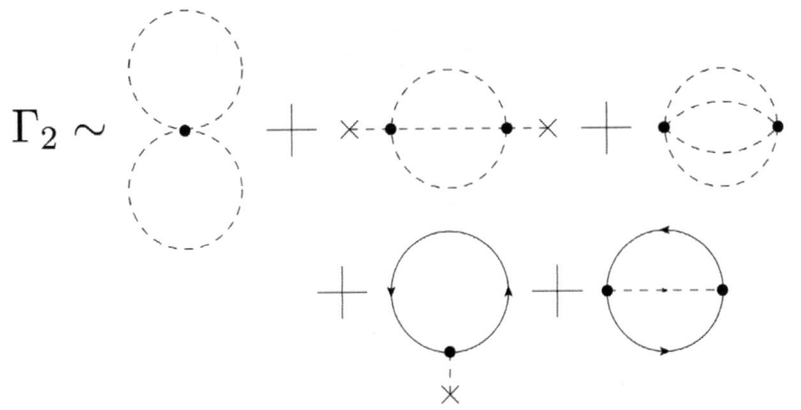

Figure 3. Approximation for the 2PI part of the effective action: the upper line shows bosonic diagrams (Hartree, sunset and basketball), where dashed lines stand for boson propagators and external lines with a cross represent mean fields. The lower line shows interactions between bosons and fermions, here a solid line stands for a fermion propagator.

In a first attempt we extend our classical set of equations (2), (4) by including also the Vlasov equation for the phase-space distribution functions of σ and π mesons in a Hartree-like approximation, where only the local part of Γ_2 is considered. Such a procedure requires the computation of self-consistently derived effective mass terms, which follow from the gap equation of the propagator. Starting from the general expression for the bosonic propagator (7) and inserting the self-energy expressions from the Hartree diagrams, leads to the following self-consistent equations

$$\begin{aligned} M_\sigma^2(x) &= \lambda\left(3\sigma^2 + 3\pi^2 - \nu^2\right) + 3\lambda G_{\sigma\sigma} + 3\lambda G_{\pi\pi}\,, \\ M_\pi^2(x) &= \lambda\left(\sigma^2 + 5\pi^2 - \nu^2\right) + \lambda G_{\sigma\sigma} + 5\lambda G_{\pi\pi}\,, \end{aligned} \tag{8}$$

where the bosonic loop integrals are given by

$$G_{\phi\phi}(t,\vec{x}) = \frac{1}{2}\int \frac{d^3p}{(2\pi)^2}\frac{1+2N_\phi(t,\vec{x},\vec{p})}{\sqrt{\vec{p}^2 + M_\phi^2(t,\vec{x})}} \quad \text{with} \quad \phi \in \{\sigma,\pi\}\,. \tag{9}$$

The Vlasov equations for the phase-space distributions of σ and π mesons are then given by

$$\left[\partial_t + \frac{p}{E_\phi(t,\vec{x},\vec{p})}\cdot\nabla_{\vec{x}} - \nabla_{\vec{x}}E_\phi(t,\vec{x},\vec{p})\,\nabla_{\vec{p}}\right] f_\phi(t,\vec{x},\vec{p}) = 0\,, \tag{10}$$

where the force term depends on the self-consistent mass expressions from (8) as gradient of

$$E_\phi(t,\vec{x},\vec{p}) = \sqrt{\vec{p}^2 + M_\phi^2(t,\vec{x})}\,. \tag{11}$$

The Vlasov equations of the chiral field components are solved as differential equations without using a test particle ansatz. Including also the tadpole contribution of Γ, we end up with the following mean-field equations

$$\begin{aligned} \partial_\mu\partial^\mu\sigma + \lambda\left(\sigma^2 + \vec{\pi}^2 - \nu^2 + 3G_{\sigma\sigma} + 3G_{\pi\pi}\right)\sigma - f_\pi m_\pi^2 + g\left\langle\bar{\psi}\psi\right\rangle = 0\,, \\ \partial_\mu\partial^\mu\vec{\pi} + \lambda\left(\sigma^2 + \vec{\pi}^2 - \nu^2 + G_{\sigma\sigma} + 5G_{\pi\pi}\right)\vec{\pi} + g\left\langle\bar{\psi}i\gamma_5\vec{\tau}\psi\right\rangle = 0\,. \end{aligned} \tag{12}$$

The system of equations (4), (10) and (12) serves as a basis to study nonequilibrium effects. Nevertheless, it does not account for dissipation as well as $q, \bar{q}$ creation and annihilation processes. In an upcoming work we will implement a dissipation kernel [13, 14], which arises from the sunset diagram (see Fig. 3) by deriving Γ_2 with respect to the mean fields.

4. Conclusion and outlook

In this proceeding we presented an improved set of equations of motion to study systems out of equilibrium, which is included in a numerical simulation. In comparison to the phenomenological work [2], here we use the 2PI effective action of the linear sigma model with constituent quarks. In the current version the set consists of mean-field equations for the components of the chiral field, the Vlasov equation for the phase-space distribution functions of σ and π mesons as well as the Vlasov equation for the quarks, which are treated as test particles.

In a further study we plan to implement a dissipation kernel for the mean-field equations, which results from the sunset diagram. For a consistent treatment it requires also the inclusion of a collision term between soft and hard modes for the Vlasov equation of chiral partners. Furthermore, an interaction between σ, π and quarks on the particle level has to be implemented to account for chemical equilibration in a dynamical and consistent way. Because of numerical challenges such a direct Boltzmann-like transport approach will be primarily suited for the study of time-dependent fluctuations in homogeneous systems. High order cumulants of conserved quantities like baryon number density offer a possibility to observe such fluctuations. Nevertheless, it is an open question whether fluctuations can also arise on the time scale of a heavy ion collision and survive the hadronization, that needs further investigation.

Acknowledgments

This work has been supported by the German Federal Ministry of Education and Research (BMBF Förderkennzeichen 05P12RFFTS). A. M. and C. W. acknowledge financial support from the Helmholtz Research School for Quark Matter Studies (H-QM) and HIC for FAIR.

References

[1] Scavenius O, Mocsy A, Mishustin I N and Rischke D H 2001 *Phys. Rev.* C **64**, 045202
[2] van Hees H, Wesp C, Meistrenko A and Greiner C 2013 Dynamics of the chiral phase transition *Preprint to be published in the proceedings of the XXXI Max Born Symposium and HIC for FAIR Workshop, 14-16 June 2013, Wroclaw, Poland*
[3] Baym G 1962 *Phys. Rev.* **127**, 1391
[4] van Hees H and Knoll J 2002 *Phys. Rev.* D **65**, 025010
[5] van Hees H and Knoll J 2002 *Phys. Rev.* D **65**, 105005
[6] Ivanov Y B, Knoll J and Voskresensky D N 1999 *Nucl. Phys.* A **657**, 413
[7] Ivanov Y B, Knoll J and Voskresensky D N 2000 *Nucl. Phys.* A **672**, 313
[8] Knoll J, Ivanov Y B and Voskresensky D N 2001 *Annals Phys.* **293**, 126
[9] Ivanov Y B, Knoll J and Voskresensky D N 2003 *Phys. Atom. Nucl.* **66**, 1902
[10] Baym G and Grinstein G 1977 *Phys. Rev.* D **15**, 28972912
[11] van Hees H and Knoll J 2002 *Phys. Rev.* D **66**, 025028
[12] Pilaftsis A and Teresi D 2013 *Nucl. Phys.* B **874**, 594
[13] Greiner C and Muller B 1997 *Phys. Rev.* D **55**, 1026
[14] Rischke D H 1998 *Phys. Rev.* C **58**, 2331

How spinodal decomposition influences observables at FAIR energies

Christoph Herold[1,2,3], Marlene Nahrgang[3,4], Igor Mishustin[3,5], Marcus Bleicher[2,3]

[1] School of Physics, Suranaree University of Technology, 111 University Avenue, Nakhon Ratchasima 30000, Thailand
[2] Institut für Theoretische Physik, Goethe-Universität, Max-von-Laue-Str. 1, 60438 Frankfurt am Main, Germany
[3] Frankfurt Institute for Advanced Studies (FIAS), Ruth-Moufang-Str. 1, 60438 Frankfurt am Main, Germany
[4] Department of Physics & CTMS, Duke University, Durham, NC 27708, USA
[5] Kurchatov Institute, National Research Center, 123182 Moscow, Russia

E-mail: `herold@fias.uni-frankfurt.de`

Abstract. The FAIR facility will make the region of high net-baryon densities experimentally accessible, where a first-order phase transition is conjectured. We investigate the dynamics of chiral symmetry breaking and the onset of confinement during a heavy-ion collision at large baryochemical potentials within a nonequilibrium chiral fluid dynamics model including effects of dissipation and noise. The order parameters are explicitly propagated and coupled to a fluid dynamically expanding medium of quarks. We demonstrate that the coupled system is strongly influenced by the nonequilibrium dynamics, leading only to a weak growth of the correlation length near the critical point. At the first-order phase transition, spinodal instabilities create domains in the order parameters and large spatial fluctuations in the baryon density within single events. As a consequence we find a clear enhancement of higher flow harmonics in coordinate space at the first-order phase transition in comparison with transitions through the crossover or critical point.

1. Introduction

The phase diagram of quantum chromodynamics (QCD) is investigated experimentally in heavy-ion collisions. The nature of the QCD phase transition has up to now only been revealed in the region of small baryochemical potential, where the transition between a hadron gas and a conjectured quark-gluon-plasma is not a phase transition but rather an analytic crossover [1, 2]. The existence of a critical point (CP) which may be inferred from mean-field effective models [3] or functional approaches [4, 5] could not yet be confirmed in experiment. This will be especially interesting for the upcoming FAIR facility, where the region of large net-baryon densities will be accessible. If there is a CP and the system is near equilibrium, then one should find a clear enhancement in event-by-event fluctuations of conserved quantities like particle multiplicities or transverse momentum [6]. Nevertheless, we know that an ideal thermalized system is impossible to create during the rapid dynamics of a particle collision.

Fluctuation signals are also important at a first-order phase transition adjacent to the CP. Here it has been demonstrated that due to spinodal decomposition spatial inhomogeneities can

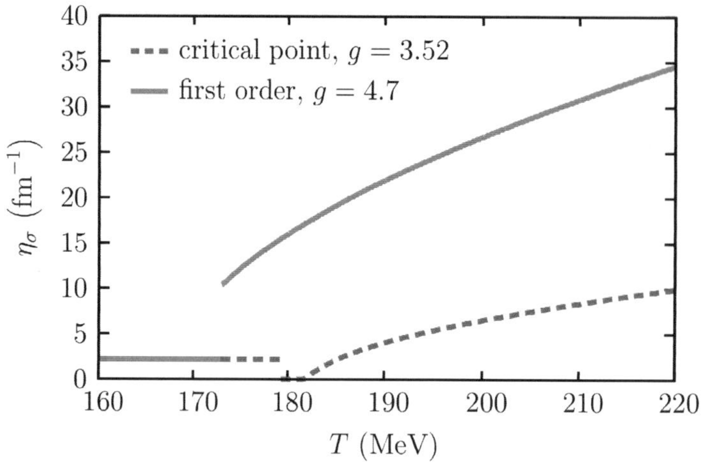

Figure 1. Damping coefficient for the sigma field as a function of temperature for a first-order and a CP scenario. We see that only around the CP, the damping vanishes. Here, the strength of the transition is tuned via the quark-meson coupling constant g. Figure from [11].

be amplified [7, 8], leading to nonstatistical fluctuations within single events. So the problems we aim to study within our dynamical model are twofold: first, whether signals or a CP can develop during the nonequilibrium dynamics of a heavy-ion collision and second, whether nonequilibrium effects are strong enough to enable the dynamical phase fragmentation at the first-order phase transition, creating some relevant signal.

2. Coupling of a quark fluid to chiral fields and Polyakov loop

Our model consists of two sectors: First, we have an ideal fluid of quarks and antiquarks which provides a locally equilibrated medium, as presumably created after the collision of two heavy nuclei. Second, we explicitly propagate fluctuations through the nonequilibrium equations of motion for the relevant order parameters, the sigma field and the Polyakov loop.

We start from the Polyakov-Quark-Meson model which exhibits the desired phase structure with a CP and a first-order phase transition on the mean-field level [9] and beyond [5]. The Lagrangian density reads

$$\mathcal{L} = \bar{q}\left[i\left(\gamma^\mu\partial_\mu - ig_s\gamma^0 A_0\right) - g\sigma\right]q + \frac{1}{2}\left(\partial_\mu\sigma\right)^2 - U(\sigma) - \mathcal{U}(\ell,\bar{\ell}) \,. \tag{1}$$

The dynamics of the chiral order parameter, the sigma field, has been derived from first principles within the two-particle irreducible effective action. This leads to additional damping and noise terms in the equation of motion [10]

$$\partial_\mu\partial^\mu\sigma + \eta_\sigma(T)\partial_t\sigma + \frac{\partial V_{\rm eff}}{\partial\sigma} = \xi_\sigma \,. \tag{2}$$

The physical process leading to the temperature-dependent damping coefficient $\eta_\sigma(T)$ is the decay of one sigma into two quarks. Therefore, around the CP where the mass of sigma vanishes, also $\eta_\sigma(T)$ is equal to zero, see Fig. 1. The stochastic noise term is assumed to be Gaussian and white, it is connected to the damping coefficient via a dissipation-fluctuation relation. In Eq. 2, $V_{\rm eff}$ denotes the effective thermodynamic potential

$$V_{\rm eff} = U(\sigma) + \mathcal{U}(\ell) + \Omega_{q\bar{q}} \,, \tag{3}$$

where the contribution $\Omega_{q\bar{q}}$ stems from integrating out the quarks degrees of freedom in the path integral formulation of the grand canonical partition function.

For the Polyakov loop, we use a phenomenological equation of motion, because the real-time dynamics is of ℓ is not yet understood, this quantity is defined only in Euclidean space-time. We interpret ℓ as an effective field and propose its nonequilibrium equation of motion as [11]

$$\eta_\ell \partial_t \ell + \frac{\partial V_{\mathrm{eff}}}{\partial \ell} = \xi_\ell \ . \tag{4}$$

Although a rigorous argument for the fixing of the damping coefficient η_ℓ cannot be given, we tested different values and found that results are in general not sensitive to this. We finally set $\eta_\ell = 5/\mathrm{fm}$. The noise term ξ_ℓ is also Gaussian and white, with a dissipation-fluctuation relation connecting it to η_ℓ [11].

As we may assume that the quarks equilibrate on a significantly smaller time scale than the order parameters, we describe their evolution via an ideal energy-momentum tensor $T_{\mathrm{q}}^{\mu\nu} = (e+p)u^\mu u^\nu - pg^{\mu\nu}$. Pressure, energy density and quark density n are obtained from the standard thermodynamic relations

$$p(\sigma, \ell, T, \mu) = -\Omega_{q\bar{q}} \ , \tag{5}$$

$$e(\sigma, \ell, T, \mu) = T\frac{\partial p}{\partial T} + \mu\frac{\partial p}{\partial \mu} - p \ , \tag{6}$$

$$n(\sigma, \ell, T, \mu) = \frac{\partial p}{\partial \mu} \ . \tag{7}$$

Energy and momentum of the fields are not conserved if the temperature changes like it does in an expanding system. However, we can account for conservation of energy and momentum in the coupled system by introducing source terms for σ and ℓ respectively, which are equal to the divergence of $T_{\mathrm{q}}^{\mu\nu}$,

$$\partial_\mu T_{\mathrm{q}}^{\mu\nu} = S_\sigma^\nu + S_\ell^\nu \ . \tag{8}$$

We could demonstrate in [11, 12] that this ensures conservation of the total energy throughout the numerical simulations. For systems at finite chemical potential, we also have to ensure quark number conservation which is achieved by the continuity equation

$$\partial_\mu n^\mu = 0 \ . \tag{9}$$

3. The correlation length in an expanding medium

The correlation length ξ is the key quantity for a CP as it is directly related to the strength of event-by-event fluctuations, the proposed observable in heavy-ion experiments. However, certain effects will clearly weaken the growth of ξ in a nonequilibrium dynamical situation. The simplest aspect is the finite size of the small system which poses a natural limit to the correlation length. Furthermore, one has to consider critical slowing down. If a system out of equilibrium approaches a CP, relaxation times become infinitely large. This has been studied in [13] within a simple phenomenological model, where the growth of ξ near the QCD CP has been estimated to a maximum of $1.5 - 2.5$ fm.

For our simulation we initialize an ellipsoidal droplet of quark matter at a high temperature and afterwards let the system expand according to the equations of motion for fields and fluid. We then extract the correlation length from an averaging procedure over the inverse mass of the system in the central region of the fireball. One might also think of other procedures how to calculate ξ as has been discussed in [14]. The results are shown in Fig. 2, for both a CP

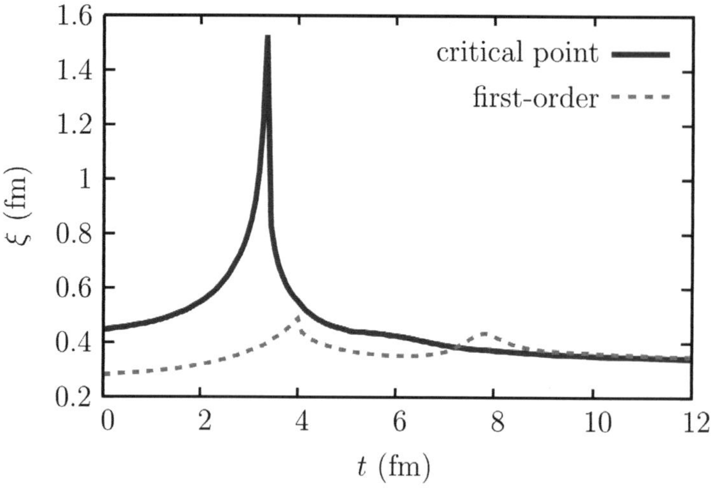

Figure 2. Correlation length as a function of time for a CP and a first-order phase transition scenario. Although we find a clear peak at the critical temperature, the growth is clearly limited to 1.5 fm. Figure from [11].

scenario and an evolution through the first-order phase transition. We find a clear peak around the critical temperature, where the correlation length grows up to 1.5 fm. For the first-order phase transition, it remains small at around $0.2 - 0.4$ fm with only two minor peaks when the system is near the transition temperature, the first time as a result of the cooling and the second time due to reheating. We see that although we are at the CP, the expected signals might not be as clearly visible as they would be in equilibrium. Furthermore, the significant enhancement is limited to a finite amount of time, so depending on where the system's freeze out happens, one might not obtain a measurable effect.

4. Anisotropic flow at the first-order phase transition

We focus now on the behavior of the system which evolves through the first-order phase transition at large net-baryon densities. We initialize a droplet of quark matter with appropriate initial conditions and let it expand and cool through the phase boundary. In [15], we found that this leads to strong non-statistical multiplicity fluctuations which do not occur at the CP or crossover. Due to spinodal instabilities that occur during the evolution, the high temperature phase dynamically fragments into small droplets, creating fluctuations in the azimuthal baryon number distribution. In contrast to the event-by-event fluctuations at the CP, this is clearly a single event signal. For a set of many events, one may average the flow harmonics in coordinate space v_n and obtain a clear enhancement in the higher harmonics compared to the CP evolution, see Fig. 3.

Although we have shown that interesting dynamical effects can be expected at the first-order phase transition, we want to investigate in the future how this might produce signals in the detector. At the moment, our model clearly lacks a hadronic phase for the low temperature regime. This could be implemented via a hadronic afterburner or a simple hadronic freeze-out. Then it would be interesting to study flow in momentum space and see if some effect can also be found here. For signals of a CP, we then aim at performing an event-by-event analysis in order to provide some realistic estimate of the expected signal.

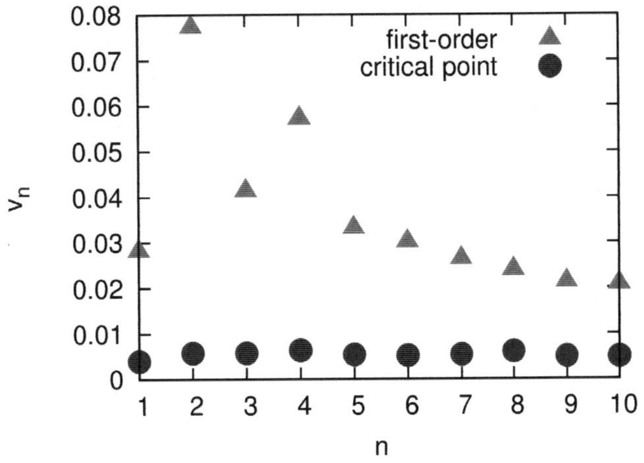

Figure 3. Average values of the harmonic coefficients in coordinate space from the azimuthal net-baryon number distribution. At the first-order phase transition we find a strongly anisotropic expansion, while small coefficients indicate a homogeneous spherical shape at the CP.

Acknowledgements
The authors thank Stefan Leupold and Carsten Greiner for fruitful discussions and Dirk Rischke for providing the SHASTA code. This work was supported by the Hessian LOEWE initiative Helmholtz International Center for FAIR. I. M. acknowledges partial support from the grant NSM-215.2012.2 (Russia).

References
[1] Aoki Y, Endrodi G, Fodor Z, Katz S D and Szabo K K *Nature* **443** (2006) 675
[2] Borsanyi S *et al.* [Wuppertal-Budapest Collaboration] *JHEP* **1009** (2010) 073
[3] Scavenius O, Mocsy A, Mishustin I N and Rischke D H *Phys. Rev.* C **64** (2001) 045202
[4] Fischer C S and Luecker J *Phys. Lett.* B **718** (2013) 1036
[5] Herbst T K, Pawlowski J M and Schaefer B *Phys. Rev.* D **88** (2013) 014007
[6] Stephanov M A, Rajagopal K and Shuryak E V *Phys. Rev.* D **60** (1999) 114028
[7] Scavenius O and Dumitru A *Phys. Rev. Lett.* **83** (1999) 4697
[8] Steinheimer J and Randrup J *Phys. Rev. Lett.* **109** (2012) 212301
[9] Schaefer B, Pawlowski J M and Wambach J *Phys. Rev.* D **76** (2007) 074023
[10] Nahrgang M, Leupold S, Herold C and Bleicher M *Phys. Rev.* C **84** (2011) 024912
[11] Herold C, Nahrgang M, Mishustin I N and Bleicher M *Phys. Rev.* C **87** (2013) 014907
[12] Nahrgang M, Herold C, Leupold S, Mishustin I N and Bleicher M *J. Phys.* G **40** (2013) 055108
[13] Berdnikov B and Rajagopal K *Phys. Rev.* D **61** (2000) 105017
[14] Nahrgang M, Herold C and Bleicher M *Nucl. Phys.* A **904-905** (2013) 899c
[15] Herold C, Nahrgang M, Mishustin I N and Bleicher M arXiv:1304.5372 [nucl-th].

Kurtoses and high order cumulants: Insights from resummed perturbation theory

Sylvain Mogliacci

Faculty of Physics, University of Bielefeld, D-33615 Bielefeld, Germany

E-mail: sylvain@physik.uni-bielefeld.de

Abstract. Cumulants of conserved charges provide important information about the physics of the quark-gluon plasma around the phase transition region, as they are by construction sensitive to changes in the degrees of freedom of the system. In this brief proceedings contribution, I report on recent results for two such quantities from two different improved perturbative frameworks, as well as discuss their relevance for heavy ion experiments.

1. Introduction

The weak coupling expansion of the QCD pressure shows few signs of convergence, and only at extremely high temperatures (see e.g. [1]). It is therefore crucial to try to improve the situation by attempting to continue the perturbative results as close to the phase transition region as possible. This is highlighted by the fact that the quark-gluon plasma (QGP) is argued [2] to reach a temperature of only a few times the pseudo-critical temperature of the deconfinement transition, $T_c = 154 \pm 9$MeV [3], where the plasma is clearly not extremely weakly coupled.

When the quark chemical potentials μ_f are turned on to study the phase diagram of the theory (see [4, 5] for present and [6, 7] for future collider experiments), lattice Monte Carlo simulations become inapplicable due to the so-called sign problem of QCD [8]. There have been many attempts to find a solution to this problem, but so far the most successful approach amounts to merely Taylor expanding the pressure in powers of μ_f/T, giving access to only moderate densities. The corresponding coefficients are related to cumulants of conserved quantum numbers in the system, and give information about their correlations and fluctuations, being practical probes of changes occurring in the degrees of freedom of the system across the transition region.

The limitation of lattice methods on the T-μ plane is one of the main reasons to attack the problem of fluctuations and correlations of conserved charges — and ultimately the equation of state — from a resummed perturbation theory point of view, where there is no sign problem. These quantities can indeed be extended to very large values of the ratio μ_f/T, hence giving information about how the QGP behaves, as it becomes more and more dense.

2. Statistical QCD and cumulants

Given the Hamiltonian $H_{\rm QCD}$ of Quantum Chromodynamics, the corresponding partition function $Z_{\rm QCD}(T, \mu_f; V)$ is written in the form $Z_{\rm QCD} \equiv {\rm Tr}\ \exp[-(H_{\rm QCD} - \sum_f \mu_f\, Q_f)/T]$. Here, Q_f and μ_f denote respectively conserved charges and the corresponding chemical potentials. In the present work, we will mainly consider the up, down and strange quark conserved charges (with their chemical potentials $\mu_{\rm u}$, $\mu_{\rm d}$, $\mu_{\rm s}$), but one can equivalently express the partition function in term of the baryon, electric charge and strangeness conserved numbers (with their chemical potentials $\mu_{\rm B}$, $\mu_{\rm Q}$, $\mu_{\rm S}$). In addition, $Z_{\rm QCD}$ gives access to a host of equilibrium thermodynamic quantities such as the pressure and entropy, following relations such as $p_{\rm QCD} = \partial(T \log Z_{\rm QCD})/\partial V \xrightarrow[V\to\infty]{} T/V \log Z_{\rm QCD}$ or $S_{\rm QCD} = \partial(T \log Z_{\rm QCD})/\partial T$.

From the partition function, it is easy to see that the mean and (co)variance of two conserved charges, i.e. $\langle Q_f \rangle$ and $\langle (Q_f - \langle Q_f \rangle) \cdot (Q_g - \langle Q_g \rangle) \rangle$, can be expressed in terms of derivatives with respect to the corresponding chemical potentials. Thus, we can write

$$\langle Q_f \rangle = T \frac{\partial}{\partial \mu_f} \log Z_{\rm QCD} \, , \quad \langle (Q_f - \langle Q_f \rangle) \cdot (Q_g - \langle Q_g \rangle) \rangle = T^2 \frac{\partial^2}{\partial \mu_f \partial \mu_g} \log Z_{\rm QCD} \, , \tag{1}$$

defining quantities called the first and second order cumulants of the related conserved charges.

Based on the above, it is clear that when considering the covariance of two different charges, such as $Q_{\rm u}$ and $Q_{\rm d}$, one is probing the correlation between the two related flavors, while when considering the variance of the same charge, one is probing its fluctuations. The former quantity gives information about possible bound-state survival [9], while the latter is related to how the system reacts to small increases of density. It also follows that equation (1) relates the cumulants to the equation of state of the system via derivatives with respect to chemical potentials. These quantities, typically referred to as susceptibilities, are denoted by

$$\chi_{{\rm u}_i\,{\rm d}_j\,{\rm s}_k\,\ldots}(T, \{\mu_f\}) \equiv \frac{\partial^n p_{\rm QCD}(T, \{\mu_f\})}{\partial \mu_{\rm u}^i \, \partial \mu_{\rm d}^j \, \partial \mu_{\rm s}^k \ldots} \quad \text{with} \quad n = i + j + k + \ldots \, , \tag{2}$$

where instead of quark numbers one may also consider any other conserved charge.

The existence and location of a possible critical point on the phase diagram of QCD is often investigated through the behavior of cumulants. Indeed, the variance of the baryon number $\chi_{{\rm B}_2}(T, \mu_{\rm B})$ is expected to display a critical-like behavior with a sharp peak at the critical point, as it is highly sensitive to changes of density in the medium. This quantity can be estimated via a Taylor expansion containing higher order cumulants at vanishing chemical potential (see e.g. [10] for discussion and results in the two flavor case). For some recent reviews on heavy ion collisions and in particular the use of cumulants in statistical QCD, the reader is referred to [2].

3. Resummed perturbative QCD

3.1. Dimensional Reduction inspired resummation

It is well known [11] that dimensional reduction allows the dynamics of the energy scales of order gT and smaller to be properly accounted for through an effective field theory, which for QCD is called Electrostatic QCD (EQCD) [12, 13]. This effective theory turns out to account for all of the infrared divergences typically encountered in naive weak coupling expansions [14]. Thus, EQCD not only provides a way out of the so-called infrared catastrophe, but in addition allows for weak coupling calculations to be carried out to high orders in perturbation theory upon a careful matching procedure with the full theory.

From the above fact, one can write the QCD pressure $p_{\rm QCD}$ as the sum of $p_{\rm hard}(g)$ and $T\,p_{\rm EQCD}(m_{\rm E}, g_{\rm E}, \lambda_{\rm E}, \zeta)$, where the function $p_{\rm hard}$ gives the contribution of the hard scale ($\sim T$) and is available through a strict loop expansion in QCD, while $p_{\rm EQCD}$ gives the contribution of the soft ($\sim gT$) and the ultrasoft ($\sim g^2T$) scales, being accessible from the EQCD partition function. The EQCD parameters $m_{\rm E}(g), g_{\rm E}(g), \lambda_{\rm E}(g)$ and $\zeta(g)$ are functions of the temperature and quark chemical potentials, and admit expansions in powers of the QCD gauge coupling g.

Typically, when computing a physical quantity, one expands the EQCD parameters in powers of g. However, as first suggested in [1] and later successfully applied to the case of the pressure at zero chemical potential in [15], one can simply consider both $p_{\rm hard}$ and $p_{\rm EQCD}$ functions of the EQCD parameters, and not expand them in g. This resums certain higher order contributions to the pressure, containing all correct contributions up to and including order $g^6 \log g$ [16, 17] when expanded in powers of g, and leads to a considerable decrease in the theoretical uncertainties through a substantial reduction of the renormalization scale dependence. Thus it also improves the convergence properties of the series. We will denote these results as the 'DR ones'.

More details about the implementation of the resummation at finite chemical potentials can be found in [18], while general details about dimensional reduction and EQCD can be found in the original references [12, 13] as well as in [17, 19, 20] for a generalization to finite density.

Journal of Physics: Conference Series **503** (2014) 012005 doi:10.1088/1742-6596/503/1/012005

3.2. Hard-thermal-loop perturbation theory

Another way of resumming important higher order contributions to thermodynamic quantities is to make use of a variationally improved perturbation theory framework (see [21, 22]), for which the basic idea is to introduce a physically relevant (usually mass) term that is added and subtracted from the Lagrangian density of the original theory. Treating the added piece with the non-interacting part of the action while adding the subtracted one to the interaction terms of it, one ends up interpolating between the original theory and a theory having dressed propagators and vertices. The procedure is of course such that working consistently to a given order in a perturbative expansion, the final results correspond to the correct original theory.

In a gauge field theory such as QCD the situation is somewhat more complicated, as one cannot simply add and subtract a local mass term for the gauge fields without breaking the gauge invariance. Instead, one can make use of the well-known hard-thermal-loop effective action [23, 24], which plays the role of a non-local mass term. This approach was first applied in [25, 26, 27], and is known as Hard-Thermal-Loop perturbation theory (HTLpt).

HTLpt amounts to reorganizing the Lagrangian density of QCD such that $\mathcal{L}_{\rm HTLpt} = [\mathcal{L}_{\rm QCD} + (1-\delta)\,\mathcal{L}_{\rm HTL}]|_{g\to\sqrt{\delta}g} + \Delta\mathcal{L}_{\rm HTL}$. $\mathcal{L}_{\rm HTL}$ is a gauge invariant HTL improvement term given by the HTL effective action, δ is a formal expansion parameter set to one after having Taylor expanded the path integral up to some order in it, and $\Delta\mathcal{L}_{\rm HTL}$ is a counter term necessary to cancel further ultraviolet divergences introduced by the resulting reorganization of the perturbative series. Essentially, this reorganization amounts to shifting the ground state of the expansion from an ideal gas of massless particles to an ideal gas of massive quasiparticles [28, 29].

More details about the implementation of the HTLpt procedure at finite chemical potentials can be found in [30, 18] as well as in [28, 31]. For related developments, see also [32].

3.3. Technical details

Working at four- and one-loop levels in the DR and HTLpt schemes, respectively, we must assign values to parameters. Following [33], we choose to apply the Fastest Apparent Convergence (FAC) scheme to the NLO expression of $g_{\rm E}$. We obtain $\bar{\Lambda}_{\rm central} = 1.445 \times 2\pi T$ and $1.291 \times 2\pi T$ for the central values with two and three flavors [30, 18], and then vary these values by a factor of two in order to estimate the sensitivity of the results with respect to this choice. Regarding the running of the coupling, we use a two-loop expression in the DR case and one-loop running in the HTLpt result. To fix the value of $\Lambda_{\rm QCD}$ we use the recent lattice result $\alpha_{\rm s}(1.5\ {\rm GeV}) = 0.326$ from [34], demanding that our one- and two-loop couplings agree with it for $\bar{\Lambda} = 1.5$ GeV. This respectively gives $\Lambda_{\rm QCD} = 176$ MeV and 283 MeV for $N_{\rm f} = 3$, as well as $\Lambda_{\rm QCD} = 204$ MeV and 324 MeV for $N_{\rm f} = 2$. These values are then varied by $\pm$ 30 MeV.

In the plots appearing in the following section, the bands corresponding to our perturbative results are composed by varying all of the parameters mentioned here in their respective ranges.

4. Results

4.1. Kurtoses

Roughly speaking, the kurtosis is a measure of how strongly peaked a given quantity is, and is defined as the ratio of the corresponding fourth and second order cumulants, the latter one squared. In figure 1, we display this quantity for the (light) quark and baryon numbers, further multiplying it by the second order cumulants for the sake of comparison with lattice data. Such a quantity can be used as a measure of how fast a phase transition is (see e.g. [35]).

In figure 1 (left), we plot our DR and HTLpt results together with lattice data from [36, 37, 38], which at temperatures around $T = 300 - 400$ MeV seem to agree nicely with the one-loop HTLpt band, but soon start approaching the DR prediction. It should be noted that the latter reproduces the overall trend of the lattice data better. On the right figure, we see similar trends in both predictions, with the difference that the DR one seems to converge to the Stefan Boltzmann limit much faster than in the case of the light quark kurtosis. This trend is seen to be in accordance with the displayed lattice data from [37, 39], and agrees with the expectation that the medium is much less sensitive to the hadronic degrees of freedom in this range of temperatures. Moreover, in figure 1 (right), we display the corresponding three-loop HTLpt prediction of the same quantity computed in [28]. Although the band is relatively large at low temperature, this result is also seen to reproduce the qualitative trend of the lattice data, furthermore agreeing quite well with the DR one. From the one-loop to the three-loop HTLpt prediction, we notice a reasonably good convergence of the series for this quantity.

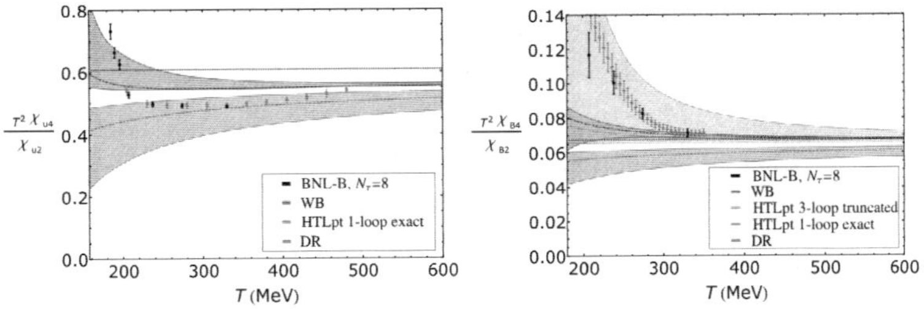

Figure 1. Ratios of cumulants related to the light quark (left) and baryon (right) numbers, with lattice data taken from [36, 37] (BNL-B) and [38, 39] (WB), as well as three-loop HTLpt results from [28]. The dashed curves correspond to the central values of the renormalization and QCD scales, while the straight dashed lines denote the Stefan Boltzmann limits.

4.2. Sixth order cumulants

In figure 2, we next display sixth order cumulants for the light quark number in the cases of three (left) and two (right) quark flavors. Such higher order cumulants are expected to be very sensitive probes of the hadronic freeze-out [40]. For these quantities, the two first orders of perturbation theory vanish, and the weak coupling expansions only start at order g^3, rendering the properties of the weak coupling series much poorer than for the lower order cumulants. Indeed, we see from both the left and right figures that our DR predictions are positive for all displayed temperatures, while the HTLpt results are consistently negative. The latter apparently agree better with the (unfortunately not continuum extrapolated) lattice data of [10, 41]. However, it should be recalled our HTLpt predictions are of one-loop order, and based on the very good agreement between our DR results and the recent three-loop HTLpt calculations of [28] (see [18] for a detailed comparison), it is probable that the forthcoming three-loop HTLpt prediction of [42] for the sixth order cumulant will also turn out to be positive.

It is interesting to note that the leading plasmon contribution to the weak coupling expansion of the sixth order cumulant is negative, while it can be seen by truncating our DR results at various orders that the sign of the result is turned positive by the higher order contributions. Noting that lattice studies seem to favor negative values for sixth order cumulants [10, 40, 41] above the transition region, we suspect that including further perturbative orders to the current DR result, it will eventually change in the temperature range considered. This will, however, likely not be verified in the near future, as the technical challenges involved in e.g. a full five-loop determination of the quantity are formidable.

5. Conclusion

We first considered kurtoses regarding the light quark and baryon number operators for three flavor QCD. We saw how resummed perturbation theory provides a very successful description of lattice data down to 2 − 3 T_c, which is clearly an indication that a weakly interacting quasiparticles picture is well suited for the fermionic sector of the theory, even at that range of temperature. Also, information on the relevant degrees of freedom was highlighted by the qualitative difference of fluctuations between light quark and baryon conserved charges. Next, we looked at sixth order cumulants in the cases of two and three flavor QCD, encountering quantities less successfully described within resummed perturbation theory. This could be understood from the fact that the weak coupling expansions only start at a relatively high order.

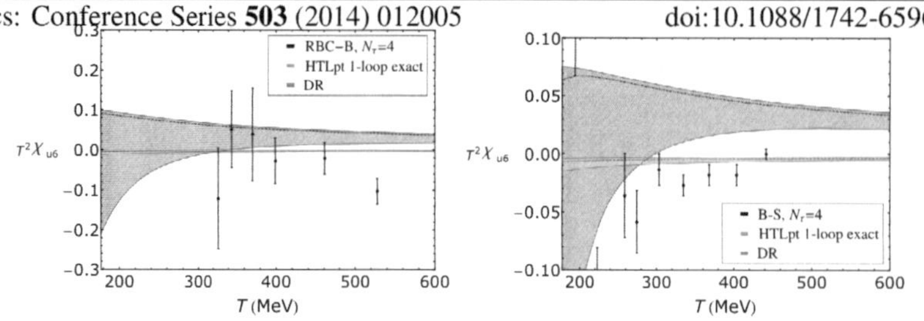

Figure 2. The sixth order cumulant for the light quark number, evaluated for $N_\mathrm{f} = 3$ (left) and $N_\mathrm{f} = 2$ (right), with lattice data from [41] (left) and [10] (right). The dashed curves inside the bands correspond to the central values of the renormalization and QCD scales.

Such quantities giving crucial information about the QGP degrees of freedom, I hope these results will be of practical use in the present and future analysis of heavy ion collision data.

References

[1] Blaizot J P , Iancu E and Rebhan A ; 2003; *Phys. Rev.* D **68** 025011 (*Preprint* hep-ph/0303045)
[2] Satz H ; 2013; *Int. J. Mod. Phys.* A **28** 1330043 (*Preprint* arXiv:1310.1209)
[3] Bazavov A *et al.* ; 2012; *Phys. Rev.* D **85** 054503 (*Preprint* arXiv:1111.1710)
[4] Tannenbaum M J ; 2012; *Preprint* arXiv:1201.5900
[5] Müller B , Schukraft J and Wyslouch B ; 2012; *Ann. Rev. Nucl. Part. Sci.* **62** 361 (*Preprint* arXiv:1202.3233)
[6] Heuser J M, CBM collaboration; 2013; *Nucl. Phys.* A **904-905** 941c
[7] Kekelidze V *et al.*; 2013; *Nucl. Phys.* A **904-905** 945c
[8] de Forcrand P ; 2009; *PoS* **LAT2009** 010 (*Preprint* arXiv:1005.0539)
[9] Ratti C *et al.*; 2012; *Phys. Rev.* D **85** 014004 (*Preprint* arXiv:1109.6243)
[10] Allton C R *et al.*; 2005; *Phys. Rev.* D **71** 054508 (*Preprint* hep-lat/0501030)
[11] Appelquist T and Pisarski R D ; 1981; *Phys. Rev.* D **23** 2305
[12] Braaten E and Nieto A ; 1995; *Phys. Rev.* D **51** 6990 (*Preprint* hep-ph/9501375)
[13] Kajantie K *et al.*; 1996; *Nucl. Phys.* B **458** 90 (*Preprint* hep-ph/9508379)
[14] Linde A D ; 1980; *Phys. Lett.* B **96** 289
[15] Laine M and Schröder Y ; 2006; *Phys. Rev.* D **73** 085009 (*Preprint* hep-ph/0603048)
[16] Kajantie K *et al.*; 2003; *Phys. Rev.* D **67** 105008 (*Preprint* hep-ph/0211321)
[17] Vuorinen A ; 2003; *Phys. Rev.* D **68** 054017 (*Preprint* hep-ph/0305183)
[18] Mogliacci S , Andersen J O , Strickland M , Su N and Vuorinen A ; 2013; *Preprint* arXiv:1307.8098
[19] Hart A , Laine M and Philipsen O ; 2000; *Nucl. Phys.* B **586** 443 (*Preprint* hep-ph/0004060)
[20] Ipp A , Kajantie K , Rebhan A and Vuorinen A ; 2006; *Phys. Rev.* D **74** 045016 (*Preprint* hep-ph/0604060)
[21] Kneur J -L and Neveu A ; 2013; *Phys. Rev.* D **88** 074025 (*Preprint* arXiv:1305.6910)
[22] Karsch F , Patkós A and Petreczky P ; 1997; *Phys. Lett.* B **401** 69 (*Preprint* hep-ph/9702376)
[23] Frenkel J and Taylor J C ; 1990; *Nucl. Phys.* B **334** 199
[24] Braaten E and Pisarski R D ; 1990; *Nucl. Phys.* B **337** 569
[25] Andersen J O , Braaten E and Strickland M ; 2000; *Phys. Rev.* D **61** 014017 (*Preprint* hep-ph/9905337)
[26] Andersen J O , Braaten E and Strickland M ; 2000; *Phys. Rev.* D **61** 074016 (*Preprint* hep-ph/9908323)
[27] Andersen J O *et al.*; 2002; *Phys. Rev.* D **66** 085016 (*Preprint* hep-ph/0205085)
[28] Haque N , Andersen J O , Mustafa M G , Strickland M and Su N ; 2013; *Preprint* arXiv:1309.3968
[29] Andersen J O *et al.*; 2011; *J. High Energy Phys.* **1108** 053 (*Preprint* arXiv:1103.2528)
[30] Andersen J O , Mogliacci S , Su N and Vuorinen A ; 2013; *Phys. Rev.* D **87** 074003 (*Preprint* arXiv:1210.0912)
[31] Haque N , Mustafa M G and Strickland M ; 2013; *J. High Energy Phys.* **1307** 184 (*Preprint* arXiv:1302.3228)
[32] Blaizot J P , Iancu E and Rebhan A ; 2001; *Phys. Rev.* D **63** 065003 (*Preprint* hep-ph/0005003)
[33] Kajantie K *et al.*; 1997; *Nucl. Phys.* B **503** 357 (*Preprint* hep-ph/9704416)
[34] Bazavov A *et al.*; 2012; *Phys. Rev.* D **86** 11403 (*Preprint* arXiv:1205.6155)
[35] Stephanov M A ; 2011; *Phys. Rev. Lett.* **107** 052301 (*Preprint* arXiv:1104.1627)
[36] Schmidt C, BNL-Bielefeld collaboration; 2013; *Nucl. Phys.* A **904-905** 865c (*Preprint* arXiv:1212.4278)
[37] Schmidt C ; 2013; *J. Phys. Conf. Ser.* **432** 012013 (*Preprint* arXiv:1212.4283)
[38] Borsányi S ; 2013; *Nucl. Phys.* A **904-905** 270c (*Preprint* arXiv:1210.6901)
[39] S. Borsányi *et al.*; 2013; *Phys. Rev. Lett.* **111** 062005 (*Preprint* arXiv:1305.5161)
[40] Friman B , Karsch F , Redlich K and Skokov V ; 2011; *Eur. Phys. J.* C **71** 1694 (*Preprint* arXiv:1103.3511)
[41] Petreczky P *et al.*, RBC-Bielefeld collaboration; 2009; *PoS* **LAT2009** 159 (*Preprint* arXiv:0911.0196)
[42] Haque N , Andersen J O , Mustafa M G , Strickland M and Su N ; in preparation

In-medium QCD sum rules for D mesons: A projection method for higher order contributions

T Buchheim, T Hilger and B Kämpfer

Helmholtz-Zentrum Dresden-Rossendorf, PF 510119, D-01314 Dresden, Germany
Technische Universität Dresden, Institut für Theoretische Physik, D-01062 Dresden, Germany

E-mail: t.buchheim@hzdr.de

Abstract. D mesons serve as excellent probes of hot and/or dense strongly interacting matter. They can provide insight into the restoration of chiral symmetry. The chiral condensate as well as other chirally odd condensates, such as certain four-quark condensates, are linked to order parameters of spontaneous chiral symmetry breaking. Thus, the evaluation of these higher order condensate contributions in the framework of QCD sum rules is of high interest. We present a general method for projecting Lorentz indices of ground state expectation values providing a crucial step towards a comprehensive calculation of higher order corrections to the operator product expansion of hadrons, especially D mesons, in a strongly interacting medium.

1. Introduction

Among the central issues of hadron physics is chiral symmetry, its breaking in vacuum and its restoration in medium. Chiral symmetry is explicitly broken due to non-zero quark masses leaving the mass term of the Lagrangian not invariant under chiral transformations. If only light quark flavors (up, down) are under consideration the assumption of massless quarks is justified. However, chiral symmetry is broken spontaneously, because the QCD ground state is not invariant under chiral transformations.

The chiral condensate $\langle \bar{q}q \rangle$ is linked to order parameters of the spontaneous chiral symmetry breaking. This chirally odd condensate, i.e. a condensate which is not invariant under chiral transformations, is non-zero and thus, signals spontaneous symmetry breaking, similarly to the vector and axial-vector mixing under axial transformations. Besides the chiral condensates also further condensates are connected to order parameters of spontaneous chiral symmetry breaking, e.g. chirally odd four-quark-condensates. Chiral spontaneous symmetry breaking gets apparent due to the mass splitting of chiral partner mesons. For instance, in the light-quark sector, the axial vector meson a_1 is significantly heavier than its chiral partner, the vector meson ρ [1].

In a strongly interacting medium the situation changes drastically: We have to deal with non-zero temperatures T and baryon densities n. In leading-order, non-zero temperatures are modeled by a pion gas and finite densities by ambient nucleons. The chiral condensate changes in a medium according to (cf. [2, 3])

$$\langle \bar{q}q \rangle_{T,n} = \langle \bar{q}q \rangle \left(1 - \frac{T^2}{8 f_\pi^2} - \frac{\sigma_N \, n}{m_\pi^2 f_\pi^2} \right) , \tag{1}$$

where f_π is the pion decay constant, m_π the pion mass and σ_N is the nucleon sigma term. Obviously the numerical value of the chiral condensate decreases for increasing temperature and density; chiral restoration would be accompanied by $\langle \bar{q}q \rangle_{T,n} = 0$.

The impact of further chirally odd condensates, e.g. certain four-quark condensates behaving as the chiral condensate, is therefore of utmost interest in relation to chiral restoration. First investigations have been performed in the light-quark meson sector, e.g. a significant impact of chirally odd four-quark condensates was found in the framework of rho meson sum rules [4]. We aim at extending these evaluations to Qq meson systems, containing an heavy (Q) and a light (q) valence quark.

Choosing certain chiral transformations restricted to light quarks, one can derive an invariant Lagrangian [5]. However, vector and axial-vector currents as well as scalar and pseudo-scalar currents mix under these transformations, i.e. also in the Qq sector certain chirally odd condensates determine the mass splitting of chiral partner mesons. Again, chiral symmetry restoration in medium is supposed to be accompanied by the vanishing of chirally odd condensates.

Theoretical investigations of four-quark condensates of Qq meson systems in medium is of large interest, since the envisaged experiments of the CBM and Panda collaborations at FAIR [6, 7] include the analysis of medium modifications of D mesons.

2. In-medium QCD sum rules of D mesons: operator product expansion

QCD sum rules proved to be a successful tool to extract spectral properties of hadrons in vacuum from the theory of strong interaction [8,9]. During the last two decennia an extension to hadrons in an strongly interacting environment has given insight into their medium modifications [2, 3], which are of large contemporary interest. The sum rule method utilizes the causal current-current correlator

$$\Pi(q) = i \int d^4x \, e^{iqx} \langle \mathrm{T}[J(x) J^\dagger(0)] \rangle \,, \tag{2}$$

where T[...] means time ordering and J denotes a current reflecting the quark content and quantum numbers of the hadron under consideration. The correlator (2) can be related to the hadrons spectral function via a dispersion relation which provides a link of QCD, formulated in quark degrees of freedom, and hadronic phenomenology. The evaluation of $\Pi(q)$ employs Wilson's operator product expansion (OPE) [10].

The OPE is an expansion at operator level. The operator product is expanded in an asymptotic series of local operators $\mathcal{O}_i$ with increasing mass dimension:

$$J(x) J^\dagger(0) = \sum_i C_i(x) \mathcal{O}_i(0) \,. \tag{3}$$

The coefficients C_i in this series expansion are the Wilson coefficients. QCD provides three operators as basic elements for this expansion: quark operators q, the gluonic field strength tensor $G_{\mu\nu}$ and the covariant derivative D_μ, which appear in infinitely combinations building the $\mathcal{O}_i$. We consider the expectation value of the current-operator product, thus, we need to use the expectation values of the operators of the expansion. These expectations values are the condensates, characterizing the complex QCD ground state. In vacuum they have certain numerical values, which may be universally used for sum rule evaluation.

QCD cannot be treated perturbatively for low momentum transfers and long distance phenomena due to the running strong coupling α_s. One of the beneficial features of the OPE is the separation of scales [11], i.e. long range effects are absorbed into the condensates and the Wilson coefficients can be calculated by means of perturbation theory. In the light-quark

Journal of Physics: Conference Series **503** (2014) 012006 doi:10.1088/1742-6596/503/1/012006

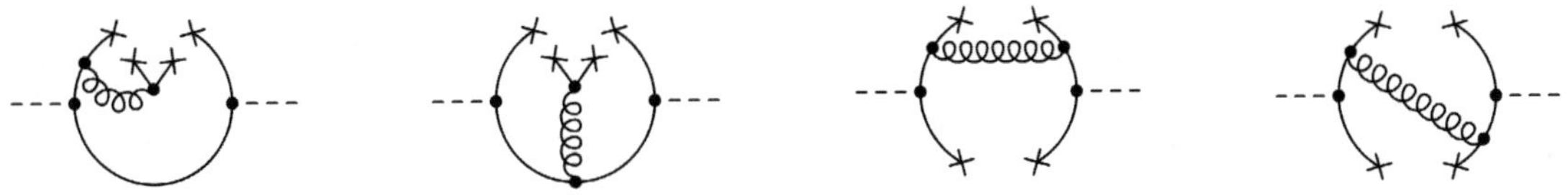

Figure 1. Diagrammatic representation of four-quark condensate contributions.

sector, all vacuum condensates up to mass dimension 6 have been identified and their Wilson coefficients have been determined, unlike the in-medium OPE for D mesons. Their coefficients are determined up to mass dimension 5 for in-medium situations [12]. The complete in-medium contributions of four-quark condensates are determined for light mesons only [13]. The three-gluon condensate contribution, also of mass dimension 6, has been calculated for vacuum situations so far [14]. This work addresses crucial computational steps towards a comprehensive calculation of Wilson coefficients of higher order condensates for in-medium situations, especially for four-quark condensates of Qq mesons. New medium specific condensates come into play, which contain heavy quark operators.

3. Four-quark condensates

Only expectation values of operators which carry the assumed symmetries of the QCD ground state are considered condensates [15]. Formally, condensates are expectation values of hermitian operator products, which are to be Lorentz and Dirac scalars as well as color singlets. They are invariant under parity and time reversal transformations. Thus, the projection of Dirac, color and Lorentz indices of the expectation values has to be performed. The projection of Dirac and color indices relies on orthogonal bases provided by the Clifford algebra and the generators and unity element of the color group, respectively [16]. Projection of Lorentz indices is more involved, in particular for in-medium situations.

In leading order, four-quark condensates contribute to D meson sum rules on tree-level in order α_s. The four-quark condensate contributions in a diagrammatic representation (cf. figure 1) are of two kinds: the meson momentum either flows through a quark or a gluon line [8]. The depicted diagrams yield condensate contributions proportional to $\langle \bar{q}_i \Gamma D_\mu D_\nu D_\lambda q_i \rangle$, $\langle \bar{q}_i \Gamma D_\mu G_{\nu\lambda} q_i \rangle$ and $\langle \bar{q} t^A \Gamma_1 q \bar{Q} t^A \Gamma_2 Q \rangle$ where q_i symbolizes either Q or q, t^A stands for a generator or the unity element of the color group and Γ, Γ_i denote basis elements of the Clifford algebra, carrying themselves up to two Lorenz indices each. The projection of these uncontracted Lorentz indices leads to 24 condensates in the Qq sector, which are invariant under parity and time reversal transformations.

4. Projection of Lorentz Indices

In contrast to Dirac and color projections, which rely on an expansion in an orthogonal basis, projecting Lorentz indices is based on given elements which carry Lorentz structures. In vacuum, these are the metric tensor $g_{\mu\nu}$ and the totally anti-symmetric pseudo-tensor $\varepsilon_{\mu\nu\lambda\sigma}$. For investigations of the temperature and/or density dependence of condensates, a further Lorentz structure is available for projection. Due to the presence of the ambient medium the Poincaré invariance is broken, and the medium velocity v_μ needs to be considered additionally for projections. It has wide ranging implications, as it allows for the projection of expectation values with an odd number of Lorentz indices which results in a multitude of new condensates.

The ansatz for an expectation value $\langle O_{\vec{\mu}_n} \rangle \equiv \langle O_{\mu_1 \ldots \mu_n} \rangle$ carrying n Lorentz indices reads

$$\langle O_{\vec{\mu}_n} \rangle = \mathbf{a} \cdot \mathbf{p}_{\vec{\mu}_n} , \tag{4}$$

where $\mathbf{p}_{\vec{\mu}_n} \equiv \mathbf{p}_{\mu_1 \cdots \mu_n}$ denotes the vector of all projection structures and $\mathbf{a}$ is the vector of the corresponding coefficients. Contracting Eq. (4) with $\mathbf{p}^{\vec{\mu}_n}$ yields

$$\langle O_{\vec{\mu}_n} \rangle \mathbf{p}^{\vec{\mu}_n} = \mathbf{p}^{\vec{\mu}_n} \left[\mathbf{p}_{\vec{\mu}_n} \cdot \mathbf{a} \right] = \left[\mathbf{p}^{\vec{\mu}_n} (\mathbf{p}_{\vec{\mu}_n})^{\mathrm{T}} \right] \mathbf{a} = P_n \mathbf{a} \,, \tag{5}$$

where we have defined the matrix $P_n = \mathbf{p}^{\vec{\mu}_n} (\mathbf{p}_{\vec{\mu}_n})^{\mathrm{T}}$. Defining a vector $\mathbf{c} = \langle O_{\vec{\mu}_n} \rangle \mathbf{p}^{\vec{\mu}_n}$, which contains the expectation values contracted with all projection structures, we obtain

$$\mathbf{a} = P_n^{-1} \mathbf{c} \,. \tag{6}$$

Thus, the projection formula for n Lorentz indices reads in compact form

$$\langle O_{\vec{\mu}_n} \rangle = \left(P_n^{-1} \mathbf{c} \right) \cdot \mathbf{p}_{\vec{\mu}_n} \tag{7}$$

or more explicitly

$$\langle O_{\vec{\mu}_n} \rangle = \left\{ \left[\mathbf{p}^{\vec{\nu}_n} (\mathbf{p}_{\vec{\nu}_n})^{\mathrm{T}} \right]^{-1} \langle O_{\vec{\kappa}_n} \rangle \mathbf{p}^{\vec{\kappa}_n} \right\} \cdot \mathbf{p}_{\vec{\mu}_n} \,. \tag{8}$$

The general projection vectors $\mathbf{p}_{\vec{\mu}_n}$ for $n \leq 5$ have the following structures:

$$\mathbf{p}_{\mu} = v_{\mu} \,, \tag{9a}$$

$$\begin{aligned} \mathbf{p}_{\mu\nu} = (&g_{\mu\nu}, \\ &v_{\mu} v_{\nu})^{\mathrm{T}} \,, \end{aligned} \tag{9b}$$

$$\begin{aligned} \mathbf{p}_{\mu\nu\lambda} = (&\varepsilon_{\mu\nu\lambda\alpha} v^{\alpha}, \\ &v_{\mu} g_{\nu\lambda}, v_{\nu} g_{\mu\lambda}, v_{\lambda} g_{\mu\nu}, \\ &v_{\mu} v_{\nu} v_{\lambda})^{\mathrm{T}} \,, \end{aligned} \tag{9c}$$

$$\begin{aligned} \mathbf{p}_{\mu\nu\lambda\sigma} = (&\varepsilon_{\mu\nu\lambda\sigma}, \\ &v_{\mu} \varepsilon_{\nu\lambda\sigma\alpha} v^{\alpha}, v_{\nu} \varepsilon_{\mu\lambda\sigma\alpha} v^{\alpha}, v_{\lambda} \varepsilon_{\mu\nu\sigma\alpha} v^{\alpha}, \\ &g_{\mu\nu} g_{\lambda\sigma}, g_{\mu\lambda} g_{\nu\sigma}, g_{\mu\sigma} g_{\nu\lambda}, \\ &v_{\mu} v_{\nu} g_{\lambda\sigma}, v_{\mu} v_{\lambda} g_{\nu\sigma}, v_{\mu} v_{\sigma} g_{\nu\lambda}, v_{\lambda} v_{\sigma} g_{\mu\nu}, v_{\nu} v_{\sigma} g_{\mu\lambda}, v_{\nu} v_{\lambda} g_{\mu\sigma}, \\ &v_{\mu} v_{\nu} v_{\lambda} v_{\sigma})^{\mathrm{T}} \,, \end{aligned} \tag{9d}$$

$$\begin{aligned} \mathbf{p}_{\mu\nu\lambda\sigma\rho} = (&v_{\mu} \varepsilon_{\nu\lambda\sigma\rho}, v_{\nu} \varepsilon_{\mu\lambda\sigma\rho}, v_{\lambda} \varepsilon_{\mu\nu\sigma\rho}, v_{\sigma} \varepsilon_{\mu\nu\lambda\rho}, \\ &g_{\mu\nu} \varepsilon_{\lambda\sigma\rho\alpha} v^{\alpha}, g_{\mu\lambda} \varepsilon_{\nu\sigma\rho\alpha} v^{\alpha}, g_{\mu\sigma} \varepsilon_{\nu\lambda\rho\alpha} v^{\alpha}, g_{\nu\lambda} \varepsilon_{\mu\sigma\rho\alpha} v^{\alpha}, g_{\nu\sigma} \varepsilon_{\mu\lambda\rho\alpha} v^{\alpha}, g_{\lambda\sigma} \varepsilon_{\mu\nu\rho\alpha} v^{\alpha}, \\ &v_{\mu} v_{\nu} \varepsilon_{\lambda\sigma\rho\alpha} v^{\alpha}, v_{\mu} v_{\lambda} \varepsilon_{\nu\sigma\rho\alpha} v^{\alpha}, v_{\mu} v_{\sigma} \varepsilon_{\nu\lambda\rho\alpha} v^{\alpha}, v_{\nu} v_{\lambda} \varepsilon_{\mu\sigma\rho\alpha} v^{\alpha}, v_{\nu} v_{\sigma} \varepsilon_{\mu\lambda\rho\alpha} v^{\alpha}, \\ &v_{\lambda} v_{\sigma} \varepsilon_{\mu\nu\rho\alpha} v^{\alpha}, \\ &v_{\mu} g_{\nu\lambda} g_{\sigma\rho}, v_{\mu} g_{\nu\sigma} g_{\lambda\rho}, v_{\mu} g_{\nu\rho} g_{\lambda\sigma}, v_{\nu} g_{\mu\lambda} g_{\sigma\rho}, v_{\nu} g_{\mu\sigma} g_{\lambda\rho}, v_{\nu} g_{\mu\rho} g_{\lambda\sigma}, v_{\lambda} g_{\mu\nu} g_{\sigma\rho}, \\ &v_{\lambda} g_{\mu\sigma} g_{\nu\rho}, v_{\lambda} g_{\mu\rho} g_{\nu\sigma}, v_{\sigma} g_{\mu\nu} g_{\lambda\rho}, v_{\sigma} g_{\mu\lambda} g_{\nu\rho}, v_{\sigma} g_{\mu\rho} g_{\nu\lambda}, v_{\rho} g_{\mu\nu} g_{\lambda\sigma}, v_{\rho} g_{\mu\lambda} g_{\nu\sigma}, \\ &v_{\rho} g_{\mu\sigma} g_{\nu\lambda}, \\ &v_{\mu} v_{\nu} v_{\lambda} g_{\sigma\rho}, v_{\mu} v_{\nu} v_{\sigma} g_{\lambda\rho}, v_{\mu} v_{\nu} v_{\rho} g_{\lambda\sigma}, v_{\mu} v_{\sigma} v_{\rho} g_{\nu\lambda}, v_{\mu} v_{\lambda} v_{\rho} g_{\nu\sigma}, \\ &v_{\mu} v_{\lambda} v_{\sigma} g_{\nu\rho}, v_{\nu} v_{\sigma} v_{\rho} g_{\mu\lambda}, v_{\nu} v_{\lambda} v_{\rho} g_{\mu\sigma}, v_{\nu} v_{\lambda} v_{\sigma} g_{\mu\rho}, v_{\lambda} v_{\sigma} v_{\rho} g_{\mu\nu}, \\ &v_{\mu} v_{\nu} v_{\lambda} v_{\sigma} v_{\rho})^{\mathrm{T}} \,, \end{aligned} \tag{9e}$$

where $\mathbf{p}_{\vec{\mu}_n}$ is listed in the following way: If possible, projection structures incorporating the totally anti-symmetric pseudo-tensor are noted first, the following lines (blocks) contain projection structures incorporating the metric tensor with increasing number of quantities v_μ (from line (block) to line (block)).

In order to separate vacuum and medium specific structures the projection is decomposed as

$$\langle O_{\vec{\mu}_n} \rangle = \langle O_{\vec{\mu}_n} \rangle^{\text{vac}} + \langle O_{\vec{\mu}_n} \rangle^{\text{med}} . \tag{10}$$

If the expectation value carries an odd number of Lorentz indices it is a purely medium specific condensate and a decomposition is obsolete. When dealing with an even number of Lorentz indices the vacuum projection $\langle O_{\vec{\mu}_n} \rangle^{\text{vac}}$ is obtained by applying the depicted method, but using $\mathbf{p}^{\text{vac}}_{\vec{\mu}_n}$ instead of $\mathbf{p}_{\vec{\mu}_n}$ containing the vacuum specific elements $g_{\mu\nu}$ and $\varepsilon_{\mu\nu\lambda\sigma}$ only. The medium projection is obtained according to $\langle O_{\vec{\mu}_n} \rangle^{\text{med}} = \langle O_{\vec{\mu}_n} \rangle - \langle O_{\vec{\mu}_n} \rangle^{\text{vac}}$.

5. Summary

Four-quark condensate contributions to the OPE of D mesons for general in-medium situations extend existing QCD sum rule evaluations. One of the major computational tasks for calculating contributions of higher order condensates, especially four-quark condensates, is the projection of Dirac, color and Lorentz indices. Generalization of the projection of Dirac and color indices is straight forward because they rely on an orthogonal basis and with regard to four-quark condensates one observes a close connection to Fierz transformations (cf. [17, 18]). However, the projection of Lorentz indices for in-medium situations is more involved. In this paper we present a general procedure which goes beyond the common itemization of explicit examples, e.g. given in [14, 15]. Our method is now being implemented in the evaluation of Wilson coefficients of those four-quark condensates which appear as α_s contributions containing terms intimately related to chiral symmetry.

Acknowledgments

This work is supported by BMBF 05P12CRGHE.

References

[1] Beringer J *et al.* (Particle Data Group) 2012 *Phys. Rev.* D **86** 010001
[2] Hatsuda T, Koike Y and Lee S H 1993 *Nucl. Phys.* B **394** 221
[3] Cohen T D, Furnstahl R J, Griegel D K and Jin X 1995 *Prog. Part. Nucl. Phys.* **35** 221
[4] Hilger T, Thomas R, Kämpfer B and Leupold S 2012 *Phys. Lett.* B **709** 200
[5] Hilger T, Kämpfer B and Leupold S 2011 *Phys. Rev.* C **84** 045202
[6] CBM Collaboration 2013 http://www.fair-center.eu/en/for-users/experiments/cbm.html
[7] PANDA Collaboration 2013 http://www-panda.gsi.de/framework/physics.php
[8] Shifman M A, Vainshtein A I and Zakharov V I 1979 *Nucl. Phys.* B **147** 385
[9] Reinders L J, Rubinstein H and Yazaki S 1985 *Phys. Rept.* **127** 1
[10] Wilson K G 1969 *Phys. Rev.* **179** 1499
[11] Shuryak E V 2004 *The QCD vacuum, hadrons and superdense matter* 2nd ed Lecture Notes in Physics (World Scientific)
[12] Hilger T, Thomas R and Kämpfer B 2009 *Phys. Rev.* C **97** 025202
[13] Thomas R, Zschocke S and Kämpfer B 2005 *Phys. Rev. Lett.* **95** 232301
[14] Nikolaev S N and Radyushkin A V 1983 *Nucl. Phys.* B **213** 285
[15] Jin X, Cohen T D, Furnstahl R J and Griegel D K 1993 *Phys. Rev.* C **47** 2882
[16] Thomas R, Hilger T and Kämpfer B 2007 *Nucl. Phys.* A **795** 19
[17] Drukarev E G, Ryskin M G, Sadovnikova V A, Gutsche T and Faessler A 2004 *Phys. Rev.* C **69** 065210
[18] Zhang T, Brauner T and Rischke D H 2010 *JHEP* **1006** 064

The $\psi(4040)$ at the future $\overline{\text{P}}$ANDA experiment

L. Zotti*[1], A. Filippi[2], S. Marcello[1], S. Spataro[1]

(1) Università degli Studi di Torino & INFN Torino
(2) INFN Torino

E-mail: `zotti@to.infn.it`

Abstract. The $\overline{\text{P}}$ANDA experiment will be carried out at the future FAIR facility, it will be a fixed target experiment, where antiproton beams, of unprecedented quality and intensity, will be used to study interactions on protons and on nuclei. $\overline{\text{P}}$ANDA will be an excellent tool to investigate those final states which include short-lived particles. Since different charmonium states can be accessed in direct formation with all the available quantum numbers in $\bar{p}p$ annihilations, the charmonium spectroscopy is one of the main goal of the experiment.
The $\overline{\text{P}}$ANDA experiment represents a unique possibility to improve both statistics and precision of existing data and to further explore the physics in the charm quark sector. Indeed, an energy scan with high precision over the full charm spectrum is still missing and it will not be delivered by future experiments currently planned as upgrade of the existing facilities. A detailed description of the possibility to reconstruct the $\psi(4040) \rightarrow D^{*+}D^{*-}$ at $\overline{\text{P}}$ANDA, together with the study of the huge hadronic background suppression will be presented. The importance of the Micro-Vertex Detector for the reconstruction of D mesons decay will be showed.

1. Introduction

The $\overline{\text{P}}$ANDA (AntiProton Annihilation at DArmstadt) experiment is one of the projects at the future Facility for Antiproton and Ion Research (FAIR) which is currently under construction as an extension of the existing GSI Laboratory (Helmholtzzentrum für Schwerionenforschung).
The FAIR facility [1] will feature a complex structure of new accelerators; primary proton and ion beams as well as secondary beams of antiprotons and radioactive ions will be provided by the FAIR accelerators with high energy, high intensity and high quality.
The $\overline{\text{P}}$ANDA spectrometer will be built as a general purpose fixed target detector that will study $\bar{p}$-p and $\bar{p}$-A annihilations [2]. It will benefit from the antiproton beams, with unprecedented intensity and quality, provided by the High Energy Storage Ring (HESR). The antiproton beam momentum will range from 1.5 GeV/c to 15 GeV/c.
The $\overline{\text{P}}$ANDA scientific program foresees charmonium spectroscopy studies as well as search for exotic hadrons, nucleon structure studies and investigation about the hadron mass in-medium modification and hypernuclear physic [2].

2. Charmonium Spectroscopy

The recent discoveries at the B-factories of unpredicted states above the $D\overline{D}$ threshold, known as X, Y and Z, have renewed the interest in the charmonium spectroscopy field. $\overline{\text{P}}$ANDA with its available energy in the center of mass system, up to $\sqrt{s} = 5.5$ GeV/c^2, will have the chance to explore deeply this mass region.

Journal of Physics: Conference Series **503** (2014) 012007 doi:10.1088/1742-6596/503/1/012007

Charmonium and open charm studies cover a large and important fraction of the $\overline{\text{P}}$ANDA physics program. Compared to past experiments, such as Fermilab E760/E835, $\overline{\text{P}}$ANDA will feature a higher momentum resolution and larger machine luminosity. The main challenge is due to the presence of the large hadronic background in $\bar{p}p$ annihilations. The ability to study charmonium states above the $D\overline{D}$ threshold is of fundamental importance in investigations of open-charm spectroscopy, search for hybrids and CP violation. In particular the reaction $\bar{p}p \to D\overline{D}$ is a benchmark channel to test the capability of background suppression. Moreover, since channels with short-lived particles in the final state will be studied, an optimal secondary vertices reconstruction is required.

3. The $\psi(4040)$ at $\overline{\text{P}}$ANDA

The $\psi(4040)$ is the 3^3S_1 excitation of charmonium. For the $\psi(4040)$, 25 decay modes have been observed but for only a few of them it was possible to fix limits on the branching fraction. The $\psi(4040)$ features a large total width (80 MeV [3]), 2 order of magnitude larger than the lower-lying 1^{--} $c\bar{c}$ states, and a weak coupling to leptons. Moreover, the decay to hidden-charm is not favored due to the availability of open-charm channels (i.e. with charmed mesons in the final states).

The reaction channel investigated in this work is:

$$\begin{aligned} p\bar{p} &\to \psi(4040) \to D^{*+}D^{*-} & \\ D^{*+(-)} &\to D^0(\overline{D^0})\pi^{+(-)} & BR: 67.7\% \\ D^0(\overline{D^0}) &\to K^-\pi^+(K^+\pi^-) & BR: 3.88\% \end{aligned}$$

where the antiprotons have a momentum of 7.71 GeV/c. The main focus of this benchmark study aims to test the capability to reconstruct secondary vertices of D-mesons, and in this way to investigate the possibility to separate the charmed signal from the huge hadronic background.

3.1. $\bar{p}p \to \psi(4040)$ Formation Cross Section

For $\bar{p}p$ annihilations the charm production cross section is so far unknown. A Breit-Wigner approach can be used to calculate the cross section for the $\psi(4040)$ resonance formation [3].

$$\sigma_R(s) = \frac{4\pi\hbar^2c^2}{(E/2)^2}\frac{2J+1}{(2S_1+1)(2S_2+1)}\frac{B_{in}B_{out}\Gamma^2/4}{(E-E_r)^2+\Gamma^2/4} \quad (1)$$

where: B_{in} is the branching ratio of $\psi(4040) \to p\bar{p}$, B_{out} is the BR of $\psi(4040) \to D^{*+}D^{*-}$, E is the center of mass energy, E_r=4039 $\pm$ 1 MeV, Γ=80 $\pm$ 10MeV [3], and S_1, S_2, and J are the spins of p, $\bar{p}$, and $\psi(4040)$, respectively.

No measurement of B_{in} is available, but it can be estimated to be $\sim 2.5 \cdot 10^{-6}$, using the branching ratio of the J/ψ resonance, and scaling the individual width of both charmonium states [2]. Using a decay branching ratio B_{out} ~33% [4] and considering the missing contribution from the continuum production, the estimated cross section is:

$$\sigma(\bar{p}p \to \psi(4040) \to D^{*+}D^{*-}) = 2.2\ nb \quad (2)$$

Taking into account the relative branching ratio of the following D-meson decays and the total inelastic cross section of $\sigma(p\bar{p} \to X)$=35 mb[3], it is possible to estimate that the $D\overline{D}$ production cross section will be suppressed at least by 11 orders of magnitude.

3.1.1. Hadronic Background The pattern of this benchmark channel features six charged tracks in the detector. Of all the 6-prong reactions, three have been investigated in detail since they could suggest the main background contamination features, and they can be considered the most effective compared to the signal. The first two contain only pions, and require a good particle identification, to be suppressed: $p\bar{p} \rightarrow 3\pi^+3\pi^-\pi^0$ has a cross section of σ = 1.3 mb, $p\bar{p} \rightarrow 3\pi^+3\pi^-$ σ = 0.21 mb. The channel that could represent the largest background source is 2K4π, since it features exactly the same signature of the considered decay channel; its cross section is σ = 0.029 mb. The contribution from this channel could be suppressed only exploiting a precise tracking which clearly identifies the secondary vertices of the delayed D-meson decays.
A two-fold approach has been followed to evaluate the background contributions and the rejection power of the applied selections. Based on the Dual Parton Model (DPM) event generator, which describes generic $p\bar{p}$ interactions at intermediate energies [5], $2.25 \cdot 10^7$ events have been generated. In a second approach the three channels mentioned above have been simulated separately.

3.2. MC Detector Setup
The detector setup used to simulate the $\psi(4040)$ benchmark channel consists of the full $\overline{\text{P}}$ANDA spectrometer, see [2]. In particular the analysis has taken advantages of only tracking devices:

- **MVD**: the Micro Vertex Detector is the innermost detector of $\overline{\text{P}}$ANDA. It is designed to measure charged particle hits as close as possible to the interaction point (IP). Its main goal is to reconstruct the decays of short-lived particles [6].
- **STT**: the Straw Tube Tracker is the central tracker of $\overline{\text{P}}$ANDA. The achievable spatial resolution is better than 150μm on x-y coordinate and it is of the order of 3mm along the z-coordinate. It will allow together with the MVD to achieve a momentum resolution of the order of 1%, for further details see [7].
- **GEM**: Gas electron multiplier disks will be placed downstream to the IP, at 1.1, 1.4 and 1.9m in order to cover the geometrical acceptance of the STT below 22° of the polar angle [2].
- **FTS**: the forward tracking system will allow to track charged particle emitted at small polar angle, in particular in the polar range below ±10° and ±5° in the horizontally and vertically direction with respect to the beam axis. Six sets of planes, will be placed before, in the middle and after the dipole magnet [2].

4. Phase Space, Acceptance Consideration

Exploiting the MC-matching routine available in the PandaRoot framework, acceptance effects, momentum and secondary vertex resolution were investigated. The $\psi(4040)$ has been studied in the decay to $D^{*+}D^{*-}$, whose momentum spans from 3.4 to 4.3 GeV/c. The subsequent decay of the $D^{*+}(D^{*-})$ mesons proceeds as a 2-body decay into a $D^0(\overline{D^0})$ meson and a $\pi^+(\pi^-)$. Because of the mass difference between the pions and the D^0-mesons, the latter carry most of the momentum boost, and their momentum ranges from ~3.1 GeV/c up to ~4.1 GeV/c. The $\pi^+(\pi^-)$ features a lower momentum distribution, as shown in Fig.1.a, and for this reason in the following it will be referred as *soft-pions*. Their low momentum and their preferred direction in the forward region are the main responsible of the low reconstruction efficiency of the studied benchmark channel. Indeed, because of the track bending due to the magnetic field, only less than 50% of the soft-pions could be properly reconstructed. The remaining four charged tracks which have to be reconstructed are coming from the weak decay of the $D^0(\overline{D^0})$ meson. Because of the mass difference between kaons and pions, the kaons will be confined in the forward direction, while pions can be found also in the backward region, as shown in Fig.1.b and 1.c.

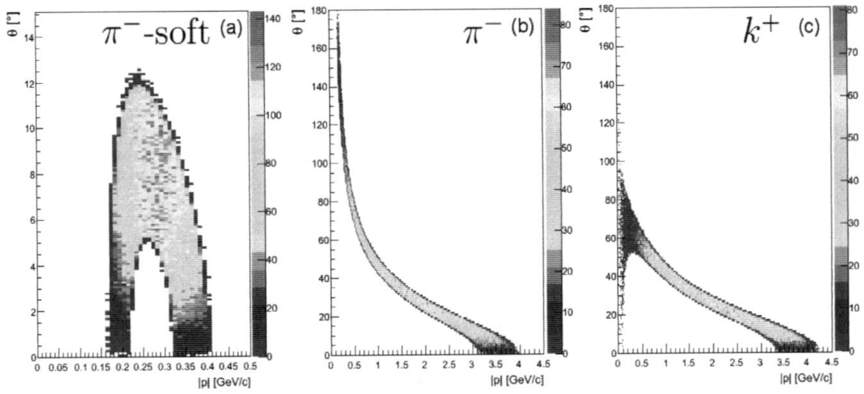

Figure 1. Phase space of the 3 particles coming from the decay of D^{*+}: respectively soft-pions (a), hard-pions (b) and kaons (c).

5. Analysis Strategy and Results

The analysis method presented in this work has been developed and tuned within the PandaRoot (rev.21574) and the FairRoot packages, April-2013.
It must be pointed out that only kinematic constraints are used. In order to suppress the expected huge hadronic background, different cuts on the momentum and mass distribution as well as kinematic and vertex fits have been applied, both on the simulated signal and on the simulated background channels. In order to take into account the detector resolution, limits on the particle momenta were set considering a range of 2σ on the momentum variable, and a tolerance interval of 10% on the angular distributions.
The method consists of four main steps:

- Build the candidates sample of six charged type of particles: K^+, K^-, π^+, π^- and two additional soft-π^+, soft-π^-. Kaons and pions feature different momentum distributions, hence a huge amount of candidates can be filtered out just imposing some general selections.
- Build the candidates sample of D^0-$\overline{D^0}$. The candidate samples are filtered out applying an initial kinematical fit which constraints the mass to the known value m_{D^0}=1.865 GeV/c^2 [3]. Afterward a mass window of $\pm 3\sigma = \pm 30$ MeV/c^2 has been choosen for a cut. Moreover, to refine the selections, further cuts on the momentum distribution have been performed, before applying a vertex fit to identify a common vertex of the Kπ pairs to get rid of the soft pions component. A cut on the $D^0(\overline{D^0})$ distance from the IP of 0.2 cm has been applied. The obtained secondary vertex resolution of neutral D-mesons is $\sim$30 μm in x-y and $\sim$110 μm along the z direction.
- Build the candidates sample of D^{*+}-D^{*-}. The candidate data samples have been filtered within a mass range of $\pm 3\sigma = \pm 36$ MeV/c^2.
- Build the cleanest data sample for $\psi(4040)$ combining the survived $D^{*+}D^{*-}$ pairs. After the candidate sample has been filtered by means of a 4C-fit, which constrains the total initial 4-momentum; in case of multiple candidates, the choice is based on the best χ^2.

The invariant mass of $D^{*+}D^{*-}$ pairs is shown in Fig.2. The peak is centered at 4040.0 MeV/c^2 and the achieved mass resolution is $\sim$1.4 MeV/c^2 , the obtained reconstruction efficiency is 3.3%.
The expected data rate can be calculated using the integrated maximum achievable luminosity of $L_{int} = 3.1 \cdot 10^{39}$cm^{-2} (considering 6 months of data taking) and the estimated cross section of Eq.2. About $6.84{\cdot}10^6$ $\psi(4040)$ should be produced, and around $\sim$4700, should decay into

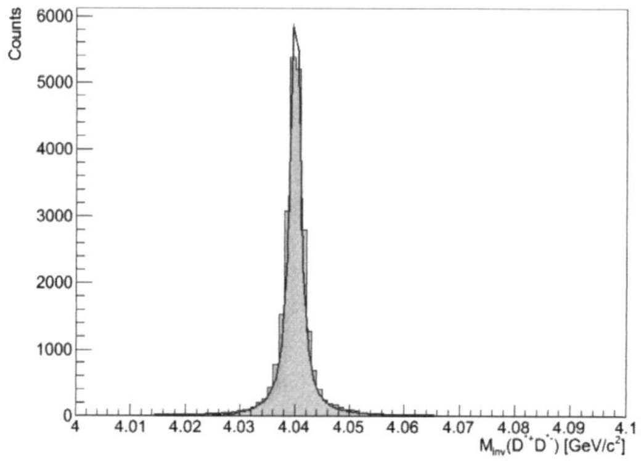

Figure 2. Invariant mass of $D^{*+}D^{*-}$ pairs.

$D^{*+}D^{*-}$. Considering the reconstruction efficiency of the $\psi(4040) \to D^{*+}D^{*-}$ channel, the number of observable $\psi(4040)$ in one year of data taking should be ~160. Using a statistical sample of $1.5{\cdot}10^7$ events for the 2K4π reactions, $2{\cdot}10^7$ for the 6π and 7π channels, and on a data sample of $2.25{\cdot}10^7$ inelastic $p\bar{p}$ reactions, the S/B ratios, reported in Tab.1, were evaluated.

Channel	S/B
2K4π	$> 2.7{\cdot}10^{-2}$
6π	$>4.9{\cdot}10^{-3}$
7π	$\geq 8{\cdot}10^{-4}$
$p\bar{p}$	$> 3.3{\cdot}10^{-5}$

Table 1. Signal to background ratio for the main background sources, a lower limit is listed when no $\psi(4040)$ was reconstructed in the data sample.

6. Conclusion

The $\psi(4040) \to D^{*+}D^{*-}$ decay chain has been investigated in detail. The reconstruction efficiency is found to be 3.3%, all the simulated backgrounds events were successfully suppressed. In particular in this work only the dominant charged decay of the D mesons has been considered; including further decay channels, such as semi-leptonic decay modes, the expected charmed signal yield should significantly improve. Furthermore, the real production cross sections of $D\bar{D}$ may be larger, which would drastically increase the charmed signal yield. The PANDA detector in its given setup will be able to contribute to investigate various physics topics in the charmonium mass region.

References

[1] http://www.fair-center.eu/
[2] PANDA Collaboration, "Physics Performance Report", http://arxiv.org/abs/0903.3905v1
[3] J. Beringer et al. (Particle Data Group), Phys. Rev. D86, 010001 (2012)
[4] G. Goldhaber et al. Phys. Lett., 69B(4), August 1977
[5] Uzhinsky et al., Cross Sections of Various Processes in Pbar P-Interactions, arXiv:hep-ph/0212369 (2002)
[6] PANDA Collaboration, Micro Vertex Detector Technical Design Report, http://arxiv.org/abs/1207.6581v2
[7] PANDA Collaboration, Straw Tube Tracker Technical Design Report, Eur. Phys. J. A (2013) 49: 25

A new method for electron momentum reconstruction in the $\bar{\text{P}}$ANDA experiment

Binsong MA on behalf of the PANDA Collaboration
Institut de Physique Nucléaire, CNRS-IN2P3 and Université PARIS-SUD, 91406 ORSAY FR
E-mail: binsong@ipno.in2p3.fr

Abstract. The Kalman Filter in existing PANDARoot framework of $\bar{\text{P}}$ANDA experiment is not optimally suited for electrons, for which the highly non-Gaussian Bremsstrahlung process yields a tail in the momentum resolution distribution. A new method was therefore developed to improve the electron momentum reconstruction with an event by event procedure. The improvements of the electron momentum resolution will be shown. The interest of the method for the electromagnetic channels studies will also be presented.

1. Introduction

The antiproton beam of FAIR (Facility for Anti-proton and Ion Research), under construction at Darmstadt and the $\bar{\text{P}}$ANDA (Anti-Proton ANnihilation at DArmstadt) detector [1] will offer an exciting opportunity for hadronic physics studies in $\bar{\text{p}}$p annihilation reactions. In particular, the $\bar{\text{P}}$ANDA experiment is well suited to investigate the electric and magnetic time-like proton form factors in the $\bar{p}p \to e^+e^-$ reaction at high q^2 , which currently lacks precise data [2].

Other electromagnetic channels of the $\bar{\text{p}}$p reaction can be used to investigate the proton structure, for example the $e^+e^-\pi^0$ [3] or $J/\Psi\pi^0$ [4] which give access to the proton-π^0 or antiproton-π^0 Transition Distribution Amplitudes (TDA).

Feasibility studies of measurements of both time-like electromagnetic form factors of the proton [5, 6, 7] and of TDA using the $e^+e^-\pi^0$ channel [8] have been performed. A major challenge of the measurements of these electromagnetic channels is the huge hadronic background, which requires excellent identification as well as an optimum angular and momentum resolutions for the reconstructed electrons.

2. $\bar{\text{P}}$ANDA sub-detectors and the tracking system

$\bar{\text{P}}$ANDA is a fixed target experiment, and its spectrometer can be divided into two parts: a forward spectrometer with covering angle lower than $10°(5°)$ in horizontal (vertical) direction and a target spectrometer at larger angles. Both parts consist of tracking detectors, detectors for charged particle identification, electromagnetic calorimeter (EMC) and muon detection systems. PANDARoot, a framework based on ROOT and Virtual Monte Carlo [9], is used for simulation and reconstruction.

The tracking system of $\bar{\text{P}}$ANDA includes three types of detectors: Straw Tube Tracker (STT), Micro Vertex Detector (MVD) and Gas Electron Multiplier (GEM) [10].

In the following part, we will focus on the algorithm of track reconstruction in the central tracking system. First, the measured points which belong to the same track are selected (pattern recognition). Then, these points are fitted by a helix parametrization (prefit). Next, a Kalman Filter [11] is applied to refine the parameters provided by the prefit. The Kalman Filter is applied in three steps: first the track parameters and the associated errors at the next plane of detector are predicted, using the knowledge at the previous plane. Then a weighted mean between the predicted and the measured values is calculated taking into account the corresponding errors. Eventually, the final parameters of the track at the vertex are deduced from the backward propagation of the particle. For the propagation of the particle in the first and the third steps, the GEANE track follower [12, 13], which takes into account energy loss by ionization and the angular deviation due to multiple scattering between two measurement planes.

3. Electron reconstruction and the Bremsstrahlung correction method

Unlike other charged particles for which the dominant causes of worsening momentum and angular resolution when going through the material of the detectors are ionization and multiple-scattering. The electron tracking resolution is strongly affected by Bremsstrahlung photon emission in the material of the tracking system.

3.1. The electron tracking problem

Simulation and reconstruction of electrons and muons with a momentum equal to 1 GeV/c with PANDARoot show that the momentum resolution distribution of electrons has a large tail and the center of the peak is also shifted to positive value (0.4%) compared to the result for muons. The proportion of events inside two standard deviations of the Gaussian fit of the resolution spectra is only about 60% in the Barrel EMC detection region ($22^\circ < \theta < 140^\circ$) and 45% in the Forward Endcap EMC detection region ($5^\circ < \theta < 22^\circ$) where the material budget is larger (Figure 1).

As the momentum resolution is a crucial parameter for the signal selection in electromagnetic channels, it is important to investigate possible improvements. The existing software with the Kalman Filter as tracking tool uses the mean energy loss and the fluctuation calculated by the track follower GEANE [12, 13]. However, electron energy loss by Bremsstrahlung has a highly non Gaussian distribution, so that the process cannot be easily treated by the Kalman Filter. Alternative methods, like the Gaussian Sum Filter [14] or Dynamic Noise Adjustment (DNA) [15] are sometimes used. As explained further in more details, we follow a different procedure based on an event by event treatment.

The scheme presented in figure 2 shows an electron produced at the target emitting a Bremsstrahlung photon when it travels through the tracking system (the STT and the MVD). Due to the energy loss, the curvature of electron track increases after the γ emission.

Most of the Bremsstrahlung photons are emitted in the MVD which is close to the target. For an electron emitted at 90° the average thickness of MVD is about 0.07 radiation length, about a factor of 4 more than for the STT. So as result, there are four to five times more photons emitted in MVD than in STT. In addition, the MVD provides only four to six points to the track, depending on the θ and ϕ direction, which is much less than the twenty five points provided by the STT. The track is therefore mostly defined by the STT points which are mainly measured after the γ emission. Therefore, in most cases, the Kalman Filter will give a reconstructed electron momentum (P_{KF}) close to the momentum after the γ emission.

3.2. Bremsstrahlung correction method

A new attempt to handle the problem of Bremsstrahlung event by event using the energy of photons measured in the EMC is proposed. In this method, for each event, a photon will be searched for in an angular window around the electron track, then the reconstructed momentum (P_{rec}) will be corrected by adding the photon energy (E_γ).

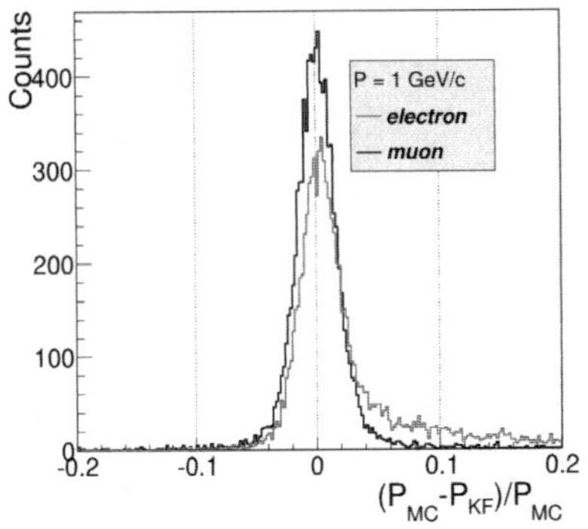

Figure 1. Comparison of electron (red) and muon (black) momentum resolution for P=1 GeV/c and θ=90° . P_{MC} is the Monte-Carlo momentum, P_{KF} is the reconstructed momentum.

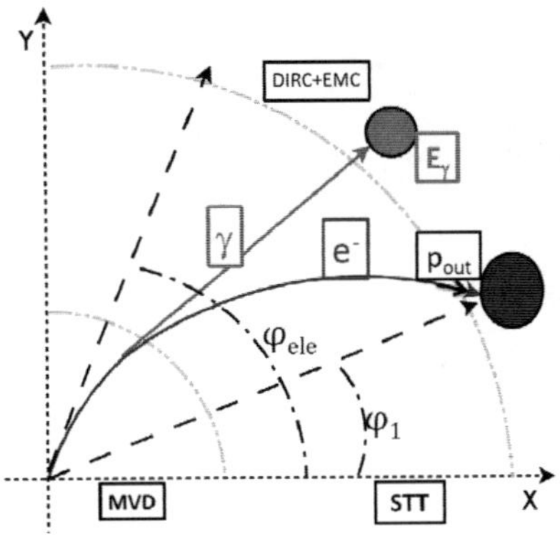

Figure 2. Scheme of an electron track with emission of photon in the $\bar{\text{P}}$ANDA central tracking system.

The existing algorithm identifies "bumps" in the distribution of energy deposits in the EMC crystals which correspond to a set of neighboring hit crystals. Photon signals are identified as those bumps which do not match with a track.

Two cases can be considered. The first one corresponds to low energy electrons (transverse momentum p_T below 1 GeV) with large curvature. If the photon is emitted at the beginning of the track, the "bumps" of electron and photon in EMC are separated. In the second case, the curvature of the electron track is not large enough, the photon bump is merged in the electron bump and the existing software can not distinguish between the two signals. For these two cases, different methods to pick up the photon energy are used.

The new algorithm first looks for photon bumps, in order to find possible Bremsstrahlung photons emitted in the tracking system. As the Bremsstrahlung photons are emitted close to the electron direction, angular information can be used. Since the magnetic field is parallel to the beam axis, only the azimuthal angle of the momentum vector of the electron is changing along the track. The photon bump has therefore to be found with a polar angle θ_γ close to the reconstructed electron one θ_{ele}. The azimuthal angle ϕ is expected to be between the reconstructed electron value ϕ_{ele} and the value at the exit of the tracking system ϕ_1 (see figure 2). The chosen limits for these angles are indicated as a rectangle in figure 3. The energy of the selected photons is then added to the reconstructed electron momentum.

To take into account the case of γ merged into the electron bumps, another algorithm based on the azimuthal distribution of the energy deposited in the EMC crystals in the electron bumps was developed. In the example of figure 4, two local maximums can be recognized in the energy distribution of a negative electron. As expected, the smaller one on the left part is due to a Bremsstrahlung photon and its energy can be extracted and added to the electron reconstructed momentum.

3.3. Improvement of the electron momentum resolution

As the geometry of EMC is not the same for the Barrel and the Forward Endcap, this method is tested separately for these two regions.

Figure 5 shows the new electron momentum resolution after correction for a transverse momentum p_T =1 GeV/c and 2 GeV/c. The result obtained when only the separated γ and electron bumps are used is also displayed as a blue line in figure 5. It can be noticed that

Journal of Physics: Conference Series **503** (2014) 012008 doi:10.1088/1742-6596/503/1/012008

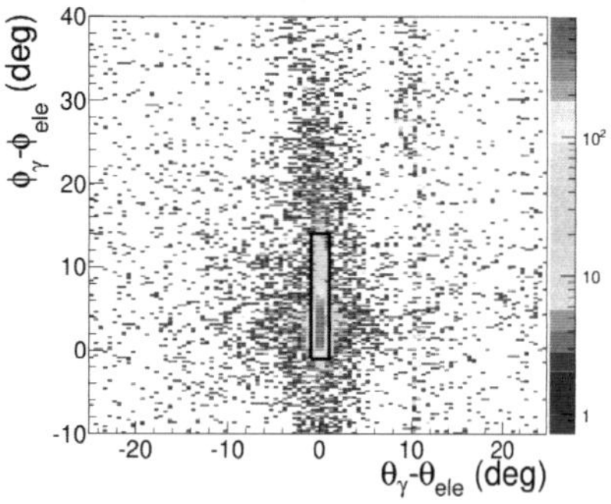

Figure 3. Correlation between the difference azimuthal and polar angles of the γ bumps and the electron track for an electron of $p_T = 1$ GeV/c. The rectangle shows the limits for the selection of Bremsstrahlung photon.

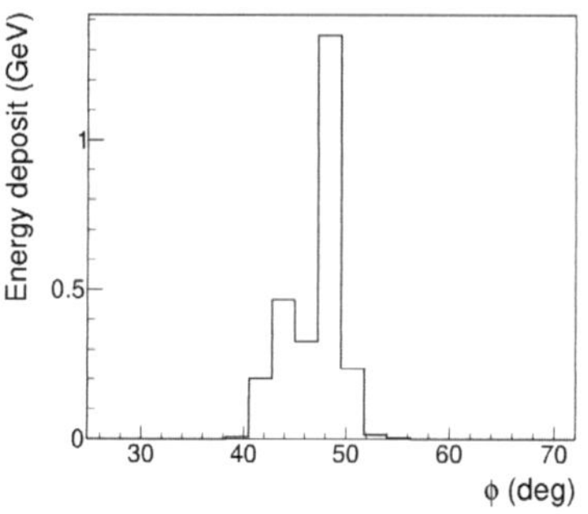

Figure 4. An example of the azimuthal distribution of electron energies deposited in the EMC crystals for an e^- bump with p_T=1 GeV/c.

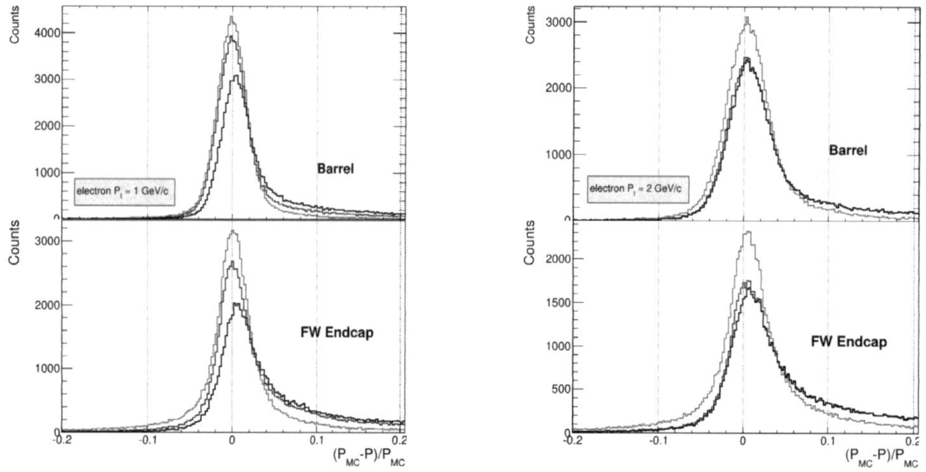

Figure 5. Comparison of the electron momentum resolution before (black) and after (red) using the Bremsstrahlung correction method. The blue line shows the result when only the separated γ bumps are used.

the analysis of the shapes is most important, especially for the large p_T. The method with combination of the two cases of photon detection (separated and merged photon and electron bumps) can strongly reduce the tail of the resolution peak, and the shift of the peak can also be fixed. After the correction, the proportion of events inside 2σ of the Gaussian fit is increased by a factor of 1.33 in the Barrel region and of 1.8 in the Forward Endcap region.

4. Application of the method on electromagnetic channels

For the proton structure studies, an important issue is the rejection of the hadronic background. Both Particle Identification (PID) and kinematical cuts are used. A larger efficiency of the kinematical cuts can be achieved by using the Bremsstrahlung correction method.

Two examples are taken to show the advantage of the method.

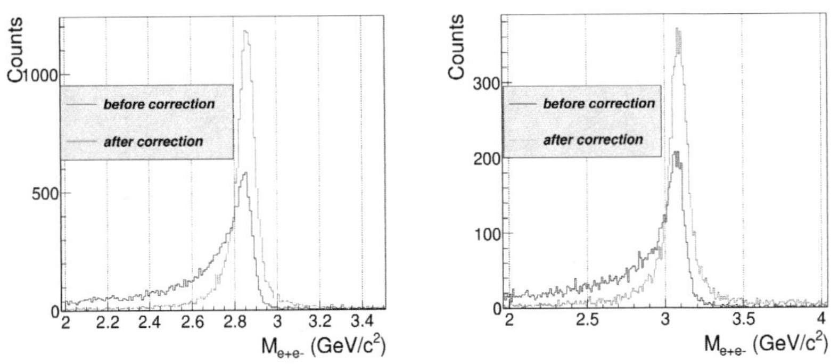

Figure 6. Distribution of the e^+e^- invariant mass before (blue) and after (red) using the method for $\bar{p}p \to e^+e^-$ (left) and $\bar{p}p \to J/\Psi\pi^0$ (right) reactions.

The first one is the proton time-like form factor measurement in the $\bar{p}p \to e^+e^-$ reaction with an antiproton beam momentum of 3.3 GeV/c. Another one is the Transition Distribution Amplitude measurement in the $\bar{p}p \to J/\Psi\pi^0$ reaction with an antiproton beam momentum of 5.51 GeV/c.

Figure 6 shows the distribution of e^+e^- invariant mass after simulation and reconstruction with PANDARoot. For the $\bar{p}p \to e^+e^-$ reaction, a selection of e^+e^- invariant mass larger than 2.72 GeV/c^2 (corresponding to $\sqrt{s} - m_{\pi_0}$) will be applied to suppress the background from the $\bar{p}p \to \pi^+\pi^-\pi^0$ reaction, in addition to PID cuts. The efficiency of this cut on the signal ($\bar{p}p \to e^+e^-$) selection is increased by a factor 1.7 when using the new method.

The second one is TDA measurement by using the $\bar{p}p \to J/\Psi\pi^0$ reaction, the e^+e^- invariant mass is used for the J/Ψ reconstruction. When using the new method, the signal yield within two standard deviations is also increased by a factor of 1.6.

5. Conclusion

The proton structure studies with $\bar{\text{P}}$ANDA need an optimized electron momentum resolution to limit the huge hadronic background. A new method to correct the Bremsstrahlung effect based on an event by event treatment and using the photon detection in the EMC is proposed. The test shown for the electromagnetic form factors and TDA measurements demonstrate that the method is well suited to handle the problem. This method will be very useful for the signal selection and background rejection in the electromagnetic channels.

[1] Physics Performance Report for PANDA, arXiv:0903.3905, hep-ex, (2009).
[2] J. Lees *et al.*, Phys.Rev. **D87**, 092005 (2013).
[3] J. Lansberg *et al.*, Phys.Rev. **D86**, 114033 (2012).
[4] B. Pire *et al.*, Phys.Lett. **B724**, 99 (2013).
[5] M. Sudol *et al.*, Eur.Phys.J. **A44**, 373 (2010).
[6] J. Boucher, Ph.D. thesis, University of Paris-Sud, France, (2011).
[7] A. Dbeyssi, Ph.D. thesis, University of Paris-Sud, France, (2013).
[8] M. Mora-Espi, Ph.D. thesis, Mainz university, Germany, (2012).
[9] I. Hřivnáčová *et al.*, Proc. of Computing in High Energy and Nuclear Physics, (2003).
[10] W. Erni *et al.*, Eur.Phys.J. **A49**, 25 (2013).
[11] R. E. Kalman, Journal of Basic Engineering 35 (1961).
[12] V. Innocente, M.Maire, and E.Nagy, GEANE, CERN Program Library, (1991).
[13] A. Fontana *et al.*, J.Phys.Conf.Ser. **119**, 032018 (2008).
[14] W. Adam *et al.*, J. Phys G: Nucl. Part. Phys. 31, 9 (2005).
[15] M. Elsing *et al.*, ATLAS Detector Paper Back-Up Note: Electrons and Photons, ATLAS NOTE, (2008).

QCD phase diagram: overview of recent lattice results

Gergely Endrődi

Institute for Theoretical Physics, Universität Regensburg, D-93040 Regensburg, Germany.

E-mail: gergely.endrodi@physik.uni-regensburg.de

Abstract. Two parameters that have a strong influence on the finite temperature QCD transition, and play an important role in various physical scenarios are the quark density and the external magnetic field. The effect of these parameters on the thermal properties of QCD is discussed, and an overview of the latest lattice results is given.

1. Introduction

Quantum Chromodynamics (QCD) is the theory of the strong interactions. Its elementary particles – quarks and gluons – cannot be observed directly at low energies, as they are confined to be constituents of bound states (hadrons). However, QCD is also an asymptotically free theory, which implies that at high energies quark and gluons are deconfined, and through a phase transition the quark-gluon plasma (QGP) phase is formed. This transition between the hadronic and QGP phases can be represented on the phase diagram of QCD. A conjecture of how it might look is depicted in Fig. 1.

To trigger deconfinement – and, thus, to probe the high-energy regime of QCD – an extreme environment is necessary, realized by high temperatures T or baryon densities ρ. The characteristic values where the transition happens correspond to $T_c \approx 10^{12}$ K and $\rho_c \approx 10^{15}$ g/cm^3. Temperatures of this strength were present in the early phase of the evolution of the universe, at about $1\mu s$ after the Big Bang. On the other hand, densities exceeding the above critical value are also conjectured to be present in the interior of compact, dense stellar objects, neutron stars. In both cases, strongly interacting matter enters its quark-gluon plasma phase and, thus, the properties of the transition between the hadronic and QGP phases are essential for the description of these systems. Furthermore, the quark-gluon plasma is also routinely reproduced in contemporary heavy-ion collisions. One relevant parameter here is the center-of-mass energy $\sqrt{s}$, which controls the temperature and the density of the created fireball. The Relativistic Heavy Ion Collider (RHIC) and the Large Hadron Collider (LHC) operate at high energies,

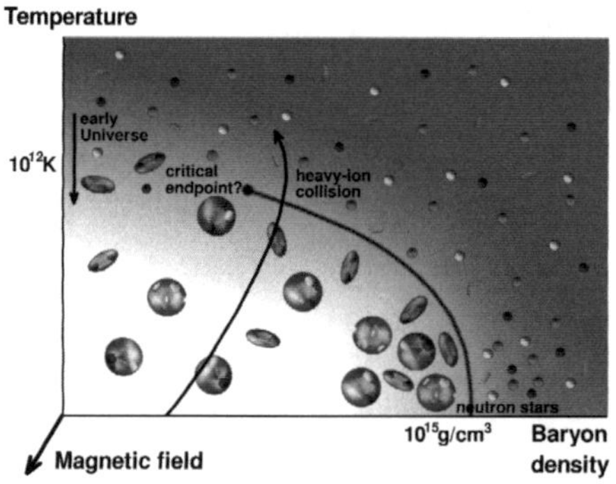

Figure 1. Conjectured QCD phase diagram.

where the initial excess of quarks over antiquarks is negligible compared to the total number of created particles, resulting at low net baryon densities. On the other hand, various experiments at lower energies aim to study the system at large net densities, for example RHIC II or the future Facility for Antiproton and Ion Research (FAIR). One of the most important objectives in the latter experiments is to identify a possible critical endpoint on the phase diagram, beyond which the transition becomes first order. Developing our theoretical understanding of the QCD phase diagram can prove highly beneficial for designing these next generation experiments.

Besides the temperature and the baryon density, there are additional parameters, which may be relevant for the above mentioned systems. Such a parameter is a background magnetic field B, which represents an important concept for early cosmology [1], for a certain class of neutron stars called magnetars [2] and for non-central heavy-ion collisions [3]. Model descriptions of such collisions show that the induced magnetic field for typical events at RHIC or at the LHC may be of the order of several m_π^2. A magnetic field of this strength competes with the strong interactions and induces several new phenomena, for example it affects chiral symmetry and its restoration and (de)confinement, it introduces a spatial asymmetry and, thus, induces Lorentz-symmetry breaking expectation values and it also changes the hadron spectrum considerably. For recent reviews on the subject, see, e.g., Refs. [3, 4]. For the phase diagram, the magnetic field introduces a third axis and defines the three-dimensional phase diagram of QCD, see Fig. 1.

Our current knowledge about the phase diagram is still very limited. In fact, most of the information illustrated in Fig. 1 is based on descriptions of strongly interacting matter in terms of low energy hadronic models (low temperatures, densities or magnetic fields) or perturbation theory (high temperatures, densities or magnetic fields). However, in order to study the transition region, where the hadronic description is no longer valid, but perturbation theory is not yet feasible due to the large QCD coupling, one needs to take into account the non-perturbative nature of QCD. The best non-perturbative and systematically adjustable approach to QCD is lattice field theory, where the field degrees of freedom are discretized on a four-dimensional space-time grid. Through the lattice approach we can study the strong dynamics solely from first principles – i.e., from the QCD Lagrangian.

The primary quantity in this approach is the QCD partition function (in the grand canonical formalism, where the density is controlled by a chemical potential μ), which is given by the functional integral over the gluon fields U,

$$\mathcal{Z} = \int \mathcal{D}U e^{-S_g} \det\left[\not{D}(B,\mu) + m\right], \tag{1}$$

where S_g is the gluonic action and $\not{D}$ the Dirac operator. The lattice regularization defines $\mathcal{Z}$ through the limiting procedure of considering finer and finer grids. This amounts to increasing the number N_s of lattice points to infinity and taking the lattice spacing a to zero in a manner that the physical size $N_s a$ is fixed. This is called the continuum limit, which requires simulations of bigger and bigger lattices, increasing the computational costs drastically. Another, computationally demanding aspect is the tuning of the bare quark masses m to their physical values. This procedure is performed such that the lattice measurements of physical observables – e.g. hadron masses at zero temperature – coincide with their experimental values. Tuning m in this manner makes the condition number of the matrix $\not{D} + m$ large and, accordingly, the simulation very expensive. Nevertheless, performing the continuum limit and using physical quark masses in the simulation are necessary ingredients to produce reliable precision results that can be compared to experiments or models.

In this talk, I will present recent lattice results about the phase diagram, concentrating first on nonzero chemical potentials, then turning to the case of nonzero magnetic fields.

2. Phase diagram at zero and low chemical potentials

The most important properties of the transition separating the hadronic and QGP phases are the characteristic temperature T_c and the order of the transition. To determine these, an observable sensitive to the transition is necessary. Such observables are the chiral condensate $\bar{\psi}\psi = \partial \log \mathcal{Z}/\partial m$ and the chiral susceptibility $\chi = \partial^2 \log \mathcal{Z}/\partial m^2$. The latter is similar to a specific heat and exhibits a peak-like structure in the transition region. The transition temperature is determined as the position of the peak, whereas the order of the transition can be extracted via a finite size scaling analysis of the height h of the peak: $h \propto V^\alpha$, where $\alpha = 1$ indicates a first-order, $0 < \alpha < 1$ a second-order phase transition, and $\alpha = 0$ an analytic crossover, which involves no singularity. Such a finite size scaling analysis was performed in Ref. [5] revealing the transition at $\mu = B = 0$ to be an analytic crossover. One implication of this is that T_c is not unique but can depend on the observable used for its definition. The transition temperature defined via various observables was determined using different discretizations of the fermionic action [6, 7, 8] [9], giving values scattered around $T_c \approx 160$ MeV.

The determination of these characteristics at $\mu > 0$ is, unfortunately, much more difficult. The lattice approach relies on the importance sampling of the partition function of Eq. (1) based on the weight $e^{-S_g} \det[\not{D} + m]$. This weight, however, becomes complex for any $\mu > 0$, rendering standard Monte-Carlo methods useless. Several methods have been developed to alleviate this complex action problem (see, e.g., Refs. [10, 11]), still, they are only applicable at low μ. One possibility is to perform a Taylor-expansion in the chemical potential at $\mu = 0$. The leading-order dependence on the chemical potential is encoded in the quark number susceptibilities $\chi_2^f = \partial^2 \log \mathcal{Z}/\partial \mu_f^2$, which are closely related to fluctuations of conserved charges and, as such, can be used for a comparison with heavy-ion collisions (f=u,d,s labels quark flavors). The left panel of Fig. 2 shows the continuum limit of χ_2^u in the transition region, as obtained in Ref. [12]. The dependence of T_c on μ can also be extracted via correlators of certain observables with χ_2^u. The expansion in the chemical potential reads $T_c(\mu) = T_c(0) \cdot [1 - \kappa \cdot \mu^2/T_c^2(0)]$. Due to the crossover nature of the transition, the coefficient κ also depends on the observable used for its definition. Using $\bar{\psi}\psi$ and χ_2^s we obtained the values [13] corresponding to the phase diagram shown in the right panel of Fig. 2. The results are in agreement with Ref. [14], where a different fermionic discretization was used. Note that equation of state-related observables (pressure, energy density, etc.) [15] tend to give somewhat larger values for the curvature [16].

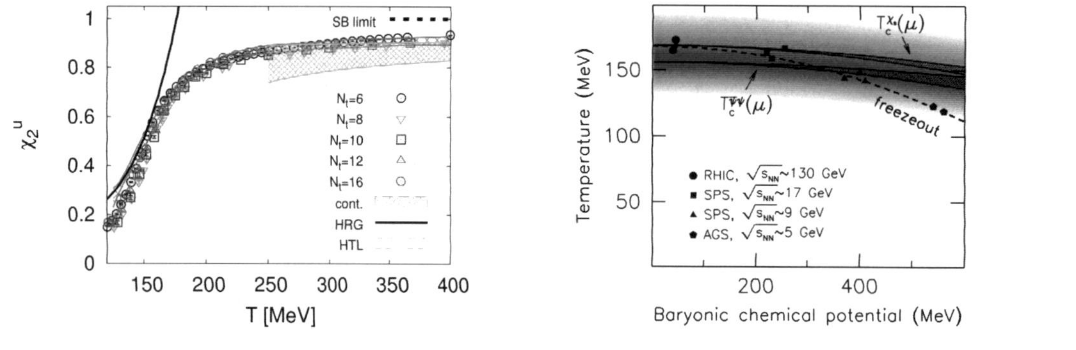

Figure 2. Left panel: continuum limit of the up quark number susceptibility compared to a hadronic model (solid line) and perturbation theory (blue dashed region). Right panel: phase diagram at low μ with two definitions of T_c, compared to the experimental freezeout curve [17].

In Ref. [13] we also determined the strength of the transition to leading order, implying a mild weakening of the transition in the low-μ region. This suggests to exclude the possibility of a critical endpoint at these low values of the chemical potential.

3. Phase diagram at nonzero magnetic fields

Next, I turn to present recent results about the phase diagram at $\mu = 0$ and $B > 0$. Here we consider a constant external magnetic field, i.e. the dynamics of B is neglected, nevertheless, the B-dependence of the fermion determinant in Eq. (1) is taken into account in the simulation. This way the magnetic field couples directly to quarks, but it also changes the gauge backgrounds, revealing its indirect coupling to gluons. It is important to stress here that – contrary to the case of nonzero chemical potentials – the magnetic field does not lead to a complex action problem, therefore, the full $T - B$ phase diagram is accessible for lattice simulations.

A pronounced effect of the magnetic field on QCD dynamics is the *magnetic catalysis* of the quark condensate, which implies $\bar{\psi}\psi$ to increase with growing B at zero temperature. Magnetic catalysis at $T = 0$ is a well-established phenomenon; it was observed in various effective theories and models, and confirmed by lattice simulations (see the recent review [18]). On the other hand, contrary to most of the low-energy models, our continuum extrapolated lattice simulations at physical quark masses have also shown that $\bar{\psi}\psi$ exhibits a non-monotonic dependence on the magnetic field around T_c, with a certain region, where it actually decreases with growing B [19]. This behavior – which we named *inverse magnetic catalysis* – operates in a way that the inflection point of the condensate (which we use to define T_c here) is shifted towards smaller temperatures (see left panel of Fig. 3). As a result, the transition temperature $T_c(B)$ is reduced as the magnetic field increases, see right panel of Fig. 3, where the phase diagram is shown.

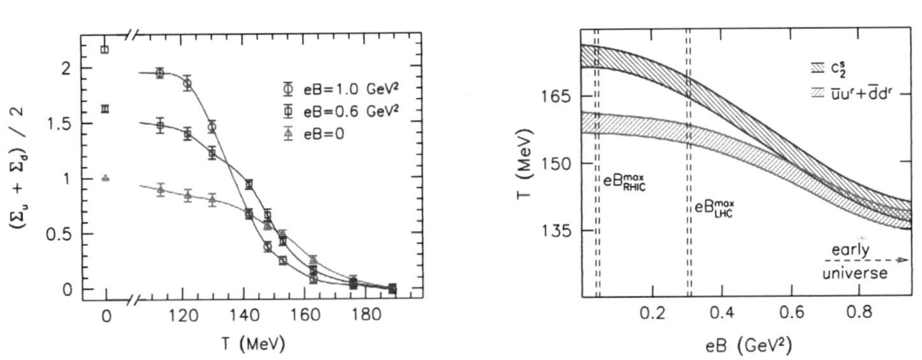

Figure 3. Left panel: catalysis of $\bar{\psi}\psi$ at low temperatures and inverse catalysis around T_c. Right panel: phase diagram in the $T - B$ plane, with T_c defined as the inflection point of the condensate (red curve) or that of the strange quark number susceptibility (blue curve).

Both the inverse catalysis around T_c and the reduction of $T_c(B)$ were rather unexpected, since most low-energy effective theories and models predicted the magnetic catalysis to prevail for all temperatures, and, $T_c(B)$ to increase. This prompted us to study the mechanism behind inverse catalysis in more detail. In Refs. [19, 20] we found that this behavior is peculiar for light quarks, and it disappears for heavier-than-physical quark masses (see also Ref. [21]). Moreover, we also found the inverse catalysis to be related to the indirect coupling of the magnetic field to gluons. There is a straightforward way to distinguish between the quark (direct) and gluonic (indirect) contributions to the condensate in terms of the partition function, Eq. (1) [22]. The expectation value of $\bar{\psi}\psi$ reads

$$\langle \bar{\psi}\psi \rangle = \frac{1}{\mathcal{Z}} \int \mathcal{D}U e^{-\beta S_g} \underbrace{\det[\not{D}(B) + m]}_{\text{indirect}} \underbrace{\mathrm{Tr}[\not{D}(B) + m]^{-1}}_{\text{direct}}. \quad (2)$$

The direct part is the dependence of the operator on B, whereas the indirect part is the dependence through the determinant. We found that the direct term always increases the expectation value, whereas the indirect contribution tends to decrease it around T_c, if the quark mass m

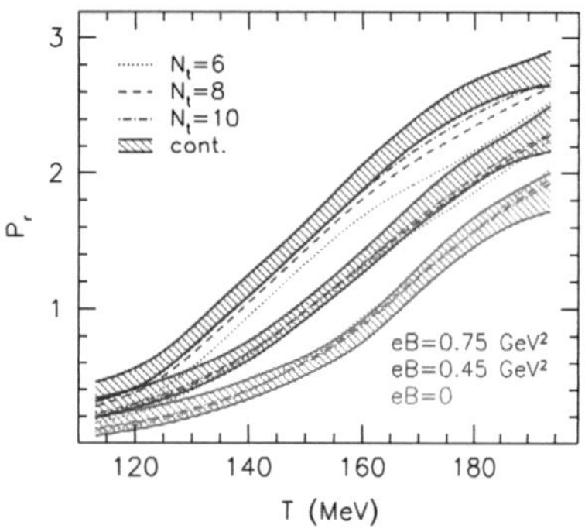

Figure 4. Continuum limit of P as a function of T for various magnetic fields.

is small [23]. It is then natural to look for the relevant degrees of freedom in the gauge background that this indirect effect couples to. The most important gauge degree of freedom around T_c is the Polyakov loop, the parallel transport winding around the temporal direction of the lattice $P = \mathrm{Tr}\exp\left[\int_0^{1/T} A_4\,\mathrm{d}t\right]$. Therefore, we proceeded by determining it in the continuum limit for various magnetic fields, see Fig. 4. The results show that P depends predominantly on the magnetic field around T_c, namely it is increased by B. At the same time, the inflection point of P is shifted to lower temperatures, revealing that around T_c, the magnetic field favors gauge configurations corresponding to the deconfined phase, i.e. with small condensate. This leads to the inverse catalysis of $\langle\bar\psi\psi\rangle$ in the transition region.

4. Summary

In this talk I discussed the effect of chemical potentials and background magnetic fields on QCD thermodynamics. The former introduces a complex action problem and prohibits direct simulation of the theory. To circumvent this problem we applied a Taylor-expansion in μ. Our continuum extrapolated results at physical quark masses reveal the phase diagram for low chemical potentials to be as depicted in the right panel of Fig. 2. On the other hand, for $\mu = 0$ and $B > 0$, the weights remain real, allowing for the determination of the whole phase diagram by conventional lattice techniques. A pronounced effect here is the inverse magnetic catalysis around T_c, and the reduction of the transition temperature with growing B, see Fig. 3.

Acknowledgments. I would like to thank the FAIRNESS 2013 organizers for the inspiring meeting. I am grateful to my colleagues with whom I collaborated on various projects presented in this talk. My work was supported by the EU (ITN STRONGnet 238353).

[1] Grasso D and Rubinstein H R 2001 *Phys.Rept.* **348** 163–266 (*Preprint* astro-ph/0009061)
[2] Duncan R C and Thompson C 1992 *Astrophys. J.* **392** L9
[3] Kharzeev D E, Landsteiner K, Schmitt A and Yee H U 2013 *Lect.Notes Phys.* **871** 1–11 (*Preprint* 1211.6245)
[4] Bali G, Bruckmann F, Constantinou M, Costa M *et al.* 2012 *PoS* **ConfinementX** 198 (*Preprint* 1301.5826)
[5] Aoki Y, Endrődi G, Fodor Z, Katz S D and Szabó K K 2006 *Nature* **443** 675 (*Preprint* hep-lat/0611014)
[6] Aoki Y, Fodor Z, Katz S and Szabó K 2006 *Phys.Lett.* **B643** 46–54 (*Preprint* hep-lat/0609068)
[7] Aoki Y, Borsányi S, Dürr S, Fodor Z, Katz S D *et al.* 2009 *JHEP* **0906** 088 (*Preprint* 0903.4155)
[8] Borsányi S *et al.* 2010 *JHEP* **1009** 073 (*Preprint* 1005.3508)
[9] Bazavov A *et al.* 2012 *Phys.Rev.* **D85** 054503 (*Preprint* 1111.1710)
[10] de Forcrand P 2009 *PoS* **LAT2009** 010 (*Preprint* 1005.0539)
[11] Aarts G 2012 *PoS* **LATTICE2012** 017 (*Preprint* 1302.3028)
[12] Borsányi S, Fodor Z, Katz S D, Krieg S, Ratti C *et al.* 2012 *JHEP* **1201** 138 (*Preprint* 1112.4416)
[13] Endrődi G, Fodor Z, Katz S D and Szabó K K 2011 *JHEP* **04** 001 (*Preprint* 1102.1356)
[14] Kaczmarek O *et al.* 2011 *Phys.Rev.* **D83** 014504 (*Preprint* 1011.3130)
[15] Borsányi S, Endrődi G, Fodor Z, Jakovác A, Katz S D *et al.* 2010 *JHEP* **1011** 077 (*Preprint* 1007.2580)
[16] Borsányi S, Endrődi G, Fodor Z, Katz S, Krieg S *et al.* 2012 *JHEP* **1208** 053 (*Preprint* 1204.6710)
[17] Cleymans J and Redlich K 1998 *Phys.Rev.Lett.* **81** 5284–5286 (*Preprint* nucl-th/9808030)
[18] Shovkovy I A 2013 *Lect.Notes Phys.* **871** 13–49 (*Preprint* 1207.5081)
[19] Bali G, Bruckmann F, Endrődi G, Fodor Z, Katz S *et al.* 2012 *JHEP* **1202** 044 (*Preprint* 1111.4956)
[20] Bali G, Bruckmann F, Endrődi G, Fodor Z, Katz S *et al.* 2012 *Phys. Rev.* **D86** 071502 (*Preprint* 1206.4205)
[21] D'Elia M, Mukherjee S and Sanfilippo F 2010 *Phys. Rev.* **D82** 051501 (*Preprint* 1005.5365)
[22] D'Elia M and Negro F 2011 *Phys. Rev.* **D83** 114028 (*Preprint* 1103.2080)
[23] Bruckmann F, Endrődi G and Kovács T G 2013 *JHEP* **1304** 112 (*Preprint* 1303.3972)

Journal of Physics: Conference Series **503** (2014) 012010 doi:10.1088/1742-6596/503/1/012010

Mesons, PANDA and the scalar glueball

Denis Parganlija

Vienna University of Technology, Institute for Theoretical Physics, Wiedner Hauptstr. 8-10/E136, 1040 Vienna, Austria

E-mail: denisp@hep.itp.tuwien.ac.at

Abstract. The non-perturbative nature of Quantum Chromodynamics (QCD) at low energies has prompted the expectation that the gauge-bosons of QCD – gluons – might give rise to compound objects denoted as glueballs. Experimental signals for glueballs have represented a matter of research for various collaborations in the last decades; future research in this direction is a main endeavour planned by the PANDA Collaboration at FAIR. Hence in this article I review some of the outstanding issues in the glueball search, particularly with regard to the ground state – the scalar glueball, and discuss the relevance for PANDA at FAIR.

1. Introduction

Approximately five decades ago a truly remarkable number of new, strongly interacting resonances was discovered leading way to a new direction in the development of nuclear physics (a non-exhaustive list of the relevant articles is presented in Ref. [1] and further work can be traced in the Particle Data Tables [2]). These discoveries coincided with the appearance of many renowned articles in theoretical physics – a selection of which can be found in Ref. [3] – that represented a basis for the development of a theory of strong interactions. The most viable approach was subsequently found to be the one where "the strong interactions are described by an unbroken gauge theory based on tricolored quarks and color octet vector gluons, with color supposed to be entirely confined, so that quarks are fractionally charged" with the suggestion that "a good name for this theory is quantum chromodynamics (QCD)" [4].

Quantum Chromodynamics has various interesting features, one of the most important of which is a scale-dependent coupling [5] of such nature that, at higher energies, it leads to weaker interactions between the relevant degrees of freedom – quarks and gluons. Conversely, at lower energies, quarks and gluons are confined into more macroscopic structures: hadrons. According to the total spin J, hadrons are divided into mesons (that possess integer spin) and baryons (that possess half-integer spin). Determination of hadron properties from Quantum Chromodynamics is a highly non-trivial task due to the inapplicability of perturbative approaches in the energy region where hadrons emerge. A circumvention of this problem is introduced by effective models containing the so-called "constituent" quarks and gluons that, in the particular case of mesons, allow for the particle classification as quarkonia ($\bar{q}q$, see Ref. [6] and references therein), tetraquarks ($\bar{q}\bar{q}qq$ [7]), hybrids (constituent quarks with excited gluon degrees of freedom [8]) and others.

The non-Abelian nature of QCD implies gluon self-interaction that is governed by the same scale-dependent coupling as in the quark sector. Hence the natural expectation is that, analogously to quarks, the gluons also bind into composite objects – the glueballs [9].

There are at least two important reasons to consider glueballs as low-energy degrees of freedom in QCD:

- Glueball mass is generated exclusively by the gluon self-interaction. The Higgs mechanism does not contribute to the mass generation of glueballs rendering them uniquely suitable for the research of the strong interaction.
- Gluons are vector particles and hence glueballs possess integer spin. This implies that glueballs belong to the meson spectrum – and therefore the meson spectrum can only be complete if the spectrum of glueballs is known.

As already noted, glueballs have long been subject to experimental and theoretical research. In general, low-energy mesons are mainly produced in these channels [6, 10]:

- antiproton-proton ($\bar{p}$-p) collisions [11]
- proton-proton (p-p) collisions [12, 13]
- electron-positron annihilation into J/ψ or $\varphi(1020)$ and their subsequent decay into further meson resonances [14]
- pion-nucleon scattering reactions [13, 15].

A glueball state is expected to be preferably produced in radiative decays and to possess a strongly suppressed $\gamma\gamma$ decay (or a strongly suppressed production in the $\gamma\gamma$ collisions).

On the theory side, there are various means suitable for the glueball identification. These range from effective models [9, 16] to lattice QCD [17] and holographic-QCD approaches [18].

However, a clear experimental signal for a glueball can only be obtained if the resonance has a strongly suppressed mixing to other states with the same quantum numbers. Otherwise the glueball search can become massively complicated, as we illustrate in the following on the example of the scalar glueball.

2. The scalar glueball

Scalar mesons are QCD degrees of freedom that possess quantum numbers $J^{PC} = 0^{++}$, where J is the total spin, P represents parity and C is the charge conjugation; mesons with isospin $I = 0$ are denoted as isoscalars.
PDG data suggest the existence of five $IJ^{PC} = 00^{++}$ states in the energy region up to 1.8 GeV: $f_0(500)$ or σ, $f_0(980)$, $f_0(1370)$, $f_0(1500)$ and $f_0(1710)$ [2]. At the level of quantum numbers, all the mentioned states are viable candidates for the scalar glueball. Then the glueball should be identified by analysing (*i*) the production mechanism and (*ii*) the decay channels.

For the production mechanism, the situation is as follows [2]:

- Produced in radiative decays: fulfilled by $f_0(980)$, $f_0(1500)$ and $f_0(1710)$.
- Not seen in $\gamma\gamma$ collisions: fulfilled by $f_0(1500)$ and $f_0(1710)$.

Then $f_0(1500)$ and $f_0(1710)$ appear to be the strongest candidates for the scalar glueball since they fulfill both of the above criteria. At this point, let us briefly comment on the above $\gamma\gamma$ data since they are of great importance for the glueball search and also since they allow us to illustrate experimental ambiguities in the f_0 sector. The main reason to include $f_0(1500)$ and $f_0(1710)$ into the above list is the fact that there was a notable absence of both of these states in the

$\gamma\gamma$ data analysed by the ALEPH Collaboration in 2000 [19]. Contrarily, an independent search by the L3 Collaboration – that approximately coincided with the ALEPH results – observed a clear signal for $f_0(1710)$ while no signal was found for $f_0(1500)$ [20]. These are currently the best, and the latest, experimental data containing negative results on photon coupling of the f_0 states, but they do not appear to be conclusive.

Hence the expectation might be that the decay channels of the f_0 resonances should provide us with more clear-cut results for the glueball identification. The reason is that, in the case of the unbroken flavour symmetry, one can calculate exact decay-width ratios valid for a glueball state without mixing to states containing quarks ("pure glueball"). Then, for example, the ratio of the pion decay width to the kaon decay width is 3/4.
Unfortunately, neither the data on $f_0(1500)$ nor those on $f_0(1710)$ follow this expectation:

- The $f_0(1500)$ resonance decays predominantly into pions (2π and 4π) whereas the kaon decay is suppressed; the ratio $\Gamma_{f_0(1500)\to\pi\pi}/\Gamma_{f_0(1500)\to KK} \simeq 4$ [2].
- The $f_0(1710)$ resonance decays predominantly into kaons and the pion decay is suppressed; the ratio $\Gamma_{f_0(1710)\to\pi\pi}/\Gamma_{f_0(1710)\to KK} \simeq (0.2 - 0.4)$, depending on collaboration [6].

These results demonstrate clearly that there is no exclusive glueball state in the scalar sector – one can only identify a resonance that is *predominantly* a glueball. Then, as already indicated, various model calculations have to be utilised in order to study mixing patterns of the pure glueball with $\bar{q}q$, $\bar{q}\bar{q}qq$ and other states [9, 16]. Lattice results [17] are also of extreme importance, especially since the inclusion of dynamical quarks has become more prominent; realistic approaches in holographic QCD can be very helpful to the cause as well [18].

However, there is a further complication: the possible existence of a new scalar state, $f_0(1790)$.

3. The $f_0(1790)$ resonance and PANDA

The BES II Collaboration claimed the existence of the new $f_0(1790)$ resonance in 2004 [21]. The mass and the decay width of the resonance were determined by the Collaboration to be $m_{f_0(1790)} = 1790^{+40}_{-30}$ and $\Gamma_{f_0(1790)} = 270^{+60}_{-30}$ MeV, respectively. The stated large decay width implies a strong overlap with $f_0(1710)$. However, there is a clear point of distinction between the two resonances: as already mentioned, $f_0(1710)$ is reconstructed predominantly in the kaon decay channels whereas $f_0(1790)$ is reconstructed predominantly in the pion decay channels.

The distinction between $f_0(1710)$ and $f_0(1790)$ is further ascertained by their production mechanism that involves J/ψ decays [21]: (*i*) $J/\psi \to \varphi K^+K^-$; (*ii*) $J/\psi \to \varphi\pi^+\pi^-$; (*iii*) $J/\psi \to \omega K^+K^-$; (*iv*) $J/\psi \to \omega\pi^+\pi^-$. The $f_0(1710)$ resonance is reconstructed from the decays (*i*) and (*iii*) whereas the decays (*ii*) and (*iv*) allow for the reconstruction of $f_0(1790)$. Then assuming $f_0(1710)$ and $f_0(1790)$ to be the same resonance leads to a contradiction: such a resonance would have to possess a pion-to-kaon-decay ratio of 1.82 ± 0.33 according to reactions (*i*) and (*ii*) and, simultaneously, the pion-to-kaon-decay ratio < 0.11 according to reactions (*iii*) and (*iv*) [21]. That is obviously not possible – and hence the BES II data suggest $f_0(1710)$ and $f_0(1790)$ to represent two distinct resonances.

Let us now discuss the importance of this statements for the PANDA experiments. The PANDA Collaboration intends to study interactions between antiprotons and fixed-target protons and nuclei in the momentum range of 1.5-15 GeV with the stated goal of exploring the spectroscopy of glueballs, multiquark and exotic states [22].

The search for glueballs is an extremely important endeavour of the Collaboration since, as

already indicated, these states offer a unique insight into the strong interaction. However, there are numerous complications along this path, exhibited for example in the case of the ground state – the scalar glueball.
This state will inevitably have strong mixing with mesons of $\bar{q}q$, $\bar{q}\bar{q}qq$ and other structures. Nonetheless, it is extremely important to identify the scalar glueball since otherwise the glueball spectrum would not be complete. In this case, however, one must bear in mind that the most basic condition for the identification is to ascertain *the number* of the scalar states. In the light of the BES II data, it is then of extreme importance for the PANDA Collaboration to search for, and confirm – or disprove – the existence of, the $f_0(1790)$ resonance since, if this resonance exists, it will inevitably overlap with the scalar glueball.

4. Conclusions

Glueball search is one of the most interesting challenges for the future PANDA experiments. These particles are expected to be generated by the strong interaction and therefore the determination of their spectrum is of extreme importance for our understanding of low-energy QCD. The determination of the ground, scalar, state is imperative for the completeness of the glueball spectrum. Due to the mixing of the pure glueball state with other mesons that have the same quantum numbers, the scalar glueball can only be reliably identified if all scalar mesons (f_0 resonances) are known. At present, it is not clear whether this is the case: the existence of a new resonance denoted as $f_0(1790)$ has been suggested by the BES II data locating $f_0(1790)$ exactly in the energy region where the scalar glueball is expected. Hence it is very important for the existence of $f_0(1790)$ to be confirmed or disproved – and the PANDA Collaboration is in a unique position to perform that task.

Acknowledgments

I am very grateful to D. Bugg, F. Giacosa, S. Janowski, P. Kovacs, A. Rebhan, D. H. Rischke and Gy. Wolf for extensive discussions regarding meson spectroscopy.

References

[1] L. D. Roper, Phys. Rev. Lett. **12**, 340 (1964); G. T. Hoff, Phys. Rev. Lett. **12**, 652 (1964); J. H. Christenson, J. W. Cronin, V. L. Fitch and R. Turlay, Phys. Rev. Lett. **13**, 138 (1964). H. C. DeStaebler, E. F. Erickson, A. C. Hearn and C. Schaerf, Phys. Rev. **140**, B336 (1965); B. H. Bransden, P. J. O'Donnell and R. G. Moorhouse, Phys. Rev. **139**, B1566 (1965); L. Bertanza, A. Bigi, R. Carrara and R. Casali, Nuovo Cim. A **44**, 712 (1966); E. W. Anderson, E. J. Bleser, G. B. Collins, T. Fujii, J. Menes, F. Turkot, R. A. Carrigan and R. M. Edelstein *et al.*, Phys. Rev. Lett. **16**, 855 (1966); K. J. Foley, R. S. Jones, S. J. Lindenbaum, W. A. Love, S. Ozaki, E. D. Platner, C. A. Quarles and E. H. Willen, Phys. Rev. Lett. **19**, 397 (1967); C. M. Ankenbrandt, A. R. Clark, B. Cork, T. Elioff, L. T. Kerth and W. A. Wenzel, Phys. Rev. **170**, 1223 (1968); J. W. Lamsa, N. M. Cason, N. N. Biswas, I. Derado, T. H. Groves, V. P. Kenney, J. A. Poirier and W. D. Shephard, Phys. Rev. **166**, 1395 (1968); R. J. Miller, S. Lichtman and R. B. Willmann, Phys. Rev. **178**, 2061 (1969); S. P. Almeida, J. G. Rushbrooke, J. H. Scharenguivel, M. Behrens, V. Blobel, I. Borecka, H. C. Dehne and J. Diaz *et al.*, Phys. Rev. **174**, 1638 (1968); T. C. Bacon, F. Bomse, T. B. Borak, T. B. Cochran, W. J. Fickinger, E. R. Goza, H. W. K. Hopkins and E. O. Salant, Phys. Rev. Lett. **22**, 43 (1969); A. Shapira, O. Benary, Y. Eisenberg, E. E. Ronat, D. Yaffe and G. Yekutieli, Phys. Rev. Lett. **21**, 1835 (1968).

[2] J. Beringer *et al.* (Particle Data Group), Phys. Rev. D**86**, 010001 (2012) and 2013 partial update for the 2014 edition.

[3] M. Gell-Mann and MLevy, Nuovo Cim. **16**, 705 (1960); Y. Nambu and G. Jona-Lasinio, Phys. Rev. **122**, 345 (1961); T. H. R. Skyrme, Proc. Roy. Soc. Lond. A **260**, 127 (1961); M. Gell-Mann, Phys. Lett. **8**, 214 (1964); F. Englert and R. Brout, Phys. Rev. Lett. **13**, 321 (1964); P. W. Higgs, Phys. Lett. **12**, 132 (1964); G. S. Guralnik, C. R. Hagen and T. W. B. Kibble, Phys. Rev. Lett. **13**, 585 (1964); S. Weinberg, Phys. Rev. Lett. **17**, 616 (1966); M. Gell-Mann, R. J. Oakes and B. Renner, Phys. Rev. **175**, 2195 (1968); S. R. Coleman, J. Wess and B. Zumino, Phys. Rev. **177**, 2239 (1969); J. S. Bell and R. Jackiw, Nuovo Cim. A **60**, 47 (1969); R. P. Feynman, Phys. Rev. Lett. **23**, 1415 (1969); S. J. Brodsky and G. R. Farrar, Phys.

Rev. Lett. **31**, 1153 (1973); G. 't Hooft, Nucl. Phys. B **72**, 461 (1974); K. G. Wilson, Phys. Rev. D **10**, 2445 (1974); S. Weinberg, Phys. Rev. D **13**, 974 (1976); R. D. Field and R. P. Feynman, Phys. Rev. D **15**, 2590 (1977).

[4] H. Fritzsch, M. Gell-Mann and P. Minkowski, Phys. Lett. B **59**, 256 (1975).

[5] D. J. Gross and F. Wilczek, Phys. Rev. Lett. **30**, 1343 (1973); D. J. Gross and F. Wilczek, Phys. Rev. D **8**, 3633 (1973); H. D. Politzer, Phys. Rev. D **9**, 2174 (1974); H. D. Politzer, Phys. Rept. **14**, 129 (1974).

[6] D. Parganlija, arXiv:1208.0204 [hep-ph].

[7] R. L. Jaffe, Phys. Rev. D **15**, 267 (1977).

[8] T. Barnes, F. E. Close, F. de Viron and J. Weyers, Nucl. Phys. B **224**, 241 (1983).

[9] R. L. Jaffe and K. Johnson, Phys. Lett. B **60**, 201 (1976); C. Rosenzweig, A. Salomone and J. Schechter, Phys. Rev. D **24**, 2545 (1981); A. Salomone, J. Schechter and T. Tudron, Phys. Rev. D **23**, 1143 (1981); R. Konoplich and M. Shchepkin, Nuovo Cim. A **67**, 211 (1982); C. Rosenzweig, A. Salomone and J. Schechter, Nucl. Phys. B **206**, 12 (1982) [Erratum-ibid. B **207**, 546 (1982)]; A. A. Migdal and M. A. Shifman, Phys. Lett. B **114**, 445 (1982); M. Jezabek and J. Szwed, Acta Phys. Polon. B **14**, 599 (1983); H. Gomm and J. Schechter, Phys. Lett. B **158**, 449 (1985); R. L. Jaffe, K. Johnson and Z. Ryzak, Annals Phys. **168**, 344 (1986).

[10] D. Parganlija, J. Phys. Conf. Ser. **426**, 012019 (2013) [arXiv:1211.4804 [hep-ph]].

[11] C. Amsler *et al.* [Crystal Barrel Collaboration], Phys. Lett. B **291**, 347 (1992); C. Amsler *et al.* [Crystal Barrel Collaboration], Phys. Lett. B **322**, 431 (1994); D. V. Bugg, V. V. Anisovich, A. Sarantsev and B. S. Zou, Phys. Rev. D **50**, 4412 (1994); C. Amsler *et al.* [Crystal Barrel Collaboration], Phys. Lett. B **340**, 259 (1994); C. Amsler *et al.* [Crystal Barrel Collaboration], Phys. Lett. B **355**, 425 (1995); A. Bertin *et al.* [OBELIX Collaboration], Phys. Lett. B **408**, 476 (1997); C. Amsler *et al.* [Crystal Barrel Collaboration], Phys. Lett. B **639**, 165 (2006).

[12] T. Akesson *et al.* [Axial Field Spectrometer Collaboration], Nucl. Phys. B **264**, 154 (1986); A. Breakstone *et al.* [Ames-Bologna-CERN-Dortmund-Heidelberg-Warsaw Collaboration], Z. Phys. C **48**, 569-576 (1990); F. Antinori *et al.* [WA91 Collaboration], Phys. Lett. B **353**, 589 (1995); D. Alde *et al.* [GAMS Collaboration], Phys. Lett. B **397**, 350 (1997); D. Barberis *et al.* [WA102 Collaboration], Phys. Lett. B **453**, 316 (1999) [arXiv:hep-ex/9903043]; R. Bellazzini *et al.* [GAMS Collaboration], Phys. Lett. B **467**, 296 (1999).

[13] T. A. Armstrong *et al.* [WA76 Collaboration], Z. Phys. C **51**, 351 (1991).

[14] C. Edwards, R. Partridge, C. Peck, F. Porter, D. Antreasyan, Y. F. Gu, W. S. Kollmann, M. Richardson *et al.*, Phys. Rev. Lett. **48**, 458 (1982); R. M. Baltrusaitis *et al.* [MARK-III Collaboration], Phys. Rev. D **33**, 1222 (1986); T. Bolton *et al.*, Phys. Lett. B **278**, 495 (1992); R. R. Akhmetshin *et al.* [CMD-2 Collaboration], Phys. Lett. B **462**, 380 (1999) [arXiv:hep-ex/9907006]; M. Ablikim *et al.* [BES Collaboration], Phys. Lett. B **598**, 149 (2004) [hep-ex/0406038]; M. Ablikim *et al.* [BES Collaboration], Phys. Lett. B **607**, 243 (2005) [arXiv:hep-ex/0411001]; F. Ambrosino *et al.* [KLOE Collaboration], Eur. Phys. J. C **49**, 473 (2007) [hep-ex/0609009].

[15] B. Hyams, C. Jones, P. Weilhammer, W. Blum, H. Dietl, G. Grayer, W. Koch, E. Lorenz *et al.*, Nucl. Phys. B **64**, 134 (1973); H. Becker *et al.* [CERN-Cracow-Munich Collaboration], Nucl. Phys. B **151**, 46 (1979); D. Alde *et al.* [Serpukhov-Brussels-Los Alamos-Annecy(LAPP) Collaboration], Phys. Lett. B **182**, 105 (1986); D. Alde *et al.* [GAMS Collaboration], Eur. Phys. J. A **3**, 361 (1998); E. M. Aitala *et al.* [E791 Collaboration], Phys. Rev. Lett. **86**, 770 (2001) [hep-ex/0007028].

[16] P. Minkowski and W. Ochs, Eur. Phys. J. C **9**, 283 (1999) [arXiv:hep-ph/9811518]; D. Parganlija, F. Giacosa and D. H. Rischke, Phys. Rev. D **82**, 054024 (2010) [arXiv:1003.4934 [hep-ph]]; S. Janowski, D. Parganlija, F. Giacosa and D. H. Rischke, Phys. Rev. D **84**, 054007 (2011) [arXiv:1103.3238 [hep-ph]]; D. Parganlija, P. Kovacs, G. Wolf, F. Giacosa and D. H. Rischke, Phys. Rev. D **87**, 014011 (2013) [arXiv:1208.0585 [hep-ph]].

[17] C. Morningstar and M. J. Peardon, AIP Conf. Proc. **688**, 220 (2004) [arXiv:nucl-th/0309068]; M. Loan, X. Q. Luo and Z. H. Luo, Int. J. Mod. Phys. A **21**, 2905 (2006) [arXiv:hep-lat/0503038]; E. B. Gregory, A. C. Irving, C. C. McNeile, S. Miller and Z. Sroczynski, PoS **LAT2005**, 027 (2006) [arXiv:hep-lat/0510066]; Y. Chen *et al.*, Phys. Rev. D **73**, 014516 (2006) [arXiv:hep-lat/0510074].

[18] C. Csaki, H. Ooguri, Y. Oz and J. Terning, JHEP **9901**, 017 (1999) [hep-th/9806021]; R. C. Brower, S. D. Mathur and C. -I Tan, Nucl. Phys. B **587**, 249 (2000) [hep-th/0003115]; G. F. de Teramond and S. J. Brodsky, Phys. Rev. Lett. **94**, 201601 (2005) [hep-th/0501022]; N. Evans, J. P. Shock and T. Waterson, Phys. Lett. B **622**, 165 (2005) [hep-th/0505250]; H. Boschi-Filho, N. R. F. Braga and H. L. Carrion, Phys. Rev. D **73**, 047901 (2006) [hep-th/0507063].

[19] R. Barate *et al.* [ALEPH Collaboration], Phys. Lett. B **472**, 189 (2000) [hep-ex/9911022].

[20] M. Acciarri *et al.* [L3 Collaboration], Phys. Lett. B **501**, 173 (2001) [hep-ex/0011037].

[21] M. Ablikim *et al.* [BES Collaboration], Phys. Lett. B **607**, 243 (2005) [arXiv:hep-ex/0411001].

[22] M. F. M. Lutz *et al.* [PANDA Collaboration], arXiv:0903.3905 [hep-ex].

Microscopic-macroscopic method for studying single-particle level density of superheavy nuclei

A.N. Bezbakh[1], **T.M. Shneidman**[1,2], **G.G. Adamian**[1], **N.V. Antonenko**[1]

[1]Joint Institute for Nuclear Research, Dubna 141980, Russia
[2] Institute of Theoretical Physics, CAS, Beijing 100190, China

E-mail: bezbakh@theor.jinr.ru

Abstract. The intrinsic level densities of superheavy nuclei in the α-decay chains of $^{296,298,300}120$ nuclei are calculated using the single-particle spectra obtained with the modified two-center shell model. The level density parameters are extracted and compared with their phenomenological values used in the calculations of the survival of excited heavy nuclei. The dependences of the level density parameters on the mass and charge numbers as well as on the ground-state shell corrections are studied.

1. Introduction

An extension of the nuclear chart to the elements with $Z > 100$ and the study of their properties provide an important insight into the determination of an island of stability. The experimental trend of nuclear properties (Q_α-values and half-lives) as well as the production cross sections of superheavy elements (SHE) reveal an increasing stability of nuclei approaching $N = 184$, and indicate quite a large shell effects behind $Z = 114$ [1-6]. This means that the predictions of relativistic and nonrelativistic mean-field models [7-11] seem to be valid. In accordance with these self-consistent models a center of the island of stability is expected at Z=120-126 and N=172 or 184. In the (N, Z)-plane, the line, along which all new SHE were discovered in the actinide-based reactions with ^{48}Ca beam, just approaches this region [1].

In Refs. [12, 13] we modified the two-center shell model (TCSM) [14] to describe the quasi-particle structure of well-studied heavy nuclei in the best way. Based on the TCSM single-particle potential and shape parametrization we calculated the ground-state binding energies and transition energies Q_α for α decays of nuclei with $105 \leq Z \leq 126$. The calculated Q_α are found to be in a good agreement with available experimental data. The dependence of Q_α on the neutron number at given Z indicates the strong shell effects at N= 184. Also our calculations result in the stronger shell effects at $Z \geq 120$ than at Z=114.

In this work we use the single-particle spectra calculated with the modified TCSM to investigate the nuclear level densities (NLD) of superheavy nuclei. The NLD is an important characteristic both for the nuclear reaction calculations and for the investigation of the stability of the new supearheavy elements. At low excitations the single-particle level densities are closely related to the shell structure and could indicate the position of the island of stability of heaviest nuclei. The level density, as a function of excitation energy, is required to calculate the survival probability and, correspondingly, the production cross section of heavy nucleus. For nuclei lighter than Cf, the systematic study of NLD was performed in Refs. [15, 16].

In this work we performed the study of the intrinsic level densities of nuclei with $100 \leq Z \leq 130$ whose α-decay chains contain $^{296,298,300}120$ which are of interest for future experiments.

2. A few words about performed calculations.

In our work we use the expression for the nuclear single-particle level density obtained in the superfluid model [17, 18]. Employing the saddle-point method, the intrinsic level density $\rho(U)$ of nucleus with Z protons, N neutrons, and excitation energy U is expressed as

$$\rho = \frac{\exp[S(\beta, \lambda_Z, \lambda_N)]}{(2\pi)^{3/2}\sqrt{D}}. \tag{1}$$

Here, S is the entropy, $\beta = T^{-1}$ is the inverse temperature, λ_Z and λ_N are the chemical potentials for protons and neutrons, respectively, and D is the determinant of the matrix comprised of the second derivatives of the entropy.

In the superfluid model adopted here, the entropy S is expressed as

$$S = 2 \sum_{k=Z,N} \sum_{\nu} \left\{ \ln[1 + \exp(-\beta E_{k\nu})] + \frac{\beta E_{k\nu}}{1 + \exp(\beta E_{k\nu})} \right\}, \tag{2}$$

where proton ($k = Z$) and neutron ($k = N$) quasi-particle energies $E_{k\nu} = \sqrt{(\varepsilon_{k\nu} - \lambda_k)^2 + \Delta_k^2}$ are calculated using the single-particle levels $\varepsilon_{k\nu}$ obtained with the TCSM. The details of the calculations with the TCSM as well as the parameters used are described in Refs. [12, 13]. The Fermi energies λ_k and correlation functions Δ_k ($k = N, Z$), are calculated at the thermodynamic equilibrium of nucleus determined by solving the system of equations

$$\begin{aligned} Z &= \sum_{\nu} \left(1 - \frac{\varepsilon_{Z\nu} - \lambda_Z}{E_{Z\nu}} \tanh[\frac{1}{2}\beta E_{Z\nu}]\right), \; N = \sum_{\nu} \left(1 - \frac{\varepsilon_{N\nu} - \lambda_N}{E_{N\nu}} \tanh[\frac{1}{2}\beta E_{N\nu}]\right), \\ \frac{2}{G_Z} &= \sum_{\nu} \frac{\tanh[\beta E_{Z\nu}/2]}{E_{Z\nu}}, \; \frac{2}{G_N} = \sum_{\nu} \frac{\tanh[\beta E_{N\nu}/2]}{E_{N\nu}}, \end{aligned} \tag{3}$$

where G_Z and G_N are the constants of pairing interaction. In Eqs. (3), the sums run over all the single-particle levels considered.

In Fig. 1, the calculated level densities $\rho_m(U)$ are presented for well-deformed nuclei ^{162}Dy, ^{166}Er together with the experimental data [19]. The agreement with experimental data is rather good, which provides us some confidence in the predictions of $\rho(U)$ in the region of SHE.

3. Dependence of level density parameter on shell corrections

Intrinsic level density is strongly influenced by nuclear shell structure. For magic or nearly magic nuclei, the level density is smaller than for mid-shell nuclei at the same excitation energy. This effect is related to the large single-particle spacings in the nuclei with the closed shell in the ground state. So, irregularities of the single-particle spectra are responsible for the shell corrections and peculiarities of the intrinsic level density.

The calculated intrinsic level densities for the superheavy nuclei are presented in Fig. 2 for two nuclei from each of the considered alpha-decay chains of the elements $^{296,298,300}120$. The solid lines correspond to the isotopes $^{296,298,300}120$ and exemplify the behavior of the level density in nuclei with quite a strong shell correction, while the dotted lines are for the isotopes $^{264,266,268}104$ with smaller shell corrections. As seen, at low excitation energies from 10 to 30 MeV the level

Journal of Physics: Conference Series **503** (2014) 012011 doi:10.1088/1742-6596/503/1/012011

densities in Z=120 isotopes are smaller within one order of magnitude than those in Z=104 isotopes due to the smaller densities of the single–particle states near the Fermi surfaces. At smaller excitation energies $U < 10$ MeV, the effects of pairing are stronger in the nuclei with weaker shell effects and the level densities of nuclei with strong and weak shell effects become comparable. At higher excitation energies, U >40–50 MeV, the shell effects are faded and the level densities of all heavy nuclei become similar.

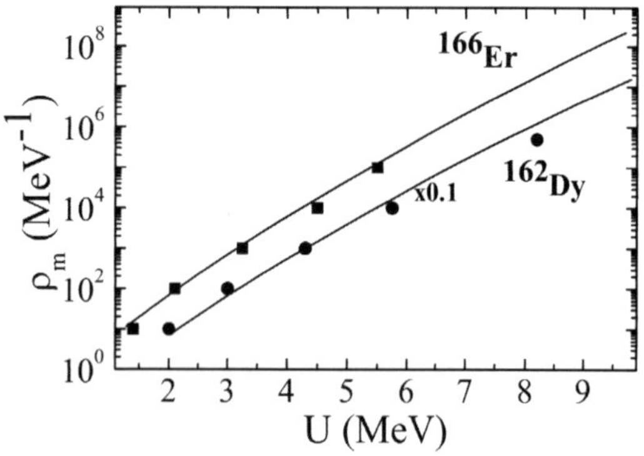

Figure 1. Calculated level densities ρ_m in ^{162}Dy and ^{166}Er as functions of excitation energy (lines) are compared with the experimental data (symbols) from Ref. [19]. For the ^{162}Dy nucleus, the experimental and theoretical results are multiplied by factor 0.1.

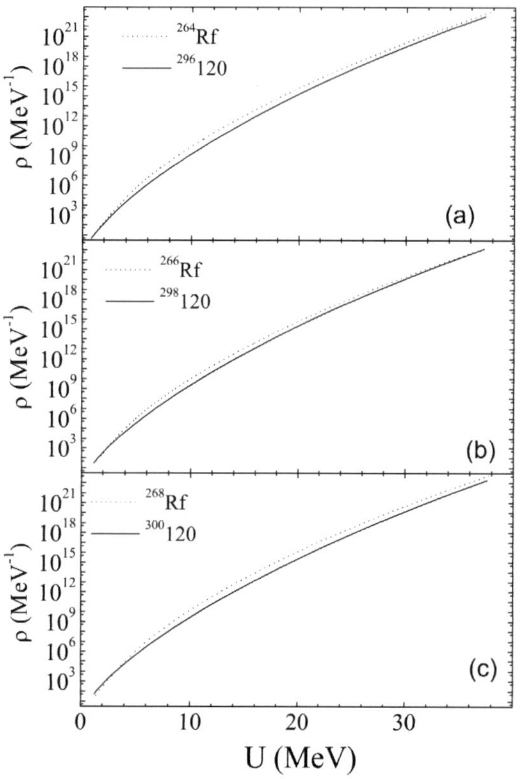

Figure 2. The calculated intrinsic level densities as functions of excitation energy for ^{264}Rf and ^{296}120 (a), ^{266}Rf and ^{298}120 (b), ^{266}Rf and ^{300}120 (c). The results for the isotopes of Rf and Z=120 nucleus are presented by dashed and solid lines, respectively.

We found that the best fit of the calculated intrinsic level density with the Fermi-gas expression is achieved if one uses the level density parameter calculated as

$$a = S^2/(4U), \tag{4}$$

where S and U are entropy and excitation energy of the nucleus, respectively. Since the thermodynamical equilibrium is assumed for each excitation energy, the level density parameter can be written as

$$a = S_N^2/4U_N + S_Z^2/4U_Z = a_N + a_Z, \tag{5}$$

where $S_N(S_Z)$ and $E_N(E_Z)$ are the entropies and excitation energies of only neutron (proton) subsystems.

The Fig. 3 presents the dependences of shell corrections δE_{sh} and level density parameters a at $U = 10$ MeV and $U = 60$ MeV excitation energies on the atomic number A for three α-decay chains containing the SHE $^{296,298,300}120$ which could be synthesized in near future with available projectiles and targets. The data of Ref. [22] are used for δE_{sh}. One can see the strong correlations between the shell corrections and level density parameter a at excitation energy $U = 10$ MeV. The larger negative shell corrections result in the decrease of the value of a with respect to the neighborhood nuclei. At higher excitation energy (U=60 MeV), the correlation between δE_{sh} and a is destroyed and (as demonstrated in upper part of Fig. 3) level density parameter becomes rather smooth function of A. Still, the shell structure slightly modifies the general trend of the level density parameter. For example, $a(^{266}\mathrm{Rf})$ is smaller slightly than $a(^{264}\mathrm{Rf})$ at U=60 MeV because of the closed neutron shell N=162 in ^{266}Rf nucleus.

Based on the study of the dependencies of a on δE_{sh}, U, and A one can use the following parametrization [20] of the level density parameter:

$$a(A,U) = \tilde{a}(A)\left[1 + \frac{1-\exp\{-U/E'_D\}}{U}\delta E_{sh}\right], \quad (6)$$

where $\tilde{a}(A)$ is the parameter smoothly depending on A. It defines a at large excitations when the shell effects are washed out. By analyzing the level density parameters with the Eq. (4), the value of the damping parameter E'_D=17.3 MeV is found. The corresponding asymptotic level density parameter $\tilde{a}(A)$ can be fitted with the following functions [20]:

$$\tilde{a}(A) = \alpha A + \beta A^2, \quad (7)$$

where the constants α=0.118 MeV^{-1} and β=-0.53×10^{-4} MeV^{-1} are found with the least square method. These values are close to those proposed in Ref. [20].

The level density parameters for the nuclei with $Z \geq 100$ are of especial interest to look for the position of the next proton and neutron shell closures beyond $Z = 82$. To separate the neutron and proton shell effects one should investigate the proton (a_Z) and neutron (a_N) level density parameters defined in Eq. (5) (see Fig. 4). At Z=108 and 120 there are minima of a_Z in all chains. This reflects quite strong proton shell effects at Z=108 and 120. At Z=120, the minima of a are the deepest and well pronounced. The similar behavior of a occurs near Z=82. The sub-shell at Z=114 exists but provides weaker shell effect than at Z=120. For nuclei with Z=124-128, the minima of a are due to the neutron shell at N=184.

4. Summary

The level density parameters were calculated for the nuclei of alpha-decay chains containing $^{296}120$, $^{298}120$, and $^{300}120$. The strong shell effects at $Z = 120$ and $N = 184$ were demonstrated that is in accordance with our previous calculations and shell-model predictions. The dependencies of the level density parameter on the shell correction and excitation energy were studied. The damping factor in the well-known expression (6) [20] was found to be E'_D=17.3 MeV. For superheavy nuclei considered, the level density parameter is approximately $A/(11-13)$ MeV at excitation energies corresponding to (3-5) neutron evaporation channels.

Acknowledgments

I express my gratitude to the organizers and sponsors of FAIRNESS'13 for opportunity to participate in the Workshop for young scientists.

References

[1] Oganessian Yu Ts 2007 *J Phys.* G **34** R165, 2010 *Phys. Rev. Lett.* **104** 142502, 2013 *Phys. Rev.* C **87** 014302
[2] Hofmann S *et al.* 2007 *Eur. Phys. J.* A **32** 251, 2009 *Lec. Notes Phys.* **764** 203
[3] Armbruster P 2008 *Eur. Phys. J.* A **37** 159

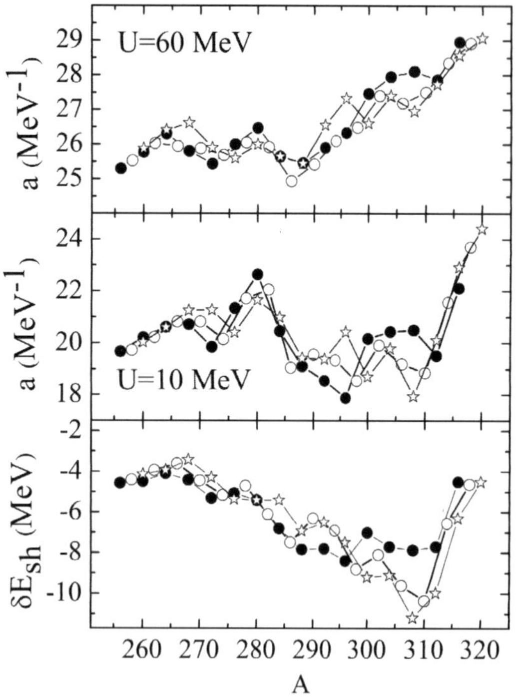

Figure 3. Calculated ground-state shell corrections δE_{sh} (lower part), the level density parameter a at U=10 (middle part) and 60 MeV (upper part), obtained with Eq. (4), as functions of mass number A. The nuclei from alpha-decay chains containing $^{296}120$, $^{298}120$, $^{300}120$ are marked by closed circles, open circles, and stars, respectively.

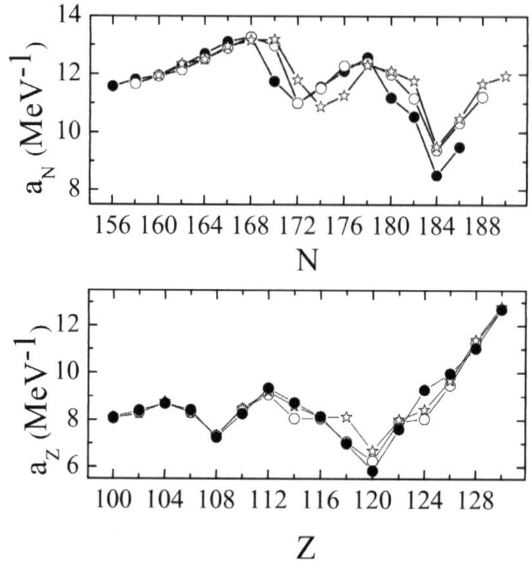

Figure 4. Calculated level density parameters a_N and a_Z at U=10 MeV as a function of neutron number N (upper part) and proton number Z (lower part), respectively. The nuclei from alpha-decay chains containing $^{296}120$, $^{298}120$, $^{300}120$ are marked by closed circles, open circles, and stars, respectively.

[4] Adamian G G, Antonenko N V and Sargsyan V V 2009 *Phys. Rev.* C **79** 054608
[5] Adamian G G, Antonenko N V, Sargsyan V V and Scheid W 2010 *Nucl. Phys.* **A834** 345c
[6] Dong J, Zuo W and Scheid W 2011 *Phys. Rev. Lett.* **107** 012501; Li Z, Sun B, Shen C H and Zuo W 2013 *Phys. Rev.* C **88** 057303
[7] Reinhard P G 1989 *Rep. Prog. Phys.* **52** 439
[8] Ring P 1996 *Prog. Part. Nucl. Phys.* **37** 193
[9] Bender M, Heenen P H and Reinhard P G 2003 *Rev. Mod. Phys.* **75** 121
[10] Meng J, Toki H, Zhou S G, Zhang S Q, Long W H and Geng L S 2006 *Prog. Part. Nucl. Phys.* **57** 470
[11] Li J J, Long W H, Margueron J and Van Giai N 2013 *arXiv: nucl-th* **1303.2765v1**
[12] Adamian G G, Antonenko N V and Scheid W 2010 *Phys. Rev.* C **81** 024320
[13] Adamian G G, Antonenko N V, Kuklin S N and Scheid W 2010 *Phys. Rev.* C **82** 054304
[14] Maruhn J and Greiner W 1972 *Z. Phys.* A **251** 431
[15] Iljinov A S *et al.* 1992 *Nucl. Phys.* **A543** 517
[16] von Egidy T and Bucurescu D 2005 *Phys. Rev.* C **72** 044311
[17] Decowski P, Grochulski W, Marcinkowski A, Siwek K and Wilhelmi Z 1968 *Nucl. Phys.* A **110** 129
[18] Adeev G D and Cherdantsev P A 1975 *Yadernaya Fizika* **21** 491
[19] Melby E, Guttormsen M, Rekstad J, Schiller A, Siem S and Voinov A 2001 *Phys. Rev.* C **63** 044309
[20] Ignatyuk A B, Smirenkin G N and Tishin A S 1975 *Yadernaya Fizika* **21** 485
[21] Shlomo S and Natowitz J B 1991 *Phys. Rev.* C **44** 2878
[22] Kuzmina A N, Adamian G G, Antonenko N V and Scheid W 2012 *Phys. Rev.* C **85** 014319

Journal of Physics: Conference Series **503** (2014) 012012 doi:10.1088/1742-6596/503/1/012012

Production of heavy meson pairs in $p\bar{p}$ collisions within a double handbag approach

A T Goritschnig[1a], **B Pire**[2b] **and W Schweiger**[1c]

[1] Institute of Physics, University of Graz, Universitätsplatz 5, 8010 Graz, Austria

[2] Centre de Physique Théorique, École Polytechnique, CNRS, 91128 Palaiseau cedex, France

E-mail: [a]alexander.goritschnig@uni-graz.at

E-mail: [b]bernard.pire@cpht.polytechnique.fr

E-mail: [c]wolfgang.schweiger@uni-graz.at

Abstract. We study the pair-production of heavy mesons in proton-antiproton annihilations within a perturbative QCD-motivated framework. In particular we investigate $p\bar{p} \to \overline{D^0}D^0$ within a double handbag approach, where a hard subprocess factorizes from soft hadronic matrix elements. The soft matrix elements can be parametrized by transition distribution amplitudes, which are off-diagonal in flavor space. The transition distribution amplitudes are modelled as overlaps of light-cone wave functions. We obtain rather robust model results for $p\bar{p} \to \overline{D^0}D^0$ cross sections, which are expected to be measured at the future $\bar{\text{P}}$ANDA detector at GSI-FAIR.

1. Introduction

In Ref. [1] we have studied the process $p\bar{p} \to \overline{D^0}D^0$ which has become of particular interest due to the planned physics program of the $\bar{\text{P}}$ANDA detector [2] at FAIR-GSI. A part of the FAIR project with its $\bar{\text{P}}$ANDA detector is devoted to the measurement of exclusive channels in proton-antiproton collisions where heavy hadron pairs are produced. Therefore theoretical input is most welcome. To investigate $p\bar{p} \to \overline{D^0}D^0$ we use the framework of perturbative QCD where we apply a generalization of the so-called "handbag approach". A handbag mechanism has been first developed to study deeply virtual Compton scattering and meson production [3]; in Ref. [4] it has been extended to the flavor off-diagonal case, which comes across here. For our studies of $p\bar{p} \to \overline{D^0}D^0$ we go along similar lines of argumentation as for $p\bar{p} \to \Lambda_c^+\bar{\Lambda}_c^-$ in Ref. [4]. Under physically plausible assumptions we argue that our process under consideration factorizes into a hard partonic subprocess amplitude and soft hadronic transition matrix elements. In doing so the hard energy scale justifying this splitting and the treatment of the partonic subprocess within perturbation theory is set by the heavy charm-quark mass m_c. In our investigations we only take the valence Fock states of the hadrons into account and consider the proton in a quark-diquark picture. The soft hadronic transition matrix elements can in principle be parametrized by transition distribution amplitudes [5]. We here represent them as an overlap of hadron light-cone wave functions (LCWFs) analogously as in Ref. [6].

2. The Double Handbag Mechanism

Before we turn to the description of $p\bar{p} \to \overline{D^0}D^0$ within a double-handbag approach we would like to specify the kinematics chosen to describe the process. The assignment of particle momenta

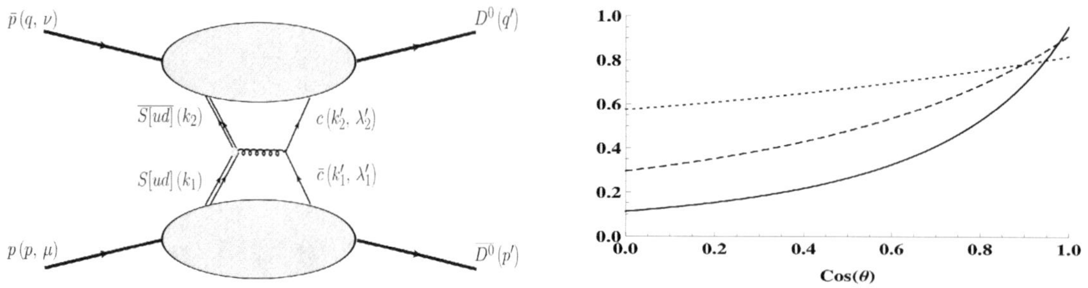

Figure 1. *Left:* The double handbag mechanism for the $p\bar{p} \to \overline{D^0}D^0$ scattering amplitude. Blobs represent the soft $p \to \overline{D^0}$ and $\bar{p} \to D^0$ contribustions. *Right:* The $\bar{x}$-moment of the wave function overlap appearing in the square brackets in Eq. (5) versus $\cos\theta$ for $s = 15, 20, 30$ GeV^2 (corresponding to the dotted, dashed and solid line, respectively).

and helicities can be seen in Fig. 1. M and m denote the masses of the $\overline{D^0}$ and the proton, respectively. We work in a center-of-momentum system (CMS) where the z axis is parallel to the 3-vector $\bar{\mathbf{p}}$, with $\bar{p} := (1/2)(p' + p)$, and the transverse momentum transfer $\mathbf{\Delta}_\perp := (\Delta^1, \Delta^2)$, with $\Delta := p' - p$, is symmetrically shared between the incoming and outgoing hadron momenta. In light-cone coordinates[1], which are ideally suited for the description of scattering processes at high energies, the proton and the $\overline{D^0}$ momentum read:

$$p = \left[(1+\xi)\bar{p}^+, \frac{m^2 + \mathbf{\Delta}_\perp^2/4}{2(1+\xi)\bar{p}^+}, -\frac{\mathbf{\Delta}_\perp}{2}\right], \quad p' = \left[(1-\xi)\bar{p}^+, \frac{M^2 + \mathbf{\Delta}_\perp^2/4}{2(1-\xi)\bar{p}^+}, +\frac{\mathbf{\Delta}_\perp}{2}\right]. \quad (1)$$

The antiproton momentum q and the D^0 momentum q' are obtained from p and p' by interchanging the plus and minus components and changing the signs of the transverse components. In Eq. (1) we have introduced the so-called "skewness" ξ which is defined as $\xi := (-\Delta^+)/(2\bar{p}^+)$. It parametrizes the relative momentum transfer between the proton and the $\overline{D^0}$ in the light-cone plus direction.

Considering $p\bar{p} \to \overline{D^0}D^0$ in the double-handbag mechanism means that the process amplitude can be split up into a nonperturbative part describing the soft hadronic $p \to \overline{D^0}$ and $\bar{p} \to D^0$ transitions and into a perturbative subprocess on the partonic level. In the latter only the minimal number of hadronic constituents which are necessary to convert the initial $p\bar{p}$ into the final $\overline{D^0}D^0$ pair take part actively. In our investigations we only take the valence Fock-state contributions into account. Furthermore, the proton is treated within a quark-diquark picture, where only scalar diquarks are considered. Thus it is the subprocess $S[ud]\overline{S[ud]} \to \bar{c}c$ which contributes on the partonic level. The virtuality of the intermediate gluon in the subprocess has to be high enough to produce the heavy $\bar{c}c$ pair. This means, the heavy charm-quark mass sets a natural hard scale in our process, such that the process $S[ud]\overline{S[ud]} \to \bar{c}c$ can be treated by means of perturbative QCD. In order to argue that we can factorize the hard and the soft parts in this way we follow a similar line of argumentation as in Refs. [4, 7], where one poses physically motivated assumptions on the parton momenta. In our framework the $p\bar{p} \to \overline{D^0}D^0$

[1] A four vector $v = (v^0, v^1, v^2, v^3)$ written in light-cone coordinates reads $v = [v^+, v^-, v^1, v^2]$ where $v^\pm := \frac{1}{\sqrt{2}}(v^0 \pm v^3)$.

amplitude then reads:

$$
\begin{aligned}
M_{\mu\nu} &= \frac{1}{4(\bar{p}^+)^2} \sum_{\lambda'_1,\,\lambda'_2} \int d\bar{x}_1 \int d\bar{x}_2\, H_{\lambda'_1,\,\lambda'_2}(\bar{x}_1,\,\bar{x}_2)\, \frac{1}{\bar{x}_1 - \xi}\, \frac{1}{\bar{x}_2 - \xi} \\
&\times \bar{v}(k'_1,\,\lambda'_1)\gamma^+\bar{p}^+ \int \frac{dz_1^-}{2\pi} e^{\imath\bar{x}_1\bar{p}^+z_1^-} \langle \overline{D^0} : p' \mid \Psi^c(-z_1^-/2)\Phi^{S[ud]}(+z_1^-/2) \mid p : p,\,\mu\rangle \\
&\times \bar{p}^+ \int \frac{dz_2^+}{2\pi} e^{\imath\bar{x}_2\bar{p}^+z_2^+} \langle D^0 : q' \mid \Phi^{S[ud]\,\dagger}(+z_2^+/2)\overline{\Psi}^c(-z_2^+/2) \mid \bar{p} : q,\,\nu\rangle\gamma^- u(k'_2,\,\lambda'_2)\,.
\end{aligned} \tag{2}
$$

$H_{\lambda'_1,\lambda'_2}(\bar{x}_1,\,\bar{x}_2)$ denotes the hard $S[ud]\overline{S[ud]} \to \bar{c}c$ amplitude, whereas the Fourier transforms of the hadronic matrix elements

$$
\bar{p}^+ \int \frac{dz_1^-}{2\pi} e^{\imath\bar{x}_1\bar{p}^+z_1^-} \langle \overline{D^0} : p' \mid \Psi^c(-z_1^-/2)\Phi^{S[ud]}(+z_1^-/2) \mid p : p,\,\mu\rangle \tag{3}
$$

and

$$
\bar{p}^+ \int \frac{dz_2^+}{2\pi} e^{\imath\bar{x}_2\bar{p}^+z_2^+} \langle D^0 : q' \mid \Phi^{S[ud]\,\dagger}(+z_2^+/2)\overline{\Psi}^c(-z_2^+/2) \mid \bar{p} : q,\,\nu\rangle \tag{4}
$$

incorporate the soft $p \to \overline{D^0}$ and $\bar{p} \to D^0$ transitions, respectively. Eq. (2) now is a convolution integral between the hard and the soft parts with respect to the average momentum fractions $\bar{x}_1 := \bar{k}_1^+/\bar{p}^+$ and $\bar{x}_2 := \bar{k}_2^-/\bar{p}^+$, where $\bar{k}_i := (1/2)(k_i + k'_i)$, $i = 1, 2$.

In the hard part of Eq. (2) the annihilation of the $S[ud]\overline{S[ud]}$ and the creation of the $\bar{c}c$ pair is separated by a lightlike distance (i.e., happens at the same light-cone time) and the active parton momenta can be approximated as being proportional to their parent hadron momenta. The soft hadronic transitions (3) and (4) are hadronic matrix elements with respect to the incoming (anti)proton and the outgoing $\overline{D^0}$ (D^0) states of appropriate c-quark and $S[ud]$-diquark field operators. The action of the field operators in Eq. (3) is that $\Phi^{S[ud]}(+z_1^-/2)$ emits an $S[ud]$ diquark from the proton into the partonic subprocess, whereas $\Psi^c(-z_1^-/2)$ reinserts a $\bar{c}$ quark into the remnants of the proton in order to give the final $\overline{D^0}$. The emission and the reinsertion (i.e., the arguments of the field operators) are again separated by a lightlike distance. An analog interpretation holds for the hadronic matrix element in Eq. (4). It should be noted that as in Ref. [4] we have used appropriate projection techniques in deriving Eq. (2), such that only contributions of the field operators survive which are dominant in our kinematical situation.

3. The soft and hard contributions of $p\bar{p} \to \overline{D^0}D^0$

In order to model the soft $p \to \overline{D^0}$ and $\bar{p} \to D^0$ transitions, Eqs. (3) and (4) respectively, we will exploit the advantages of light-cone quantum field theory and derive an overlap representation in terms of hadronic LCWFs [6]. In the following we will only discuss the $p \to \overline{D^0}$ transition, since one can proceed in complete analogy for the $\bar{p} \to D^0$ one. Within the light-cone quantum field theoretical framework we apply the Fourier representation of the field operators as well as the Fock-state decomposition of the hadron states in Eq. (3). Each Fock-state component, $\mid S[ud]u\,\rangle$ for the proton and $\mid \bar{c}u\,\rangle$ for the $\overline{D^0}$, comes with a LCWF as a "prefactor". Furthermore, we only take into account LCWFs where we neglect partonic orbital angular momenta, such that the helicities of the partons add up to the parent hadron helicity. Thus we need only two LCWFs: one for the proton and one for the $\overline{D^0}$, which we call ψ_p and ψ_D, respectively. Such hadronic LCWFs only depend on the relative parton momenta with respect to the parent hadron momentum, but not on the total hadron momentum.

Before stating the result for the $p\bar{p} \to \overline{D^0}D^0$ amplitude after applying the overlap representation, we first take into account the fact that according to Refs. [8, 9] the D-meson wave function is strongly peaked around $x_0 \approx m_c/M$ with respect to its momentum fraction dependence. This behavior is also inherited by the LCWF overlap such that it notably weights momentum fraction close to the peak position in the hard scattering amplitude. It is therefore possible to replace the momentum fractions inside the hard-scattering amplitude by their peak position and to take the subprocess amplitude out of the convolution integral. Then the $p\bar{p} \to \overline{D^0}D^0$ amplitude takes on the simpler form

$$\begin{aligned} M_{\mu\nu} =& 2\mu\nu\, H_{-\mu,-\nu}(x_0, x_0) \Big[\int d\bar{x} \frac{1}{\sqrt{\bar{x}^2 - \xi^2}} \int \frac{d^2\bar{k}_{1\perp}}{16\pi^3} \\ & \times \psi_D(\hat{x}'(\bar{x}_1, \xi), \hat{\mathbf{k}}'_\perp(\bar{\mathbf{k}}_{1\perp}, \bar{x}_1, \xi))\, \psi_p(\tilde{x}(\bar{x}_1, \xi), \tilde{\mathbf{k}}_\perp(\bar{\mathbf{k}}_{1\perp}, \bar{x}_1, \xi)) \Big]^2 , \end{aligned} \quad (5)$$

with the hard subprocess amplitudes calculated within perturbative QCD

$$H_{++} = +4\pi\alpha_s(x_0^2 s)\, F_s(x_0^2 s)\, \frac{4}{9} \frac{2M}{\sqrt{s}} \cos\theta\,, \quad H_{+-} = -4\pi\alpha_s(x_0^2 s)\, F_s(x_0^2 s)\, \frac{4}{9} \sin\theta\,, \quad (6)$$

$H_{--} = -H_{++}$ and $H_{-+} = H_{+-}$. We have added the color factor 4/9 to the subprocess amplitudes, α_s is the strong coupling constant and the phenomenological form factor F_s takes care of the composite nature of the $S[ud]$ diquark at the SgS vertex, cf. Refs. [10, 11, 12].

4. Model results

In order to obtain model results for the $p\bar{p} \to \overline{D^0}D^0$ cross sections we have to specify the model LCWFs for the proton and the D meson which appear in the overlap representation. We take the wave functions as have been used in Refs. [1, 13],

$$\psi_p(\tilde{x}, \tilde{\mathbf{k}}_\perp) = N_p\, \tilde{x}\, e^{-a_p^2 \frac{\tilde{\mathbf{k}}_\perp^2}{\tilde{x}(1-\tilde{x})}} \quad \text{and} \quad \psi_D(\hat{x}', \hat{\mathbf{k}}'_\perp) = N_D\, e^{-a_D^2 \frac{\hat{\mathbf{k}}'^2_\perp}{\hat{x}'(1-\hat{x}')}}\, e^{-a_D^2 M^2 \frac{(\hat{x}'-x_0)^2}{\hat{x}'(1-\hat{x}')}} \quad (7)$$

for the proton and the $\overline{D^0}$, respectively. The arguments of the LCWFs are the relative momentum fractions $\tilde{x}$ and $\hat{x}'$ and the intrinsic transverse momenta $\tilde{\mathbf{k}}_\perp$ and $\hat{\mathbf{k}}'_\perp$ of the active constituents inside the proton and the $\overline{D^0}$, respectively, with respect to their parent hadron momenta. The $\overline{D^0}$ wave function has its peak around x_0 through its Gaussian mass exponential. The free parameters appearing in each wave function, namely the normalization constant $N_{p/D}$ and the transverse size parameter $a_{p/D}$, can be fixed by requirering a specific value for the root-mean-square of the intrinsic transverse momentum of the active constituent and the valence Fock-state probability or decay constant. This means that the model parameters can thus be associated with physical parameters. Choosing $P_p = 0.5$ and $\langle \mathbf{k}_\perp^2 \rangle_p^{1/2} = 280\,\text{MeV}$ for the proton and $P_D = 0.9$ and $f_D = 206\,\text{MeV}$ for the D meson [14] gives: $N_p = 61.8\,\text{GeV}^{-2}$, $a_p = 1.1\,\text{GeV}^{-1}$ and $N_D = 55.2\,\text{GeV}^{-2}$, $a_D = 1.1\,\text{GeV}^{-1}$.

On the right panel of Fig. 1 we show the overlap integral with respect to the $\bar{x}$ and the $\bar{\mathbf{k}}_\perp$ integration which appears within the square brackets of Eq. (5) when using the wave functions (7) with the parameters as chosen above. We have plotted it as a function of $\cos\theta$ for different values of Mandelstam s, namely $s = 15$, 20 and $30\,\text{GeV}^2$ corresponding to the dotted, dashed and solid lines, respectively. One observes that the magnitude of the overlap decreases with increasing CMS energies s and, for all values of Mandelstam s, it shows an increase when going to small scattering angles.

On the left-hand side of Fig. 2 we present our predictions for the differential $p\bar{p} \to \overline{D^0}D^0$ cross section $d\sigma_{p\bar{p}\to\overline{D^0}D^0}/d\Omega$ versus $\cos\theta$ for Mandelstam $s = 15\,\text{GeV}^2$. It is in the region of nb and it also shows the increase in magnitude for decreasing CMS scattering angle θ. This property is inherited from the behavior of the overlap. On the right-hand side of Fig. 2 we show

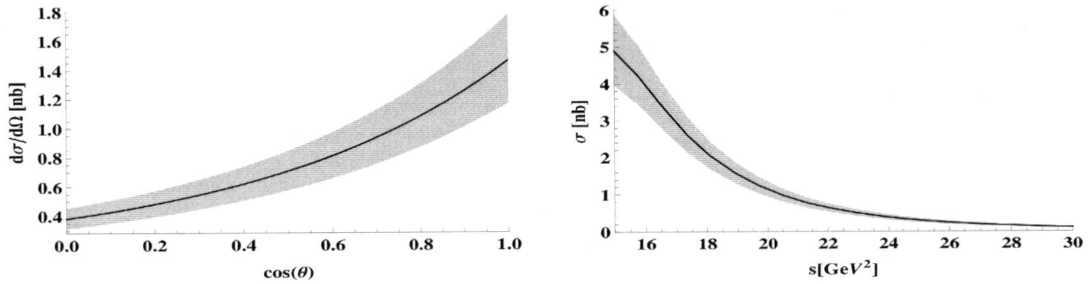

Figure 2. *Left:* The differential cross section $d\sigma_{p\bar{p}\to\overline{D^0}D^0}/d\Omega$ for $s = 15\,\mathrm{GeV}^2$ versus $\cos\theta$. *Right:* The integrated $p\bar{p} \to \overline{D^0}D^0$ cross section as a function of s.

our integrated cross-section predictions as a function of s. It is of the order of nb, which still is in the range $\bar{\mathrm{P}}$ANDA is able to measure. Our cross section predictions are in accordance with the results of Ref. [13], where also a quark-diquark model has been used, but are one order of magnitude smaller than the hadronic interaction-model calculations of Refs. [15, 16].

5. Summary

We have investigated the process $p\bar{p} \to \overline{D^0}D^0$ within a double handbag approach where the process amplitude can be factorized into a hard subprocess amplitude on the constituent level and soft hadronic $p \to \overline{D^0}$ and $\bar{p} \to D^0$ matrix elements. We have treated the hard subprocess perturbatively and modelled the soft hadronic transitions as a LCWF overlap. In doing so we have obtained predictions for the differential and integrated $p\bar{p} \to \overline{D^0}D^0$ cross sections.

Acknowledgments

We would like to thank the organizing committee of the "FAIRNESS 2013" workshop for their efforts. ATG is supported by the Austrian science fund FWF under Grant No. J 3163-N16.

References

[1] Goritschnig A T, Pire B and Schweiger W 2013 *Phys. Rev.* D **87** 014017; 2013 **Erratum-ibid** *Phys. Rev.* D **88** 079903(E)
[2] Lutz M F M et al. [PANDA Collaboration] *arXiv* hep-ex/0903.3905
[3] Mueller D, Robaschik D, Geyer B, Dittes F M and Horejsi J 1994 *Fortsch. Phys.* **42** 101; Radyushkin A V 1996 *Phys. Lett.* B **380** 417; Ji X D 1997 *Phys. Rev. Lett.* **78** 610
[4] Goritschnig A T, Kroll P and Schweiger W 2009 *Eur. Phys. J.* A **42** 43
[5] Pire B, Semenov-Tian-Shansky K and Szymanowski L 2010 *Phys. Rev.* D **82** 094030; 2011 *Phys. Rev.* D **84** 074014; 2013 *Phys. Lett.* B **724** 99; Lansberg J P, Pire B, Semenov-Tian-Shansky K and Szymanowski L 2012 *Phys. Rev.* D **85** 054021
[6] Diehl M, Feldmann T, Jakob R and Kroll P 2001 *Nucl. Phys.* B **596** 33; 2001 **Erratum-ibid** *Nucl. Phys.* B **605** 647; Brodsky S J, Diehl M and Hwang D S 2001 *Nucl. Phys.* B **596** 99
[7] Diehl M, Feldmann Th, Jakob R and Kroll P 1999 *Eur. Phys. J.* C **8** 409
[8] Körner J G and Kroll P 1992 *Phys. Lett.* B **293** 201
[9] Kniehl B A and Kramer G 2005 *Phys. Rev.* D **71** 094013; 2006 *Phys. Rev.* D **74** 037502
[10] Anselmino M, Kroll P and Pire B 1987 *Z. Phys.* C **36** 89
[11] Anselmino M, Caruso F, Kroll P and Schweiger W 1989 *Int. J. Mod. Phys.* A **4** 5213
[12] Kroll P, Pilsner T, Schürmann M and Schweiger W 1993 *Phys. Lett.* B **316** 546
[13] Kroll P, Quadder B and Schweiger W 1989 *Nucl. Phys.* B **316** 373
[14] Beringer J et al. [Particle Data Group] 2012 *Phys. Rev.* D **86** 010001
[15] Khodjamirian A, Klein C, Mannel T and Wang Y M 2012 *Eur. Phys. J.* A **48** 31
[16] Haidenbauer J and Krein G 2011 *Few Body Syst.* **50** 183

Journal of Physics: Conference Series **503** (2014) 012013 doi:10.1088/1742-6596/503/1/012013

Antihyperon-hyperon production in antiproton-proton annihilations with PANDA

Karin Schönning[a] and Erik Thomé[a] for the PANDA collaboration

[a] Dept. of Physics and Astronomy,bUppsala University, Box 516, 75120 Uppsala, Sweden

E-mail: karin.schonning@physics.uu.se

Abstract. Hyperon production is an excellent probe of the strong interaction in the confinement domain. The spin observables provide a powerful tool in understanding the underlying physical processes. Expressions of the seven non-zero polarisation parameters of the Ω baryon from decay angular distributions have recently been derived by the Uppsala group. Simulation studies show that all strange and single charmed hyperon production channels have great prospects with the PANDA detector.

1. Introduction

Hyperon production in $\bar{p}p \to \overline{Y}Y$ reaction gives important insights into strangeness and charm production. In this work, we consider single- and multi-strange and single-charmed hyperons. Their production from light-quark systems like $\bar{p}p$ implies processes where light quarks are replaced by heavier quarks. The relevant degrees of freedom of a certain process are given by its energy scale, which for strangeness production is governed by the mass of the strange quark, $m_s \approx 100$ MeV. This is close to the QCD cut-off, $\Lambda_{QCD} \approx 200$ MeV where the strong coupling constant, α_s, grows so large that perturbative QCD breaks down. As a consequence, it is unclear what the relevant degrees of freedom for processes involving strangeness production are: quarks and gluons, or hadrons? The production of strange hyperons therefore probes QCD in the intermediate region between the perturbative and the non-perturbative regime, *i.e.* the confinement domain, that we, up to now, know very little about. The scale of charm production is governed by $m_c \approx 1300$ MeV, more than ten times larger than m_s. The strong coupling constant in this region is $\alpha_s \approx 0.3$, just barely enough for a perturbative treatment to be valid. Comparing the production of strange hyperons with charmed could thus give important insights into the differences in the underlying physics at these two separate energy scales. Theoretical models describing hyperon production in $\bar{p}p \to \overline{Y}Y$ reactions are often based on the quark-gluon picture, where a light quark-antiquark pair from the initial nucleons annihilate into a gluon that splits into a heavier quark-antiquark pair ($s\bar{s}$ or $c\bar{c}$) [1]. For strange hyperons, there are also kaon exchange models, where the production of single (multiple) strangeness hyperons are modeled by the single (multiple) exchange of a t-channel kaon [2]. There have also been attempts to combine a quark-gluon approach with a kaon exchange approach [3].

Spin variables are often very powerful in discriminating between different theoretical models. The hyperon spin variables can be related to the spin of individual quarks. The Λ hyperon can be modeled by a ud spin 0 di-quark combined with an s-quark that carries the spin of the entire hyperon. A similar picture can be drawn for the Λ_c hyperon, and by comparing spin observables of Λ and Λ_c, we can learn about the role of spin in the creation of s- and c-quarks.

Another interesting aspect of hyperons is CP violation. One of the unsolved mysteries of contemporary physics is why the universe consists of matter while antimatter is almost absent. Violation of CP symmetry is one possible source that gives rise to this asymmetry [4]. CP violation has been observed in meson decays, but so far never for baryons. For hyperons, CP violation observables are accessible *via* angular distributions of hyperon decay products.

In this work, the potential of spin measurements in multi-strange and charmed hyperon production is described. First, we outline the polarisation parameters of spin $\frac{1}{2}$ and spin $\frac{3}{2}$ hyperons, the latter as recently derived by the Uppsala group. Then we present the spin observables of the full $\bar{p}p \to \bar{Y}Y$ process in the case of spin $\frac{1}{2}$ hyperons, based on previous work by *e.g* the PS185 collaboration. We conclude with the prospects for hyperon studies with PANDA, based on simulation studies performed by the Uppsala group.

2. Spin observables for spin $\frac{1}{2}$ and spin $\frac{3}{2}$ hyperons

Polarisation parameters for spin $\frac{1}{2}$ hyperons

All physics information about a quantum mechanical ensemble is contained within the density matrix ρ. In an expansion of hermitian matrices Q^L_M and polarisation parameters r^L_M [5], the density matrix of a particle of arbitrary spin j is given by

$$\rho = \frac{1}{2j+1} I + \sum_{L=1}^{2j} \frac{2j}{2j+1} \sum_{M=-L}^{L} Q^L_M r^L_M \tag{1}$$

where the first term denotes the unpolarised differential cross section I and the second the polarised part, containing the r^L_M parameters. L is the angular momentum and M its projection. In the case of spin $\frac{1}{2}$ particles, the Q^L_M are the Pauli matrices and the polarisation parameters correspond to the vector polarisations P_l, P_m and P_n. The indices are defined in the coordinate system in the left panel of Fig. 1. For particles produced in parity conserving processes, *e.g* hyperons in strong interactions, symmetries of the spin density matrix imply that $P_l = P_m = 0$, giving non-zero polarisation only perpendicular to the production plane. The angular distribution of a particle from a given decay can be expressed in terms of the spin density matrix of the mother particle: $I = T\rho_{initial}T^\dagger$, where T is the decay matrix [6]. For the case of a weak decay of a spin $\frac{1}{2}$ baryon into a spin $\frac{1}{2}$ baryon and a pseudoscalar meson, the final state can have angular momentum $L = 0$, that violates parity, or $L = 1$, that conserves parity. This gives a decay matrix with one parity violating part T_s and one parity conserving part T_p. From their products, the asymmetry parameters $\alpha = 2Re(T_s^*T_p)$, $\beta = 2Im(T_s^*T_p)$ and $\gamma = |T_s|^2 - |T_p|^2$ can be defined. By definition, they fulfill $\alpha^2 + \beta^2 + \gamma^2 = |T_s|^2 + |T_p|^2 = 1$. For example, in the $\Lambda \to p\pi^-$ decay, the angular distribution of the proton is given by

$$I(\cos\theta_p) = \frac{1}{4\pi}(1 + \alpha_\Lambda P_n \cos\theta_p). \tag{2}$$

Some hyperons decay into other hyperons, *e.g.* the Ξ baryons, as illustrated in the right panel of Fig. 1. In the $\Xi^- \to \Lambda\pi^-, \Lambda \to p\pi^-$ process, the asymmetry parameters β and γ of the Ξ^- hyperon are accessible *via* the angular distribution of the protons [7]:

$$I(\theta_p, \phi_p) = \frac{1}{4\pi}(1 + \alpha_\Xi\alpha_\Lambda \cos\theta_p + \frac{\pi}{4}\alpha_\Lambda P_\Xi \sin\theta_p(\beta_\Xi \sin\phi_p - \gamma_\Xi \cos\phi_p)). \tag{3}$$

This is valid if the reference system is defined such that the spin of the Ξ^- hyperon is along the z-axis, the momentum of the daughter Λ hyperon is in the positive xz plane and the angles

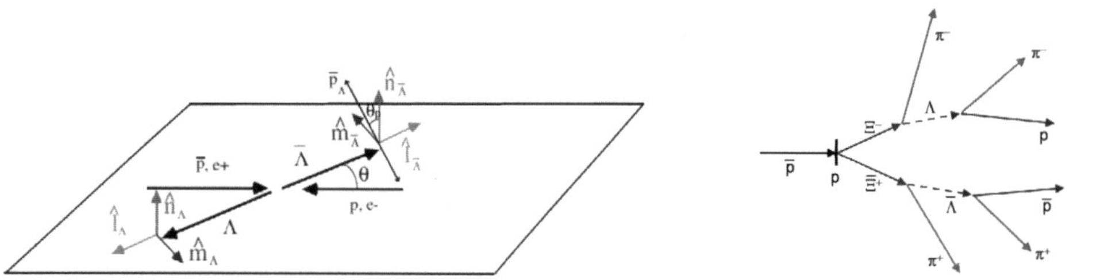

Figure 1. Left: Coord. system of $\overline{p}p \to \overline{Y}Y$. Right: The $\overline{p}p \to \overline{\Xi}\Xi, \Xi \to \Lambda\pi, \Lambda \to p\pi$ decay.

of the Ξ^- decay are integrated over. Furthermore, α_Λ is known, α_Ξ can be extracted from the angular distribution of the Λ from the Ξ^- decay and finally, $\alpha_\Xi^2 + \beta_\Xi^2 + \gamma_\Xi^2 = 1$ can be applied to extract all unknowns from Eq. 3.

Polarisation parameters for spin $\frac{3}{2}$ hyperons

For spin $\frac{3}{2}$ hyperons, the spin structure is more complicated. The L number in Eq. 1 can be 3, 2 or 1. This gives three Q_M^1, five Q_M^2 and seven Q_M^3 matrices, all in all fifteen Q_M^L matrices with fifteen corresponding r_M^L parameters. The spin density matrix was derived in Ref. [9]. Using symmetries imposed by strong interaction, eight polarisation parameters were found to be zero.

In the weak decay of a spin $\frac{3}{2}$ baryon into a spin $\frac{1}{2}$ baryon and a pseudoscalar, *e.g.* $\Omega \to \Lambda K^-$, the final state can have angular momentum $L = 1$, giving the parity conserving part of the decay matrix, or angular momentum $L = 2$, giving the parity violating part. It has been found that the asymmetry parameter α_Ω is consistent with zero [10] and in the following, it is set to exactly zero to simplify the calculations. Defining the reference system in the same way as in the spin $\frac{1}{2}$ case (see left panel of Fig. 1 but replace Λ with Ω and p with Λ) the angular distribution of the Λ from the Ω decay is given by

$$I(\theta_\Lambda, \phi_\Lambda) = \frac{1}{4\pi}(1 + \frac{\sqrt{3}}{2}(1 - 3\cos^2\theta_\Lambda)r_0^2 - \frac{3}{2}\sin^2\theta_\Lambda \cos 2\phi_\Lambda r_2^2 + \frac{3}{2}\sin 2\theta_\Lambda \cos\phi_\Lambda r_1^2) \quad (4)$$

Using the *method of moments* [8], the three polarisation parameters appearing in Eq. 4 can be obtained in terms of the moments of the sine and cosine functions, denoted within $\langle\rangle$:

$$\begin{aligned} r_0^2 &= \tfrac{15}{2\sqrt{3}}(\tfrac{1}{3} - \langle\cos^2\theta_\Lambda\rangle) \\ r_2^2 &= \tfrac{8}{3}(1 - \langle\cos^2\theta_\Lambda\rangle - 2\langle\sin^2\theta_\Lambda \sin^2\phi_\Lambda\rangle) \\ r_1^2 &= 5\langle\cos\theta_\Lambda \sin\theta_\Lambda \cos\phi_\Lambda\rangle \end{aligned} \quad (5)$$

For details, see Ref. [9]. More information about the remaining four non-zero polarisation parameters can be obtained by studying the combined angular distribution of the Λ hyperons from the Ω decay and the protons from the subsequent Λ decay, $I(\theta_\Lambda, \phi_\Lambda, \theta_p, \phi_p)$. The expression is lengthy for these proceedings but can be multiplied with certain trigonometric functions and integrated over the angles to extract the polarisation parameters [11]:

$$\begin{aligned} r_{-1}^1 &= -\frac{20\sqrt{10}\langle(3\cos\theta_\Lambda - 1)\sin\phi_p\rangle}{3\pi\alpha_\Lambda\gamma_\Omega} \\ r_{-1}^3 &= \frac{2\sqrt{5}\langle(15\cos\theta_\Lambda - 1)\sin\phi_p\rangle}{\sqrt{3}\pi\alpha_\Lambda\gamma_\Omega} \\ r_{-2}^3 &= -\frac{1024\langle\sin\phi_\Lambda\cos\phi_p\rangle}{3\pi^2\alpha_\Lambda\gamma_\Omega} \\ r_{-3}^3 &= -\frac{1}{5\sqrt{6}}(\frac{640}{\pi\alpha_\Lambda\gamma_\Omega}\langle\sin\phi_\Lambda\cos\phi_\Lambda\sin\phi_p\rangle + 4\sqrt{15}r_{-1}^3 + 3\sqrt{10}r_{-1}^1) \end{aligned} \quad (6)$$

Spin observables for $\frac{1}{2}$ hyperons in the $\overline{p}p \to \overline{Y}Y$ reaction

In the $\overline{p}p \to \overline{Y}Y$ reaction, not only the polarisation of individual hyperons are of interest, but also their correlations. In Ref. [12], the derivation of the spin observables for the spin $\frac{1}{2}$ case is performed in detail. Here we only summarise the results. The spin observables of the $\overline{p}p \to \overline{Y}Y$ reaction can be written in terms of the momentum vectors of the final state antibaryon $\overline{B}$ and baryon B from the $\overline{Y}$and Y decays:

$$I_0^{\overline{B}B} = \frac{I_0^{\overline{Y}Y}}{64\pi^3} \sum_{\mu,\nu=0}^{3} \sum_{i,j=0}^{3} \overline{\alpha}\alpha P_i^{\overline{p}} P_j^{p} \chi_{ij\mu\nu} \overline{k}_\mu k_\nu \tag{7}$$

where $I_0^{\overline{Y}Y}$ is the unpolarised angular distribution, α ($\overline{\alpha}$) denotes the asymmetry parameter of the Y ($\overline{Y}$), P_j^p ($P_i^{\overline{p}}$) the polarisation of the initial state proton (antiproton), k_ν ($\overline{k}_\mu$) the momentum vector of the final state decay baryon (antibaryon) and $\chi_{ij\mu\nu}$ the 256 spin observables. For the case of an unpolarised beam and an unpolarised target, $i = j = 0$, giving sixteen accessible variables. Due to symmetries, only six of these are non-zero: the polarisations $P_{\overline{Y},\mu} = \chi_{00\mu0}$ and $P_{Y,\nu} = \chi_{000\nu}$ and the spin correlations between the Y and the $\overline{Y}$, $C_{\mu\nu} = \chi_{00\mu\nu}$.

3. Prospects for PANDA

The foreseen PANDA experiment at FAIR opens up new possibilies in hyperon physics. The antiproton beam from the HESR storage ring, operating in an energy range between 1.5 GeV and 15 GeV, will interact with an internal hydrogen target. The PANDA experiment will have high luminosity and provide a near 4π acceptance featuring precise tracking and vertex reconstruction, sophisticated particle identification and calorimetry. For further details, see Ref. [13] and references therein. Simulation studies, described in Refs. [9, 16] show excellent prospects: high signal rate, low background rate and good detection efficiency over the full phase space for all single- and multi-strange and single-charmed hyperons. The estimated signal rates are given in Table 1. They are calculated assuming the high luminosity operating mode of the HESR that corresponds to $2{\cdot}10^{32}\text{cm}^{-2}\text{s}^{-1}$, as expected in the final version of FAIR. The only cross section that is well known here is the one for $\overline{p}p \to \overline{\Lambda}\Lambda$ at 1.64 GeV. In the remaining strangeness channels the quoted cross sections are extrapolations from data collected at other energies. For the Ω and Λ_c channels, no data exist. The quoted values of $\sigma(\overline{p}p \to \overline{\Omega}^+\Omega)$ and $\sigma(\overline{p}p \to \overline{\Lambda}_c\Lambda_c)$ are obtained from model calculations in Ref. [14] and [14, 15], respectively, and may be up to an order of magnitude different in reality. For the Λ and Ξ channels, the large foreseen statistics enables precision CP violation studies. If the cross section estimate for $\overline{p}p \to \overline{\Xi}^+\Xi^-$ is correct, the large Ξ data samples from HyperCP [17] can be exceeded in $\approx$80 days of data taking at 4 GeV. PANDA has the advantage that the $\overline{p}p$ in the initial state produces the same amount of hyperons and antihyperons. The simulation studies in Refs. [9, 16], show that spin observables of all hyperon channels can be well reconstructed in PANDA. Some examples are shown in Fig. 2.

4. Summary

The future PANDA experiment at FAIR has a unique opportunity to give new insights into the strong interaction in the confinement domain. Spin observables in antihyperon-hyperon production provide a powerful tool for this purpose. It will be possible to study the $\overline{p}p \to \overline{\Omega}^+\Omega^-$ reaction for the first time and seven non-zero polarisation parameters have recently been derived by the Uppsala group. Furthermore, high precision tests of CP violation in hyperon decay can be performed with the foreseen hyperon data samples from PANDA.

References

[1] Kohno M. and Weise W., 1986 *Phys. Lett.* B. **179** 15; Rubinstein H.R. and Snellman H. 1985 *Phys. Lett.* B. **165** 187; Furui S. and Faessler A. 1987 *Nucl. Phys.* A **486** 669;Burkardt M. and Dillig M. 1988 *Phys. Rev.* C **37** 1362; Alberg M.A. *et al.* 1988 *Z. Phys.* A **331** 207.

Journal of Physics: Conference Series **503** (2014) 012013 doi:10.1088/1742-6596/503/1/012013

Table 1. Approximate foreseen production rates of various hyperon channels [9, 16]. The cross sections marked with a * have a very large uncertainty. Details are given in the text.

Beam mom. (GeV/c)	Reaction	σ (μb)	Eff (%)	Decay	Rate at $2{\cdot}10^{32}\mathrm{cm}^{-2}\mathrm{s}^{-1}$
1.64	$\bar{p}p \to \bar{\Lambda}\Lambda$	64	11	$\Lambda \to p\pi^-$	580 s^{-1}
4	$\bar{p}p \to \bar{\Lambda}\Lambda$	≈50	23	$\Lambda \to p\pi^-$	980 s^{-1}
15	$\bar{p}p \to \bar{\Lambda}\Lambda$	≈10	14	$\Lambda \to p\pi^-$	120 s^{-1}
4	$\bar{p}p \to \bar{\Lambda}\Sigma^0$	≈40	31	$\Sigma^0 \to \Lambda\gamma$	600 s^{-1}
4	$\bar{p}p \to \bar{\Xi}^+\Xi^-$	≈2	19	$\Xi^- \to \Lambda\pi^-$	30 s^{-1}
12	$\bar{p}p \to \bar{\Omega}^+\Omega^-$	≈0.002*	≈30	$\Omega \to \Lambda K^-$	≈80 h^{-1}
12	$\bar{p}p \to \bar{\Lambda}_c^-\Lambda_c^+$	≈0.1*	≈35	$\Lambda_c \to \Lambda\pi^+$	≈25 d^{-1}

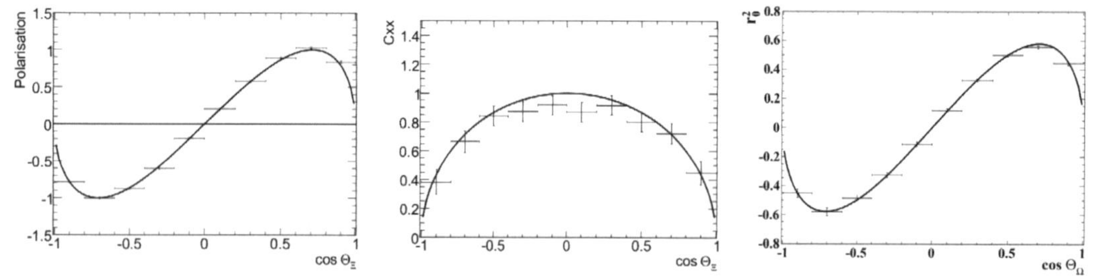

Figure 2. Left: The polarisation of the Ξ^- as a function of the Ξ^- scattering angle $\cos\theta_\Xi$. Middle: The spin correlation in the x direction (m in Fig. 1) of the $\bar{\Xi}^+$ and the Ξ^-. Right: The polarisation parameter r_0^2 of the Ω as a function of $\cos\theta_\Omega$. The points represent reconstructed values and the lines the input trigonometric functions.

[2] Tabakin F. and Eisenstein R.A. 1985 *Phys. Rev.* C **31** 1857; Kohno M. and Weise W., 1986 *Phys. Lett.* B. **179** 15; La France P. *et al.* 1988 *Phys. Lett.* B. **214** 317; Timmermans R.G.E. *et al.* 1992 *Phys. Rev.* D **45** 2288; Haidenbauer J. *et al.* 1992 *Phys. Rev.* C **46** 2516.
[3] Ortega P.G. *et al.* 2011 *Phys. Lett.* B. **696** 352.
[4] Sakharov A.D. 1967 *Pis'ma Zh. Eksp. Teor. Fiz.* **5** 32; Sakharov A.D. 1976 *JETP Lett.* **5** 24, [hep-ex/0408100].
[5] Doncel M.G. *et al.* 1972 *Nucl. Phys* B **38** 477; Doncel M.G. *et al.* 1973 *Phys. Rev.* D **7** 815.
[6] Pilkuhn H. 1967 *The interaction of hadrons* (North-Holland Publishing Company, Amsterdam)
[7] Koch W. 1968 *Analysis of scattering and decay* ed. Nikolic M. (Gordon and Breach, new York-London-Paris).
[8] Frodesen A.G. *et al.* 1979 *Probability and Statistics in Particle Physics* (Universitetsforlaget).
[9] Thomé E. 2012 *Multi-Strange and Charmed Antihyperon-Hyperon Physics for PANDA* Ph. D. Thesis, Uppsala University.
[10] Yao W. M. *et al.* [Particle Data Group] 2006 *J. Phys.* G **33** 1.
[11] Thomé E. 2013 *A Method for Measurement of Polarisation Parameters in the* $\bar{p}p \to \bar{\Omega}^+\Omega^-$ *Reaction* (in preparation).
[12] Paschke K. 2001 Ph. D. Thesis, Carnegie Mellon University.
[13] Fioravanti E. 2013 *Experimental overview of PANDA* (These proceedings).
[14] Kaidalov A.B. and Volkovitsky P.E. 1994 *Z. Phys.* C **63** 51
[15] Titov A.I. and Kämpfer B. 2008 *Phys. Rev.* C **78** 025201; Gornitschnig A.T. *et al.* 2009 *Eur. Phys. J.* A **42** 43; He J. *et al.* 2011 *Phys. Rev.* D **84** 114010; Khodjamirian A. *et al.* 2012 *Eur. Phys. J.* A **48**.
[16] Grape S. 2009 *Studies of PWO Crstals and Simulations of the* $\bar{p}p \to \bar{\Lambda}\Lambda, \bar{\Lambda}\Sigma^0$ *Reactions for the PANDA experiment* Ph.D. Thesis, Uppsala University.
[17] T. Holmstrom *et al.* 2004 *Phys. Rev. Lett.* **93** 262001.

Journal of Physics: Conference Series **503** (2014) 012014 doi:10.1088/1742-6596/503/1/012014

Electron identification in Au+Au collisions at 1.23 GeV/u using multivariate analysis

Szymon Harabasz for the HADES Collaboration
Technische Universität Darmstadt, Schlossgartenstraße 9, 64289 Darmstadt, Germany
Jagiellonian University, ul. Reymonta 4, 30-059 Kraków, Poland
E-mail: s.harabasz@gsi.de

Abstract. Au+Au collisions at a beam kinetic energy of 1.23 GeV/u have been measured by HADES in 2012. Lepton identification in this experiment has been done using a multivariate algorithm based on an artificial neural network. In the proceedings, details of the identification method and its assessment in terms of purity of the final lepton sample are presented. The obtained purity reaches 95% and the amount of identified electrons and positrons is sufficient to perform further steps of the physics analysis with e^+e^- pairs.

1. Introduction

Dilepton spectroscopy allows to investigate the properties of hadrons embedded in a strongly interacting medium. Dileptons are not subject to the strong force, which makes them able to leave a region of a nuclear reaction practically undistorted by any final-state interactions. Moreover, as leptons are radiated during all stages of nuclear collisions, they provide an insight into the hot and dense fireball as well as into first-chance interactions and the freeze-out. This is in contrast to hadrons, whose multiplicities and kinematical properites are established in the last stage, when hadronic interactions vanish. Considering that the dilepton invariant mass spectrum is also closely related to the issue of modifications of light vector meson in nuclear matter, dilepton spectoscopy offers a possibility to investigate the question of mass generation. However, production of dileptons is suppressed by a factor of the order of α^2 (α - fine structure constant), thus their measurements are highly demanding from an experimental point of view.

An enhanced yield of dileptons with masses below the vector meson (i.e. ρ^0 and ω meson) pole mass appears to be a general feature of heavy-ion reactions, from bombarding energies as low as 1 GeV/u, studied by the former DLS experiment [1] at the Bevalac, through the range of SPS energies (40–158 GeV/u) used by the CERES [2][3] and NA60 [4] experiments at CERN, up to the highest energies (with $\sqrt{s}_{NN} = 200$ GeV) employed by the PHENIX [5] and STAR [6] experiments (the latter performed also a beam energy scan [7]) at the RHIC. This enhancement is defined as the excess of the measured pair yield over the summed-up cocktail of dileptons from long-lived sources, namely the decays of π^0, η, and ω mesons. It is hence expected to probe the early phases of the collision.

The High-Acceptance DiElectron Spectrometer (HADES) operates at the GSI Helmholtzzentrum für Schwerionenforschung in Darmstadt where it has started a systematic investigation of dilepton production in the SIS/Bevalac energy regime of 1–2 GeV/u.

Journal of Physics: Conference Series **503** (2014) 012014 doi:10.1088/1742-6596/503/1/012014

One of main results is shown in Fig.1, which compares dilepton invariant mass spectra in heavy-ion and nucleon-nucleon collisions. It appears, that while the light system C+C may be treated as a superposition of individually colliding nucleons [8], this is not any longer true for heavier Ar+KCl system, where there is an additional source of dilepton radiation emerging above $M_{ee} = 150$ MeV/c^2. Current results show the scaling of this so called "excess" as $\propto A_{\mathrm{part}}^{1.4}$ with the number of nucleons participating in the interaction [9].

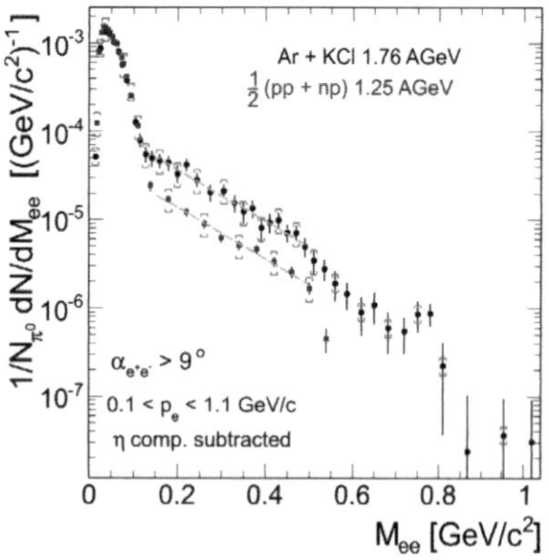

Figure 1. Invariant mass spectrum of dielectrons measured in Ar+KCl at 1.76 GeV/u compared to a reference spectrum from elementary collisions. Opening angle and single particle momentum cuts are applied and a "trivial" component of η (with $T_{1/2}$ longer than hot and dense matter stage) is identified and subtracted.

To confirm this result and to further investigate dilepton production enhancement, an experiment with the heaviest system available at SIS18, Au+Au at 1.23 GeV/u was performed in April–May 2012. In this experiment, due to a large hadron multiplicity, traditional methods of electron identification have been proved to be insufficient. In this paper, results of a multivariate lepton identification in Au+Au experiment are presented.

2. The HADES Experiment

HADES is a charged-particle detector consisting of a 6-coil toroidal magnet centered on the beam axis and six identical detection sections covering polar angles between 18° and 85° with nearly complete azimuthal coverage. The main component serving for electron and positron selection is a hadron-blind Ring-Imaging Cherenkov detector (RICH). Further particle identification power is provided by combined effect of (i) the time-of-flight (TOF) measurement in a plastic scintillator and Resistive-Plate Chamber (RPC) walls, (ii) the electromagnetic shower characteristics observed in a PreShower detector, and (iii) the energy-loss signals from the scintillators of the TOF wall as well as from the four planes of drift chambers serving as tracking stations. A detailed description of the HADES setup can be found in [10].

In the experiment discussed here, a gold beam with a kinetic energy of 1.23 GeV/u was used to bombard a 15-fold segmented gold target (for details see Tab.1).

3. Electron identification using multivariate analysis

In the data analysis, e^+ and e^- are identified by applying appropriate selection cuts to the RICH ring observables, time-of-flight, PreShower and energy loss signals. Particle momenta were obtained by tracking the charged particles through the HADES magnetic field.

In a first attempt one could impose one-/two-dimensional conditions (so called hard cuts) on various observables in order to suppress hadrons and to keep as many electrons as possible. However, to find the optimal set of cuts is very difficult. Applying or modifying a requirement on one quantity changes what would be regarded as best condition on all the others. Another difficulty is that a one- or two-dimensional cut ignores all other dimensions (other variables). This makes the selection less flexible. In addition, it is possible to accept a particle, which just

Table 1. Basic facts about April-May 2012 Au+Au beam-time.

Beam kinetic energy	1.23 GeV/u
Total beam-on-target time	557 hours
Beam intensity	$(1.2 - 1.5) \cdot 10^6$ ions/s
Trigger rate	8 kHz
Data rate	200 MB/s
Total events collected	$7.3 \cdot 10^9$
Total data collected	140 TB

slightly passes all tests and on the other hand reject a particle which meets well all conditions but one.

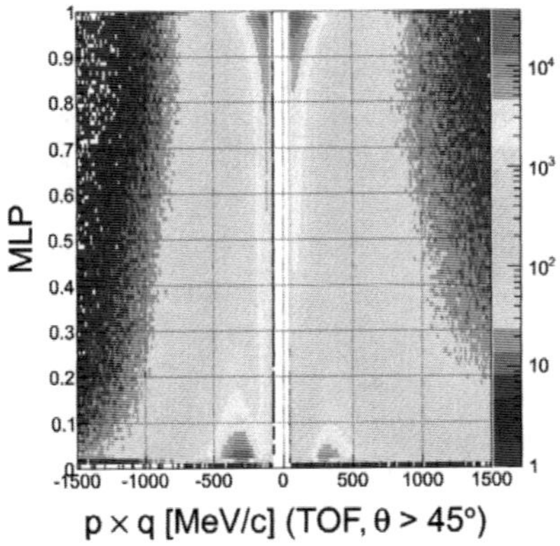

Figure 2. Neural network output (denoted as MLP for multilayer perceptron) versus particle's momentum for tracks falling into TOF scintillator wall.

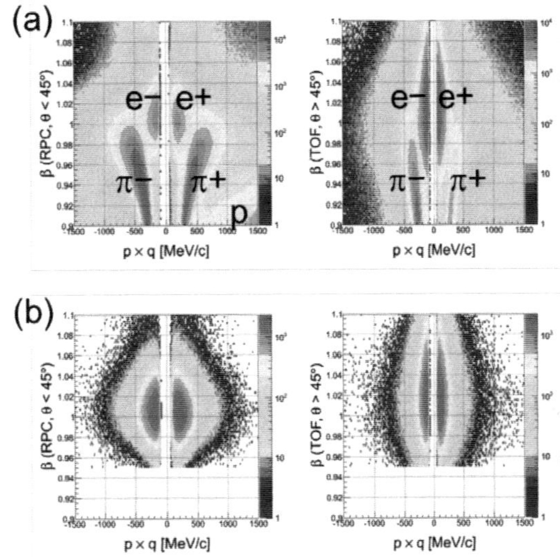

Figure 3. Particle's velocity versus momentum for (a) all particles, (b) particles, for which the neural network output function has values larger than 0.6. On the left side there are particles measured in the RPC detector, on the right side – in the TOF scintillator.

All these difficulties may be addressed by classification algorithms of multivariate analysis, which allow to create a multidimensional decision boundary optimizing it in all dimensions (variables) at once. In the current work, the implementation available as a part of the ROOT data analysis framework (TMVA [11]) was used. From the number of algorithms offered, the Multilayer Perceptron (MLP) model of artificial neural network was chosen.

The algorithm has to be trained with well defined samples of signal and background particles, to determine characteristic properties of the two and be able to classify tracks of the whole data sample later on. There are two challenges in training the neural network to give the correct response: one is the selection of pure signal and background training samples. The other one is the choice of variables with the highest separation power between hadron and lepton tracks. The signal and background samples are taken from experimental data (not from simulation), using a strong cut on the matching of track to RICH ring, which has sufficiently large separation power to generate a pure electron sample. The matching can be quantified by track-ring matching quality:

$$C = \sqrt{(\Delta\theta_{\text{track-ring}})^2 + (\Delta\phi_{\text{track-ring}} \cdot \sin\theta)^2}.$$

Electron tracks are well matched to rings and in their case C distribution is peaked around 0 (cf. Fig.4 (b)), while for any other particle species it has a broad shape. For the training of the neural network tracks with $C < 0.5^\circ$ are treated as signal, while those with $C > 7^\circ$ are regarded as background.

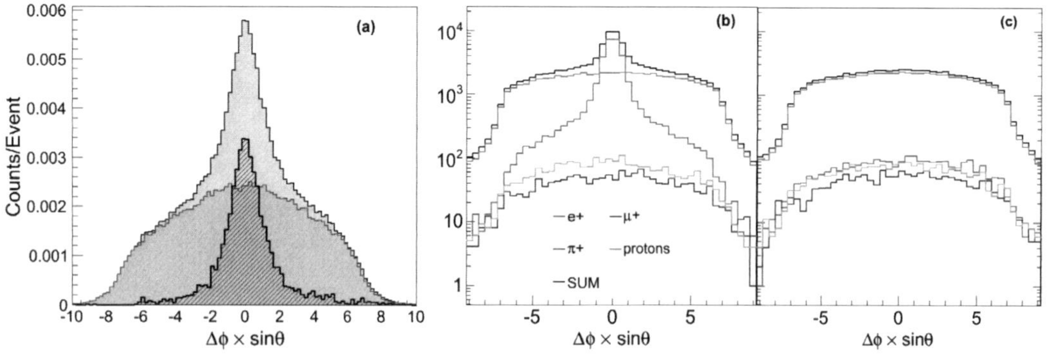

Figure 4. Spatial correlation between MDC tracks and RICH rings: (a) experimental data (green), rotated RICH (background - orange), difference of the two (signal - dark blue hatched); UrQMD simulation without (b) and with (c) rotated RICH, depicting different particle types components.

The neural network output is shown in Fig.2 as a function of particle momentum. The closer to value 1, the more certainly the particle is a lepton; events closer to 0 are related to background. Finding a boundary which separates between signal and background is a matter of compromise between purity and efficiency of the identification. In the further analysis, tracks with MLP values smaller than 0.6 have been discarded. A comparison of velocity-momentum plots before and after applying this MLP cut is presented in Fig.3. It can be seen that most of hadrons have been removed and a significant statistics of electrons has been kept.

4. Background estimations

To assess the quality of electron identification, one has to estimate the amount of background still present in the sample. This can be done by adding 60° to the azimuthal position of each RICH ring (thus moving it to the neighboring sector of the spectrometer) and then perform the usual ring-track matching procedure and subsequent analysis. True matches, corresponding to leptons, should then be destroyed, but false i.e. random matches should be statistically reproduced in the spectra. The overlay of a distribution of not rotated events with the background prepared in such a way is shown in Fig.4 (a). In Fig.4 (b) and (c) one can compare contributions of different particle species from simulation (UrQMD [12][13], Au+Au at 1.25 GeV/u, $b_{\max} = 9$ fm) analyzed in the same manner as the experimental data. The agreement of pions, protons and muons spectra between data and data with and without rotated rings confirms that the RICH rotating procedure allows to reproduce perfectly all components of background.

The difference between the original data and the reproduced background can be regarded as a signal. The purity of the electron sample may be then calculated as $P = \frac{\text{signal}}{\text{signal+background}}$. It is shown in Fig.5 for different momentum bins, compared to the best hard cut result obtained before the application of the multivariate analysis method. Global values are 84.8 % for hard cut and 94.8 % for multivariate analysis. A way to calculate the identification efficiency is currently under investigation, but comparing total number of identified leptons $2.45 \cdot 10^8$ with $1.71 \cdot 10^8$ from hard-cut analysis supports the superiority of the multivariate approach.

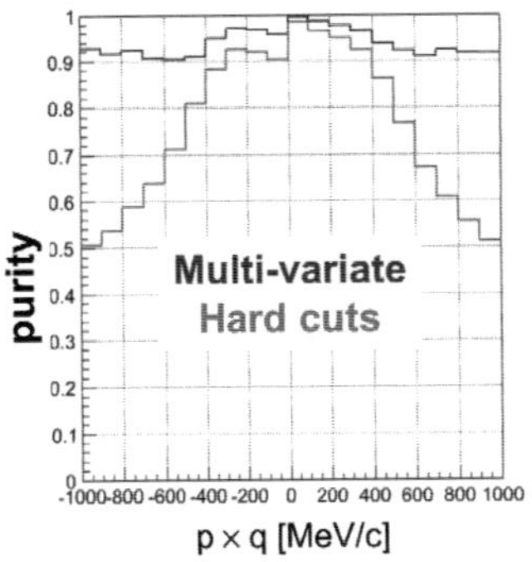

Figure 5. Lepton sample purity in momentum bins, calculated as described in text, for multi-variate analysis using neural network approach compared to the best result obtained with hard cut analysis.

5. Summary

The multilayer perceptron, which is one of the algorithms of multivariate analysis of the TMVA framework provides a unique way to obtain a pure and efficient sample of electrons in a strongly hadron dominated environment of the Au+Au at 1.23 GeV/u experiment. Such a sample is necessary to obtain reliable spectra of dileptons with properly treated calculation of the combinatorial background which has to be subtracted from the pair yield. Multivariate analysis is superior to the traditional hard cut approach and will be applied in further steps of dielectron analysis.

Acknowledgments

The collaboration gratefully acknowledges the following funding: INFN-LNS Catania (Italy); LIP Coimbra (Portugal): PTDC/FIS/113339/2009; SIP JUC Cracow (Poland): NN202198639; GSI Darmstadt (Germany): Helmholtz Alliance HA216/EMMI; TU Darmstadt (Germany): VH-NG-823, Helmholtz Alliance HA216/EMMI; HZDR, Dresden (Germany): 283286, 05P12CRGHE; Goethe-University, Frankfurt (Germany): Helmholtz Alliance HA216/EMMI, HIC for FAIR (LOEWE), GSI F&E, BMBF 06FY9100I; TU München, Garching (Germany): BMBF 06MT7180; JLU Giessen (Germany): BMBF:05P12RGGHM; University Cyprus, Nicosia (Cyprus): UCY/3411-23100; IPN Orsay, Orsay Cedex (France): CNRS/IN2P3; NPI AS CR, Rez, (Czech Republic): MSMT LG 12007, GACR 13-06759S.

References

[1] Porter R J *et al.* (DLS Collaboration) 1997 *Phys. Rev. Lett.* **79** 1229
[2] Agakichiev G *et al.* (CERES Collaboration) 2005 *Eur. Phys. J.* C **41** 475
[3] Adamova D *et al.* (CERES Collaboration) 2003 *Phys. Rev. Lett.* **91** 042301
[4] Arnaldi R *et al.* (NA60 Collaboration) 2006 *Phys. Rev. Lett.* **96** 162302
[5] Adare A *et al.* (PHENIX Collaboration) 2010 *Phys. Rev.* C **81** 034911
[6] Huang B *et al.* (STAR Collaboration) 2013 *Nucl. Phys.* A 904–905
[7] Guertz F *et al.* (STAR Collaboration) 2012 *EPJ Web of Conferences* **36**
[8] Agakishiev G *et al.* (HADES Collaboration) 2010 *Phys. Lett.* B **690** 118
[9] Agakishiev G *et al.* (HADES Collaboration) 2011 *Phys. Rev.* C **84** 014902
[10] Agakishiev G *et al.* (HADES Collaboration) 2009 *Eur. Phys. J.* A **41** 243–277
[11] Hoecker A *et al.* 2007 *PoS* **ACAT** 040 (*Preprint* `physics/0703039`)
[12] Bass S *et al.* 1998 *Prog. Part. Nucl. Phys.* **41** 255 – 369
[13] Bleicher M *et al.* 1999 *J. Phys.* G **25** 1859

Journal of Physics: Conference Series **503** (2014) 012015 doi:10.1088/1742-6596/503/1/012015

Reconstruction of π^0 and η mesons via conversion method in Au+Au at 1.23AGeV with HADES

C Behnke[1] **for the HADES Collaboration**

[1]Goethe Universität Frankfurt, Max-von-Laue-Str. 1, 60438 Frankfurt am Main

E-mail: C.Behnke@gsi.de

Abstract. Lepton pairs emerging from decays of virtual photons are one of the most promising probes of dense hadronic matter. The HADES experiment at GSI Helmholtzzentrum für Schwerionenforschung in Darmstadt studies dielectron radiation as well as strangeness production in various reactions, i.e. proton, deuteron and heavy-ion. The understanding of the corresponding experimental results calls for supporting studies from various model calculations. For a more model independent understanding of the dilepton cocktail the production cross sections of particles created during the freeze-out is crucial. In this contribution we will present results from a two photon analysis of Au+Au in 1.23 GeV/u collisions providing information on π^0 and η mesons. Therefore the capability of HADES to detect e^+e^- pairs from external conversion of real photons will be explained and first results will be shown.

1. Introduction

Heavy-ion collisions provide a unique possibility to study nuclear matter under extreme conditions in the laboratory. The most abundantly produced and recostructed particles are pions, kaons, protons and light fragments. But the most promising tool to probe the microscopic properties of strongly interacting matter under extreme temperatures and densities is electromagnetic radiation. Photons (γ) and dileptons (e^+e^-) are penetrating all stages of a heavy-ion collision without strong interactions. Lepton pairs, emerging from decays of virtual photons (γ^*) are emitted during the whole evolution of the fireball. On the other hand they will not only probe the hot and dense stage but also the first chance collisions and the freeze-out of the collision. To estimate the contribution of the first chance collisions one can measure so called reference spectrum from elementary reactions, i.e. pp and np. These reactions have been successfully measured by the HADES Collaboration [1, 2, 3].

The contribution to the dilepton invariant mass spectra from the freeze-out stage is dominated by mesons with a long lifetime (long compared to the lifetime of the fireball), i.e. the neutral π^0, η and the ω meson. It is crucial to understand the contribution of the π^0 for the normalization of dilepton invariant mass spectra. The cross section of the η is essential to determine the non trivial enhancement of low mass (M_{ee} between 0.15 and 0.55 GeV/c^2) lepton pairs, that was found by DLS [4] and HADES [5] at Bevalc/SIS, CERES [6] and NA60 [7] at CERN, and STAR[8] and Phenix [9] at RHIC. The detection of the γ from Dalitz decays (meson $\rightarrow \gamma e^+e^-$) would give opportunity to fully reconstruct the meson yield. Besides the Dalitz decays π^0 and η mesons have dominant decay into two photons (meson $\rightarrow \gamma\gamma$). The Two Arm Photon Spectrometer (TAPS) Collaboration provided a large set of measurements for this decay channel [10].

In Fig. 1 the multiplicities of the π^0 and η mesons, as functions of beam energy and system size are shown. The data were obtained with TAPS [10], the π^+ data as shown in Fig. 1 was measured by KaoS Collaboration [11]. Curves are log-log polynomial fits to the data in lighter systems. The curves for the Au+Au data points are from extrapolation of the light systems (C+C, Ca+Ca). The the Au+Au data sets were measured with much less accuracy compared to the light systems. The estimated multiplicities of the π^0 and η mesons in Au+Au at 1.23 GeV/u are indicated with red stars on a dashed line and extracted values are 6.4 ± 15% for the π^0 and 0.09 ± 50% for the η. To make a solid statement on a possible e^+e^- enhancement, one has to minimize the uncertainties on the trivial sources, in particular on η. The HADES experiment has no dedicated detector system to measure photons. Therefore photons are detected via conversion into e^+e^- pairs. Such analysis has been already preformed at the beam times Ar+KCl at 1.76 GeV/u [12] and p+Nb at 3.5 GeV/u [13] with HADES. This procedure was successfully used by other experiments [14, 15].

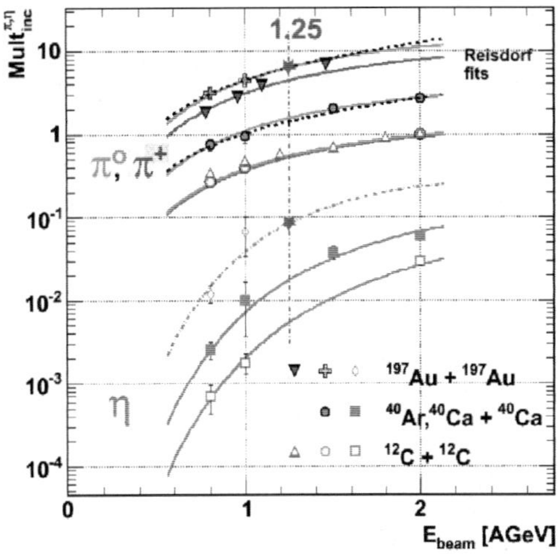

Figure 1: Excitation functions of inclusive π and η production in the C+C, Ca+Ca and Au+Au systems by TAPS Collaboration [10] and KaoS Collaboration [11]

1.1. The HADES spectrometer

To study e^+e^- pairs radiated from cold nuclear matter, hot and dense nuclear matter as well as in elementary collisions, the High Acceptance DiElectron Spectrometer (HADES) is installed at the SIS 18 (GSI Helmholtzzentrum für Schwerionenforschung, Darmstadt). It consists of 6 identical sectors that covers a full azimuthal angle and polar angle from 18° to 85°. The tracking is performed with 4 x 6 Multiwire Drift Chambers and a superconducting toroidal magnet. For the time-of-flight measurement the ToF detector and a Resistive Plate Chamber (RPC) walls are used. HADES measured the collision system Au+Au at the highest (achievable at SIS18) beam energies, E_{kin} = 1.23 GeV/u in April/May 2012. In total 7.3 x 10^9 events, that corresponds to 140 TByte of data have been collected. The trigger on hit multiplicity in ToF_{Mult} =>20 (PT3) corresponds an impact parameter $b_{max} \approx 9$ fm.

2. Conversion in the HADES spectrometer

Since HADES has no photon detector yet, the measurement of the electromagnetic decays of π^0 and η is only possible via external conversion of photons in detector material. Therefore the knowledge of the conversion probability is needed. The conversion probability can be extracted with the help of the simulation tool GEANT3. Lepton pair production of a γ can only occur when the photon passes near a massive particle such as an atomic nucleus. The recoil of the massive particle satisfies momentum conservation without carrying off a significant amount of energy, so the assumption that all of the energy of the photon goes into the electron-positron pair is a good approximation. The differential cross section for such process depends on the atomic number Z of the material in which the interaction occurs. In a compound material the element i in which the interaction occurs is chosen randomly according to the probability:

$$Prob(Z_i, E_\gamma) = \frac{n_{ati}\sigma(Z_i, E_\gamma)}{\sum_i [n_{ati} \cdot \sigma_i(E_\gamma)]}, \tag{1}$$

where Z_i stands for the Atomic Number of the material in which the photon will convert and E_γ the energy of the photon. In Fig. 2a the conversion probability inside the HADES as a function of the photon energy and the polar Θ angle is shown. Since the target is segmented into 15 vertical aligned gold discs (r =1.2 mm, thickness = 0.25mm), a lager polar angle corresponds to a longer flightpath trough target material. The dependence of Eq. 1 is reflected here. In Fig. 2b the conversion probability for the main inner detector parts are is presented. The systematic errors are in the order of 5% due to uncertainties in the material budget, for details see [13].

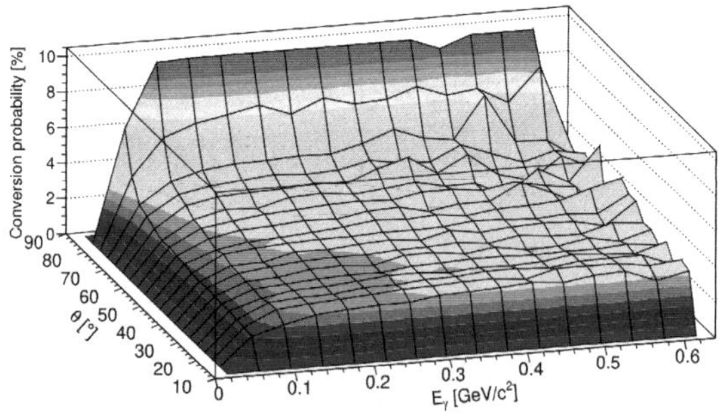

(a)

Detector component	Material	% (π^0)	% (η)
Target	Gold	0.32	0.54
Target holder	Carbon	0.026	0.045
Beam pipe	Carbon	0.26	0.48
Radiator gas	C4F10	0.59	0.91
Sum		1.2	2.0

(b)

Figure 2: (a) Conversion probability (in %) as function of the γ energy and the Θ Angle. (b) Conversion probability (in %) for different material in the center of the spectrometer for γ coming from π^0 and η decay

3. Estimations of π^0 and η multiplicities

To validate the feasibility of the two photon and photon e^+e^- analysis the expected multiplicities of the neutral mesons reconstructed from 4 leptons in finale state need to be estimated. To do that one can use the following equation:

$$N_{AuAu} = Mult_{TAPS} \cdot N_{events} \cdot BR \cdot P_{Conv} \cdot \epsilon_{reconstruction} \cdot \epsilon_{acceptance}, \tag{2}$$

where N_{AuAu} is the expected multiplicity for 4 leptons in final state for Au+Au at 1.23 GeV/u, $Mult_{TAPS}$ is the multiplicity of the mesons (as estimated from Fig. 1) , N_{events} the total number of events, BR is the decay branching ratio [16], P_{Conv} the conversion probability and $\epsilon_{reconstruction} \cdot \epsilon_{acceptance}$ the assumed efficiencies for reconstruction and geometrical acceptance. In Table 1 BR, and P_{Conv} for these channels are summed up. Note that the product of the BR and P_{Conv} is on the same order of magnitude for the Dalitz and the 2γ decay channels.

Channel	BR	P_{Conv}	Total
$\pi^0 \to \gamma\gamma \xrightarrow{conv} e^+e^-e^+e^-$	$98.8 \cdot 10^{-2}$	1.44%	$1.4 \cdot 10^{-7}$
$\pi^0 \to \gamma e^+e^- \xrightarrow{conv} e^+e^-e^+e^-$	$1.17 \cdot 10^{-2}$	1.2%	$1.4 \cdot 10^{-7}$
$\pi^0 \to e^+e^-e^+e^-$	$3.34 \cdot 10^{-5}$		$0.3 \cdot 10^{-7}$
$\eta \to \gamma\gamma \xrightarrow{conv} e^+e^-e^+e^-$	$39.3 \cdot 10^{-2}$	4%	$4.7 \cdot 10^{-7}$
$\eta \to \gamma e^+e^- \xrightarrow{conv} e^+e^-e^+e^-$	$7.0 \cdot 10^{-3}$	2%	$4.2 \cdot 10^{-7}$
$\eta \to e^+e^-e^+e^-$	$<6.9 \cdot 10^{-5}$		$2.1 \cdot 10^{-7}$

Table 1: Branching ratios and conversion probabilities into various decay channels

Based on [12, 13] reconstruction efficiency for π^0 and η has been estimated to be 10% and acceptance 1% and 2%, respectively. Summing up all 4 lepton final state decay channels the expected yield will be around $1.5 \cdot 10^4$ entries for the π^0 and 750 for the η meson.

4. Analysis of Real Data

HADES provides possibility to identify electrons using various detector observables, i.e. RICH ring properties, particle velocity, energy loss in ToF (Time of Flight) detector and MDC chambers (Multiwire Drift Chambers), etc. Leptons from γ conversion can be characterized by a small momenta and e^+e^- has small opening angles (α <2°), see Fig 3b. Due to the small opening angles they will be identified as a single ring in the RICH. Therefore for the lepton identification currently only the velocity versus momentum dependence for the particle identification is used. Algorithms to reconstruct close pairs are under development. In Fig. 3a the distribution, as reconstructed with the RPC detector (Θ <45°), is shown. Different particles can be seen clearly. The lepton selection criteria are velocity (β) between 0.97 and 1.1 and a momentum smaller than 500 MeV/c. The black box in Fig 3a indicates the identification criteria. After such selection all

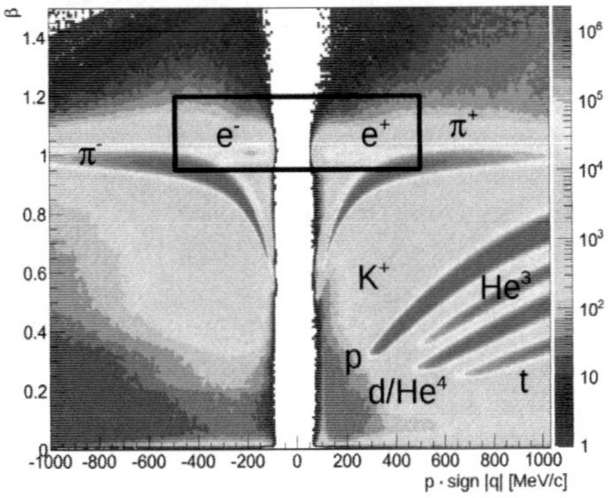

(a) Velocity versus momentum · sign$|q|$ in the RPC region of the HADES spectrometer.

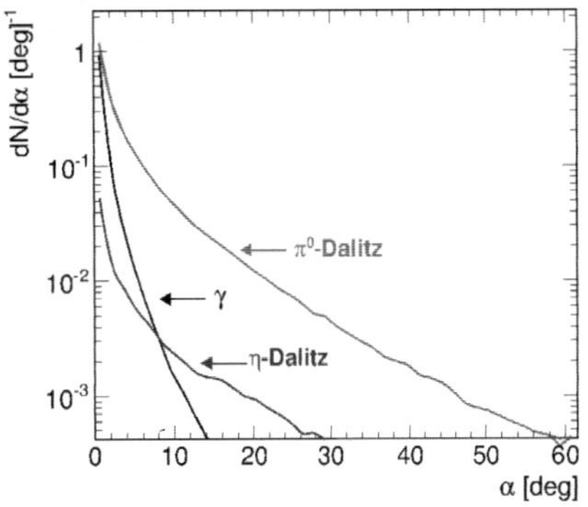

(b) Opening angle distribution for different dilepton sources.

Figure 3: Velocity versus momentum · sign$|q|$ and opening angle for different dilepton sources.

surviving lepton candidates are combined to pairs to reconstruct photons or pairs form Dalitz decay. In this first analysis topological selection criteria on opening angle are used to identify leptons from conversion
(α <2.5°) and from the Dalitz decay (α <20°) at the same time. Those cuts are based on former analysis as discussed in Ref. [12]. As it has been mentioned before, the angle between the e^+e^- pairs from conversion is small. To avoid correlation between random photons the angle between the photons (photons originating from various sources) is required to be at least 5°. In Fig. 4 the preliminary invariant mass distribution of the 4 lepton candidates is shown. At an invariant mass of 140 MeV/c^2 a clear π^0 peak is visible. The background was simulated with a 4th order polynomial function. The extracted π^0 counts on the order of 8000 counts. This result demonstrates that the π^0 can be reconstructed in Au+Au collisions via conversion method.

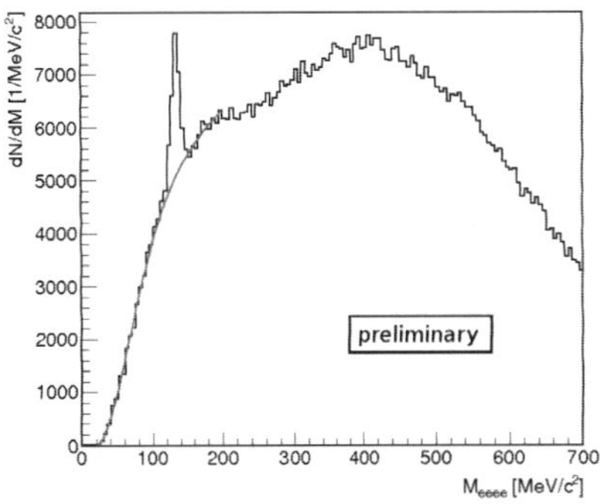

Figure 4: 4 lepton invariant mass spectrum after the topological cuts.

5. Summary and Outlook

In this work a feasibility study of the reconstruction of neutral mesons π^0 and η via photon conversion was studied. The conversion probability in the inner parts of the HADES detector was calculated. The π^0 and η multiplicities were estimated. Lepton candidates were identified using velocity and momentum and combined to 4 lepton invariant mass spectrum. The π^0 peak is visible and is about 8000 signal entries. The reconstruction of the η is ongoing.

Electron identification and topological cuts are still an issue for optimizations. Later acceptance and efficiency corrections will be extracted and in the final step the multiplicities for both mesons will be determined for Au+Au at 1.23 GeV/u.

Acknowledgments

The collaboration gratefully acknowledges the following funding: INFN-LNS Catania (Italy); LIP Coimbra (Portugal): PTDC/FIS/113339/2009; SIP JUC Cracow (Poland): NN202198639; GSI Darmstadt (Germany): Helmholtz Alliance HA216/EMMI; TU Darmstadt (Germany): VH-NG-823, Helmholtz Alliance HA216/EMMI; HZDR, Dresden (Germany): 283286, 05P12CRGHE; Goethe-University, Frankfurt (Germany): Helmholtz Alliance HA216/EMMI, HIC for FAIR (LOEWE), GSI F&E, BMBF 06FY9100I, HGS-Hire, H-QM research school TU München, Garching (Germany): BMBF 06MT7180; JLU Giessen (Germany): BMBF:05P12RGGHM; University Cyprus, Nicosia (Cyprus): UCY/3411-23100; IPN Orsay, Orsay Cedex (France): CNRS/IN2P3; NPI AS CR, Rez, (Czech Republic): MSMT LG 12007, GACR 13-06759S.

References

[1] G. Agakishiev, et al., Origin of the low-mass electron pair excess in light nucleus-nucleus collisions, Phys.Lett. B690 (2010) 118–122. arXiv:0910.5875, doi:10.1016/j.physletb.2010.05.010.

[2] G. Agakishiev, et al., Inclusive dielectron production in proton-proton collisions at 2.2 GeV beam energy, Phys.Rev. C85 (2012) 054005. arXiv:1203.2549, doi:10.1103/PhysRevC.85.054005.

[3] G. Agakishiev, et al., Inclusive dielectron spectra in p+p collisions at 3.5 GeV, Eur.Phys.J. A48 (2012) 64. arXiv:1112.3607, doi:10.1140/epja/i2012-12064-y.

[4] R. J. Porter, et al., Dielectron Cross Section Measurements in Nucleus-Nucleus Reactions at 1.0A GeV, Phys. Rev. Lett. 79 (1997) 1229–1232. doi:10.1103/PhysRevLett.79.1229.

[5] G. Agakishiev, et al., Dielectron production in Ar+KCl collisions at 1.76A GeV, Phys.Rev. C84 (2011) 014902.

[6] G. Agakichiev, et al., e+ e- pair production in Pb - Au collisions at 158-GeV per nucleon, Eur.Phys.J. C41 (2005) 475–513. arXiv:nucl-ex/0506002, doi:10.1140/epjc/s2005-02272-3.

[7] R. Arnaldi, et al., First measurement of the rho spectral function in high-energy nuclear collisions, Phys.Rev.Lett. 96 (2006) 162302. arXiv:nucl-ex/0605007, doi:10.1103/PhysRevLett.96.162302.

[8] B. Huang, Dielectron differential cross section in Au + Au collisions at different beam energies at STAR, Nucl.Phys. A904-905 (2013) 565c–568c. doi:10.1016/j.nuclphysa.2013.02.077.

[9] A. Adare, et al., Detailed measurement of the e^+e^- pair continuum in $p+p$ and Au+Au collisions at $\sqrt{s_{NN}} = 200$ GeV and implications for direct photon production, Phys.Rev. C81 (2010) 034911. arXiv:0912.0244, doi:10.1103/PhysRevC.81.034911.

[10] R. Averbeck, R. Holzmann, V. Metag, R. S. Simon, Neutral pions and eta mesons as probes of the hadronic fireball in nucleus-nucleus collisions around 1 AGeV, Phys. Rev. C 67 (2003) 024903.

[11] F. Laue, et al., Production of charged pions, kaons and anti-kaons in relativistic C + C and C + Au collisions, Eur.Phys.J. A9 (2000) 397–410. arXiv:nucl-ex/0011010, doi:10.1007/s100500070024.

[12] R. Holzmann, M. Gumberidze, SR2012-PHN-NQM-EXP-06 Inclusive π^0 and η production in the 1.76 GeV/u Ar+KCl reaction, Vol. 2013-1 of GSI Report, GSI Helmholtzzentrum für Schwerionenforschung, Darmstadt, 2013.

[13] G. Agakishiev, A. Balanda, D. Belver, A. Belyaev, J. Berger-Chen, et al., Inclusive pion and eta production in p+Nb collisions at 3.5 GeV beam energyarXiv:1305.3118.

[14] J. Adams, et al., Photon and neutral pion production in Au + Au collisions at s(NN)**(1/2) = 130-GeV, Phys.Rev. C70 (2004) 044902. arXiv:nucl-ex/0401008, doi:10.1103/PhysRevC.70.044902.

[15] B. Abelev, et al., Neutral pion and η meson production in proton-proton collisions at $\sqrt{s} = 0.9$ TeV and $\sqrt{s} = 7$ TeV, Phys.Lett. B717 (2012) 162–172. arXiv:1205.5724, doi:10.1016/j.physletb.2012.09.015.

[16] K. Nakamura, et al., Review of particle physics, J. Phys. G37 (2010) 075021. doi:10.1088/0954-3899/37/7A/075021.

Journal of Physics: Conference Series **503** (2014) 012016 doi:10.1088/1742-6596/503/1/012016

Monte-Carlo simulation of lepton pairs production in "$p\bar{p} \rightarrow e^+e^- + X$" events at PANDA experiment.

Anna Skachkova

Joint Institute for Nuclear Research, Joliot-Curie 6, 141980 Dubna, Moscow region, Russia

E-mail: `Anna.Skachkova@cern.ch`

Abstract. Electron-positron pairs production via the quark-antiquark annihilation process in collisions of antiproton beam having E_{beam} = 15 GeV with the proton target in PANDA experiment is modeled by using the PYTHIA6.4 Monte-Carlo event generator and PandaRoot framework. The considered quark level subprocess pass through the production of virtual photon which converts into the lepton pair ($q\bar{q} \rightarrow \gamma^* \rightarrow e^+e^-$). The spectrum of the final state leptons in this process of quark-antiquark annihilation obviously depends on the form of parton distributions inside colliding protons and may provide an interesting information about the quark dynamics inside the hadron. The measurement of the total transverse momentum of a lepton pair as a whole system may give an important information about the intrinsic transverse momentum k_T that appears due to the Fermi motion of quarks inside the nucleon. In the article the distributions of different kinematical variables of final state electrons/positrons are presented. The influence of the detector environment on the presence of fake leptons is demonstrated. The signal leptons registration efficiency is calculated. The background contribution is also discussed.

Introduction

The intermediate energy experiment PANDA ($E_{beam} \leq$ 15 GeV) may play an important role because it allows to study the energy range where the perturbative methods of QCD (pQCD) come into interplay with a rich physics of resonance production. A detailed and high-precision experimental study at PANDA may allow to discriminate between a large variety of existing nonperturbative approaches and models that already exist or are under the development now. Dilepton events may serve as a powerful tool to get out the information about the parton distribution functions (PDFs) in hadrons [1]. The plans to study this process are included into the "Physics book" [2] of PANDA experiment at HESR (High-Energy Storage Ring at FAIR). This study may provide an interesting information about quark dynamics inside the nucleon [3].

This paper is a continuation of the work [1], [3] done to study the lepton pairs production in the processes $p\bar{p} \rightarrow \mu^+\mu^- + X$ and $p\bar{p} \rightarrow e^+e^- + X$ at the beam energy E_{beam} = 14 GeV. Here we consider the last process but for the case of beam energy E_{beam}= 15 GeV (which corresponds to the center-of-mass energy of the $p\bar{p}$ system E_{cm} = 5.474 GeV). The simulation is done with PYTHIA6.4 [4] and the PandaRoot [5] framework which takes into account the PANDA detector geometry and fulfills the global simulation of the detector by using the implemented Geant4 [6] program. The main goal of the current work was to monitor the influence of detector environment on leptons propagation and their final distributions.

Observations and Interpretation

We have used PYTHIA6.4 to generate 10000 $p\bar{p} \rightarrow \mathrm{e^+e^-} + X$ events which include the $2 \rightarrow 2$ quark level $q\bar{q} \rightarrow \gamma^* \rightarrow \mathrm{e^+e^-}$ subprocess. This simulation has shown that one can expect to gain about $7 \cdot 10^6$ such lepton pairs production events per year for the luminosity $L = 2 \cdot 10^5 mb^{-1}s^{-1}$, which corresponds to the cross section $\sigma = 4.59 \cdot 10^3$ pb and the assumption of the full year of detector operation. Electron-positron pairs from these events were used for the further study with the program of full simulation PandaRoot with Geant4.

It was found that the number of events where one of the signal leptons is not registered in detector amount to 1285 for 10000 generated events. There are also 127 events in which no one of signal leptons was registered. That means we can expect the registration of only 85.8% of signal events. The number of signal events where both of the signal leptons satisfy the criteria $P_e > 0.2$ GeV/c and $PT_e > 0.2$ GeV/c is 62.6%.

Also we have performed the comparison of kinematical distributions of signal leptons after passing through the detector with the corresponding initial ones obtained with PYTHIA. In the following pictures the distributions of leptons obtained with the simulation by PandaRoot and Geant4 are shown by pink histograms. For comparison, distributions of the same leptons obtained at the level of the generation by PYTHIA, i.e. before passing through the detector, are shown in the same pictures by violet lines. At the final distributions, having the pink color, are shown only those electrons and positrons which belong to those events in which both signal leptons are registered. All of the statistical boxes contain the information corresponding to GEANT simulation.

The first figures show the distributions of the number of generated signal leptons N_{ev} versus the their momentum components $P_x^{\mathrm{e}^\pm}$(Fig.1), $P_y^{\mathrm{e}^\pm}$(Fig.2) and $P_z^{\mathrm{e}^\pm}$(Fig.3). Distributions over Px and Py are identical to each other and follow the initial distributions at PYTHIA level except some loss of events, denoted above. Distribution of longitudinal component shows the excess over PYTHIA results in the region of small $0.2 < P_z^{\mathrm{e}^\pm} < 0.6$ GeV/c and some reduction of number of events at the medium values of $1 < P_z^{\mathrm{e}^\pm} < 9$ GeV/c.

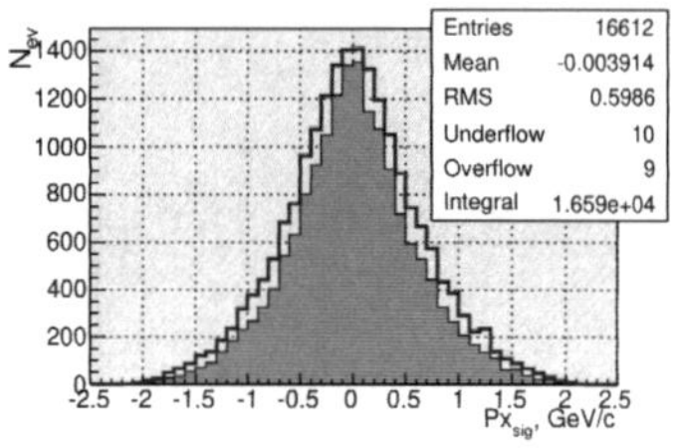

Figure 1. $P_x^{\mathrm{e}^\pm}$.

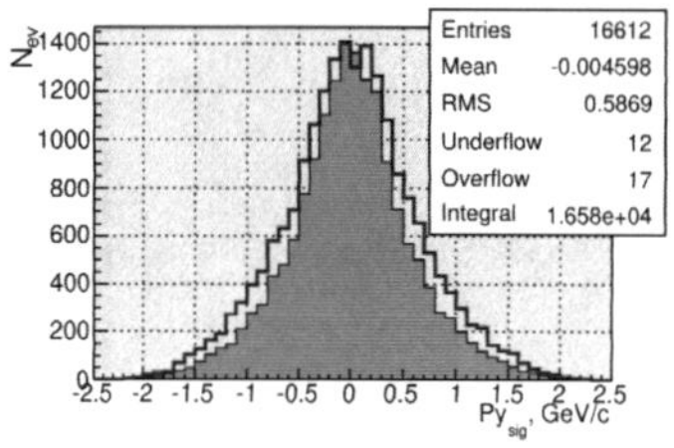

Figure 2. $P_y^{\mathrm{e}^\pm}$.

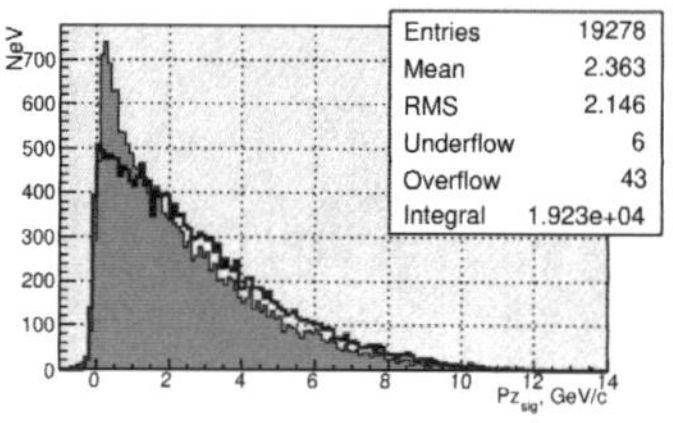

Figure 3. $P_z^{\mathrm{e}^\pm}$.

The next figures demonstrate distributions of the signal e^+/e^- over their transverse momentum $PT^{\mathrm{e}^\pm}$(Fig.4) and full momentum $P^{\mathrm{e}^\pm}$(Fig.5). One can see that PandaRoot spectra shift to the region of lower momenta as comparing to the PYTHIA ones. In the region of small transverse momentum $0 < PT^{\mathrm{e}^\pm} < 0.25$ GeV/c the number of events obtained by PandaRoot slightly exceeds the initial distributions. At the higher $PT^{\mathrm{e}^\pm} > 0.4$ GeV/c their number is reduced significantly. The distribution over full momentum P shows the excess over PYTHIA results in the region of small $0.3 < P^{\mathrm{e}^\pm} < 0.8$ GeV/c and some reduction of number of events at the medium values of $1 < P^{\mathrm{e}^\pm} < 9$ GeV/c.

The distribution of the number of generated signal leptons N_{ev} versus their polar angle $\Theta^{\mathrm{e}^\pm}$, measured from the z-axis directed along the beam line, is given at Fig.6. The distribution over the polar angle $\Theta^{\mathrm{e}^\pm}$ shows the maximum at the values about 15 degrees, whereupon it sharply falls

Journal of Physics: Conference Series **503** (2014) 012016 doi:10.1088/1742-6596/503/1/012016

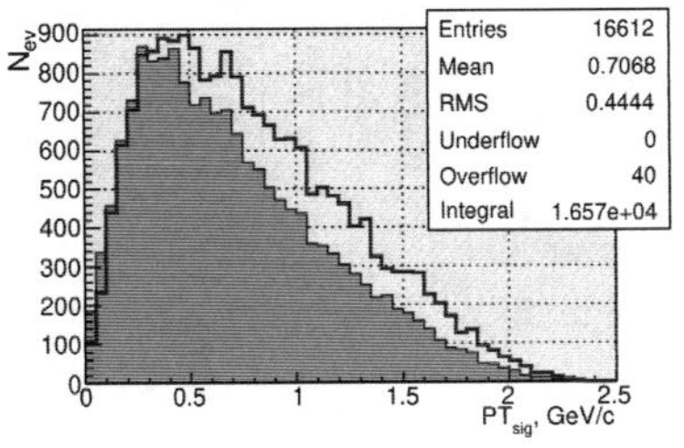

Figure 4. Transverse momentum $PT^{e^{\pm}}$.

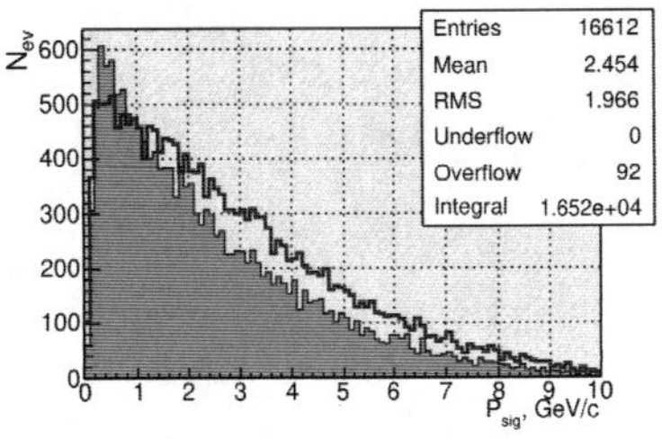

Figure 5. Full momentum $P^{e^{\pm}}$.

up to the values about 120 degrees. Fig.7 presents the distribution over the azimuthal angle $\phi^{e^{\pm}}$ which shows a relatively uniform shape. Both angular distributions, accounting for the statistical errors and decrease of the number of registered particles, repeat the initial distributions obtained with PYTHIA.

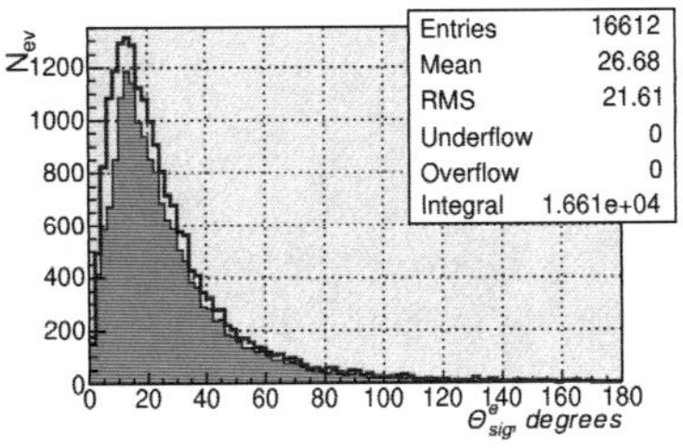

Figure 6. Polar angle $\Theta^{e^{\pm}}$.

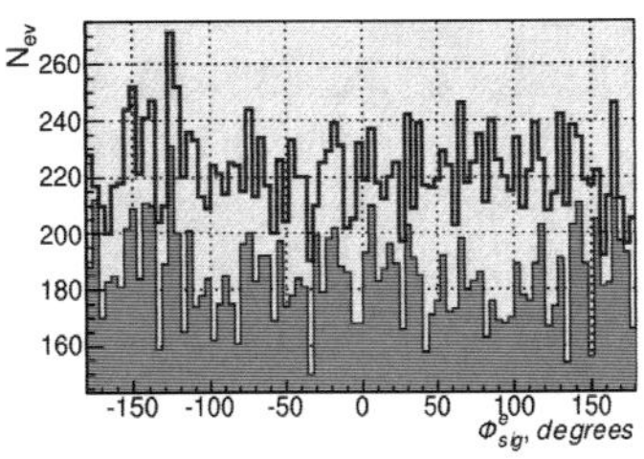

Figure 7. Azimuthal angle $\phi^{e^{\pm}}$.

Figures 8, 9 and 10 present the registration efficiencies over the electron/positron transverse momentum $Eff_{PT^{e^{\pm}}}$, full momentum $Eff_{P^{e^{\pm}}}$ and polar angle $Eff_{\Theta^{e^{\pm}}}$ correspondingly. The efficiency of registration *Eff* is calculated as a ratio of the number of leptons registered in a definite momentum (angle) region while modeling in PandaRoot to that ones which were initially generated in the same momentum (angle) region in PYTHIA.

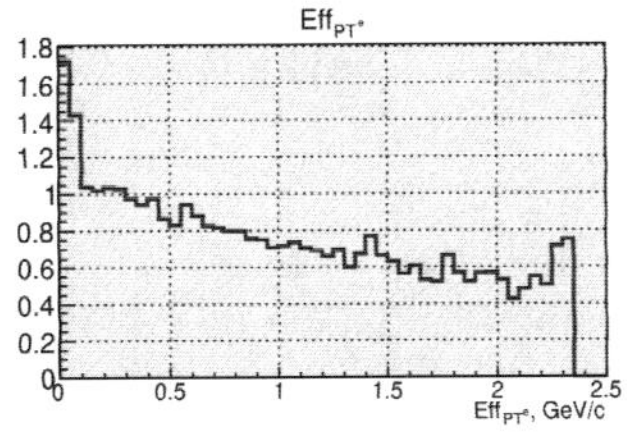

Figure 8. Transverse momentum registration efficiency $Eff_{PT^{e^{\pm}}}$.

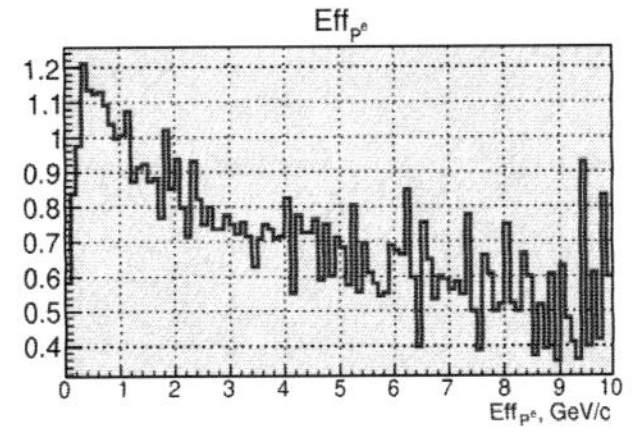

Figure 9. Full momentum registration efficiency $Eff_{P^{e^{\pm}}}$.

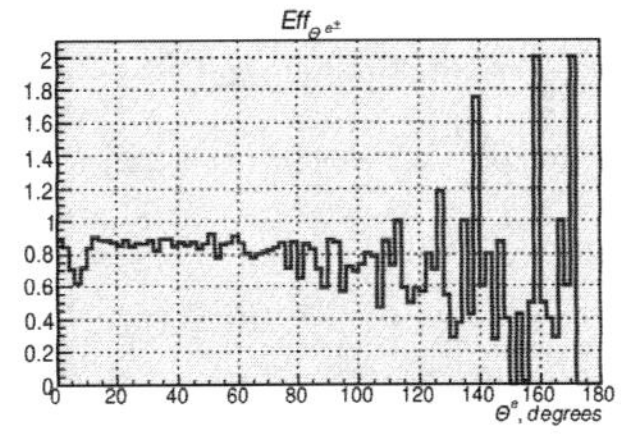

Figure 10. Polar angle registration efficiency $Eff_{\Theta^{e^{\pm}}}$.

One can see that at the low values of $PT < 0.1$ GeV/c the value of registration efficiency $Eff_{PT^{e^{\pm}}} > 1$ that can be caused by the production of some additional low energetic e^{+}/e^{-} in a result of interaction with the detector environment. At the higher transverse momentum the value of registration efficiency evenly falls linearly up to $Eff_{PT^{e^{\pm}}} \approx 0.4 - 0.6$ and the $Eff_{PT^{e^{\pm}}}$ spectrum stops at the value of $PT = 2.35$ GeV/c.

Journal of Physics: Conference Series **503** (2014) 012016 doi:10.1088/1742-6596/503/1/012016

The value of e^+/e^- full momentum registration efficiency $Eff_{P^{e\pm}}$ sharply grows at the values $P < 0.4$ GeV/c from $Eff_{P^{e\pm}} = 0.6$ to the value of 1.2, that can be explained by the reason described above. Then the distribution curve practically linearly falls up to the value of $Eff_{P^{e\pm}} = 0.5$.

From Fig.10 showing the electron/positron registration efficiency $Eff_{\Theta^{e\pm}}$ over their polar angle $\Theta^{e^\pm}$, one can see that the value of this efficiency has almost the same value $Eff_{\Theta^{e\pm}} \approx 0.83$ (that corresponds to the whole loss of the number of signal events) in the polar angle region $0 < \Theta^{e^\pm} < 80^o$ with some fall up to the value $Eff_{\Theta^{e\pm}} \approx 0.6$ in the region of $\Theta^{e^\pm} \approx 7^o$, with the following uniform (with account of statistical errors) fall to $Eff_{\Theta^{e\pm}} \approx 0.4$ at $\Theta^{e^\pm} \approx 160^o$.

Backgrounds

The main source of the background to the process of lepton pairs production are the so called minimum bias events, i.e. the events with low PT [1], [3] as well as different kind of QCD events. The total cross section of minimum-bias events is about $6 \cdot 10^6$ times higher than the cross section of the signal event, i.e. there are about 10^7 background events per one signal event. The total cross section of QCD processes is also significant, but it is about 5 order of magnitude less than the one of the minimum-bias events.

We took 10^5 background events initially simulated in PYTHIA and passed them through the detector using Geant4 in PandaRoot. The comparative kinematical distributions of the e^+/e^- obtained in PYTHIA and PandaRoot are represented further. The final state electrons in background processes in PYTHIA appear from the hadron and meson decays according to the Lund fragmentation model. In PandaRoot they are produced from the analogous decays laid in Geant program.

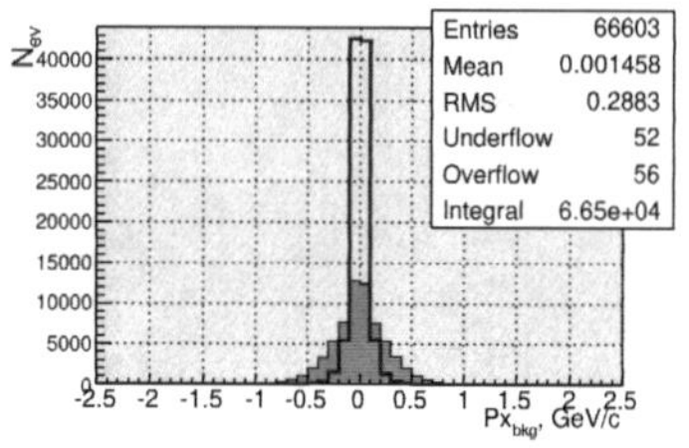

Figure 11. $Px^{e^\pm}_{bkg}$.

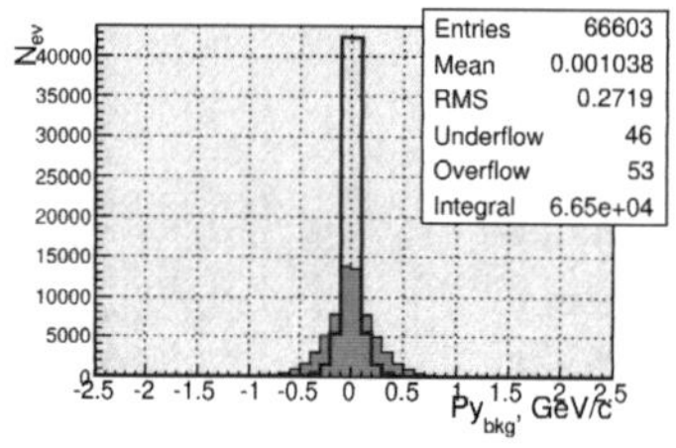

Figure 12. $Py^{e^\pm}_{bkg}$.

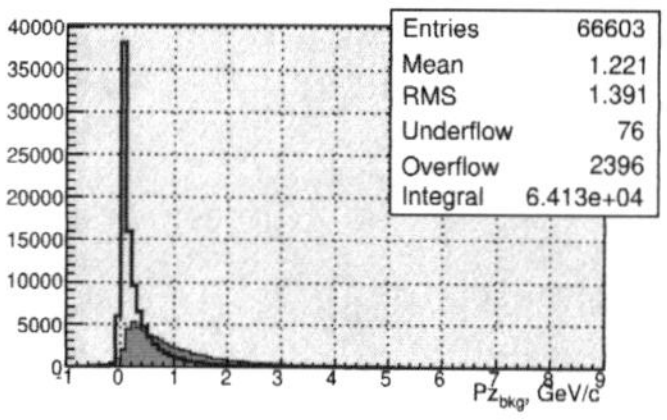

Figure 13. $Pz^{e^\pm}_{bkg}$.

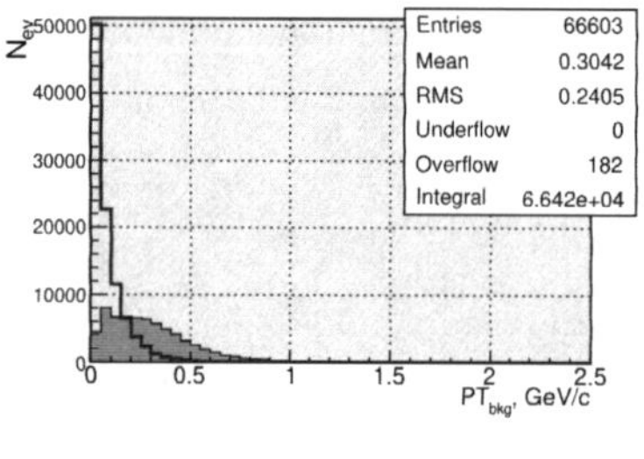

Figure 14. Transverse $PT^{e^\pm}_{bkg}$.

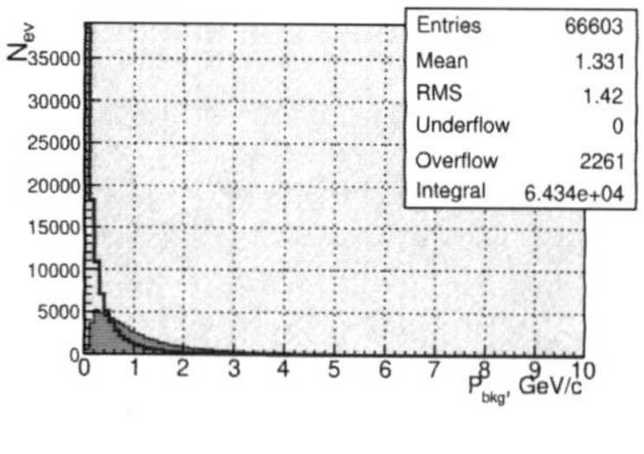

Figure 15. Full $P^{e^\pm}_{bkg}$.

As one can see from the obtained distributions, the electrons/positrons produced from decays of different particles in detector volume happen to be more energetic in comparison with analogous ones simulated in PYTHIA. On the other hand, their number is significantly reduced in comparison with PYTHIA predictions. Namely, they have in average on 0.5 GeV/c higher momentum in transverse plane (Px (Fig.11), Py (Fig.12) and PT (Fig.14)), and about 1 GeV/c higher momentum in longitudinal component (Pz (Fig.13) and correspondingly P (Fig.15)).

The distribution of the background leptons obtained by PandaRoot over the polar angle $\Theta^{e^{\pm}}$ (Fig.16) has the similar shape with electrons/positrons, obtained in PYTHIA simulation, but has some clear excess in direction of 5-35 degrees with the peaks on 3-5 and 9-11 degrees.

The distribution of the background electrons/positrons over the zenith angle $\phi^{e^{\pm}}$ (Fig.17) has evidently non uniform character with a prevalence in the region of 0 and 180 degrees, as well as with the holes close to 80 (-80) and -100 degrees.

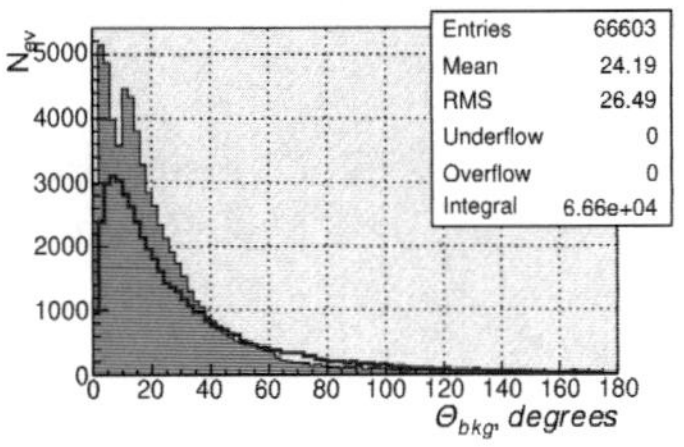

Figure 16. $\Theta^{e^{\pm}}_{bkg}$.

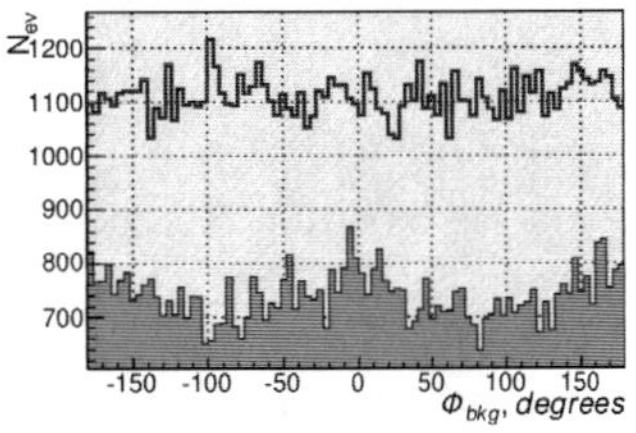

Figure 17. $\phi^{e^{\pm}}_{bkg}$.

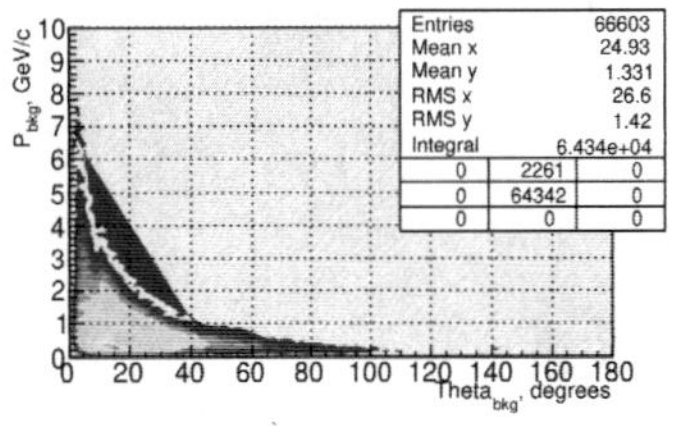

Figure 18. $P^{e^{\pm}}_{bkg}/\Theta^{e^{\pm}}_{bkg}$.

The correlation plot of electron/positron distribution over their full momentum P and the polar angle $\Theta^{e^{\pm}}$ (Fig.18) shows the maximum in the region of 3-4 GeV/c over momentum and 10-21 degrees over the polar angle.

The analysis of 100000 generated background events, discussed above, has shown that in these 10^5 background events 68803 electrons/positrons were produced. The number of background events where no one of electron/positron was registered amount to 55448, i.e. 55.4%. The number of background events including 1 electron/positron is 29311, i.e. 29.3%. Thus the number of events having 2 and more electrons/positrons is 15241, i.e. 15.2%. The number of events, which satisfy to the selection criterion – the presence of exactly 2 leptons of different charge with $P^{e^{\pm}} > 0.2$ GeV/c and $PT^{e^{\pm}} > 0.2$ GeV/c is equal to zero!

Thus the aforesaid selection criterion for signal events - the presence of exactly 2 leptons of different charge with $P^{e^{\pm}} > 0.2$ GeV/c and $PT^{e^{\pm}} > 0.2$ GeV/c provides the background suppression at least for the 5 order of magnitude (in 10^5 times), exactly as it was predicted in the previous papers based on the simulation at the level PYTHIA [1], [3]. But since the cross section difference of background and signal events is 7 orders of magnitude, the more prolonged and deep study of background events is needed with their simulation at much higher statistic.

Acknowledgments

The work was carried out with the financial support of SAEC "Rosatom" and Helmholtz Association.

References

[1] Skachkova A N, Skachkov N B and Alexeev G D Monte-Carlo simulation of lepton pair production in "p pbar → l+l- + X" events at $E_{beam} = 14$ GeV arXiv: hep-ph/0506139v2;

[2] PANDA collab. Erni W et al Physics Performance Report for PANDA: Strong Interaction Studies with Antiprotons 2007 arXiv: hep-ex/0903.3905

[3] Skachkova A N and Skachkov N B On lepton pair production in proton-antiproton collisions at intermediate energies 2009 *Phys.Part.Nucl.Lett* **6** N4(153) 504-18.

[4] Sjöstrand T, Mrenna S and Skands P 2006 *JHEP* **05**(2006) 026 hep-ph/0603175

[5] *http://panda-wiki.gsi.de/cgi-bin/view/Computing/PandaRoot*

[6] *Nuclear Instruments and Methods in Physics Research* **A 506**(2003) 250-303, *IEEE Transactions on Nuclear Science* **53** No.1(2006) 270-278. *http://geant4.web.cern.ch/geant4/*

Strange and heavy mesons in hadronic matter

Daniel Cabrera[1,0], Luciano M Abreu[2], Elena Bratkovskaya[1], Andrej Ilner[1], Felipe J Llanes-Estrada[3], Angels Ramos[4], Laura Tolos[1,5] and Juan M Torres-Rincon[5]

[1] Institute for Theoretical Physics and Frankfurt Institute for Advanced Studies, Frankfurt University, 60438 Frankfurt am Main, Germany
[2] Instituto de Física, Universidade Federal da Bahia, 40210-340. Salvador, BA, Brazil
[3] Departamento de Física Teórica I, Universidad Complutense, 28040 Madrid, Spain
[4] Departament d'Estructura i Constituents de la Matèria, Universitat de Barcelona, Diagonal 647, 08028 Barcelona, Spain
[5] Institut de Ciències de l'Espai (IEEC/CSIC), Campus Universitat Autònoma de Barcelona, Facultat de Ciències, Torre C5, E-08193 Bellaterra, Spain

Abstract. We present selected results on the properties of strange and heavy-flavoured mesons in a hot and dense nuclear medium, with emphasis in selfconsistent coupled-channel approaches based on the chiral Lagrangian. In the strangeness sector, we discuss how the enhanced reactivity of light strange vectors at FAIR conditions can be tied to in-medium effects on their predominant decay modes (e.g. $\bar{K}^* \to \bar{K}\pi$) and to the excitation of strange baryons in vector-meson nucleon interactions. In the heavy-flavour sector, we focus on recent determinations of the transport coefficients of charmed and bottomed mesons in a hadron gas at vanishing baryonic chemical potential. We comment on the role of microscopic transport simulations to establish a connection between theoretical models and experimental observables from heavy-ion collisions (HICs).

1. Introduction

The features of strongly interacting matter in a broad range of temperature and density has been a subject of great interest in the last decades, in connection with fundamental aspects of the strong interaction such as the nature of (de)confinement or the physical mechanism of chiral symmetry restoration, and with the microscopic dynamics and composition of neutron stars.

In particular, understanding the dynamics of light strange mesons in a nuclear environment has posed a challenge to theoretical models, amongst different reasons, due to the failure of the low-density approximation to describe the interaction of the $\bar{K}$ meson with the medium, as it is concluded from the phenomenology of kaonic atoms [1]. The properties of both K and $\bar{K}$ close to threshold energy have also been thoroughly investigated in HICs at SIS energies [2, 3, 4, 5]. The analysis of experimental data in conjunction with microscopic transport approaches has allowed to draw solid conclusions regarding the production mechanisms of strangeness, the freeze-out conditions exhibited by K^+ and K^- mesons and the use of K^+ as a probe of the nuclear matter equation of state at high densities. A good agreement with data has been achieved for many different observables when the in-medium properties of kaons are implemented. Still, a simultaneous description of all observables involving antikaon production is still missing, leaving room for a more elaborated description of the most relevant reactions (e.g., $\pi Y \to \bar{K}N$).

[0] Speaker

In the heavy flavour sector, the study of the mechanisms of equilibration of heavy mesons in a hadronic medium has been lately subject of intense activity. This has been partially motivated by recent developments on the theoretical description of the interaction of D and B mesons with other (lighter) hadrons, based on phenomenological approaches [6] or effective theories of QCD [7, 8, 9, 10]. The effect of these interactions in the heavy meson spectrum is quantified by the corresponding transport coefficients.

We report in this talk on recent progress on the medium modifications of strange and heavy flavoured mesons, highlighting (some new) results from chirally motivated hadronic many-body approaches in connection with microscopic transport simulations.

2. Strange mesons in a hot/dense medium

2.1. Selfconsistent determination of $\bar{K}$ and K spectral functions at FAIR conditions

The early departure of the $\bar{K}$ nuclear potential from the low-density approximation is successfully described in coupled-channel approaches implementing the exact unitarity of the scattering matrix and a selfconsistent evaluation of the kaon selfenergy [11, 12]. The onset of an attractive s-wave $\bar{K}N$ interaction at low densities is a consequence of an upper shift of the $\Lambda(1405)$ resonance induced by Pauli blocking on the nucleon states. The simultaneous dressing of all intermediate meson-baryon states in such approaches typically leads to an attractive $\bar{K}$ potential of about 40-60 MeV at normal nuclear matter density (ρ_0), a rather shallower potential than obtained within relativistic mean-field calculations [13] or required by phenomenological analysis of kaonic atom data [1].

In Ref. [14] the coupled-channel approach of [11], based on the meson-baryon chiral Lagrangian, was extended to account for finite temperature so as to match the experimental conditions at the future FAIR facility at GSI. The set of coupled equations determining the $\bar{K}(K)N$ scattering amplitudes and the $\bar{K}(K)$ selfenergies were solved selfconsistently for the lowest partial waves (schematically $\Pi_{\bar{K}(K)} = \sum_{\vec{p}} n(\vec{p}) T_{\bar{K}(K)N}$) accounting for Pauli-blocking effects, finite-temperature baryonic mean-field potentials and the meson selfenergies of pions and kaons. The model achieves a good description of vacuum low-energy scattering observables with a minimal number of parameters. The $\Lambda(1405)$ resonance is dynamically generated in the isospin I=0 channel from the leading order (Weinberg-Tomozawa) interaction, once unitarity is restored by solving the Bethe-Salpeter equation for the scattering amplitudes. A p-wave selfenergy for the kaons was implemented accounting for YN^{-1} excitations, where $Y = \Lambda, \Sigma, \Sigma^*$ (see also [15]). The role of higher partial waves (beyond $L = 0$), leading to non-isotropic in-medium cross sections, has been emphasised in the context of heavy-ion collisions [2].

The $\bar{K}$ spectral function, $S_{\bar{K}}(q_0, \vec{q}; T) = -\pi^{-1} \mathrm{Im}\, [q_0^2 - \vec{q}^{\,2} - m_{\bar{K}}^2 - \Pi_{\bar{K}}(q_0, \vec{q}; T)]^{-1}$, is shown in Fig. 1 (left) at $\rho = \rho_0, 2\rho_0$ and different temperatures. Far from staying as a narrow quasi-particle, the $\bar{K}$ exhibits a complex structure as a result of the interference of the quasi-particle peak with the $\Lambda(1405)$ and the p-wave YN^{-1} excitations. Increasing the temperature smears the Fermi surface of nucleons, reducing the attraction felt by the $\bar{K}$ and thus bringing the quasi-particle peak closer to the free position, whereas the more substantial mixing between the different YN^{-1} modes spreads the spectral function over a wider range of energies. The kaon (K), in contrast, survives as a narrow state in the nuclear medium and only at high densities and temperatures some broadening of the K peak is observed. A repulsive K mass shift of about 30 MeV was obtained in cold matter at normal density, close to the result within a low-density $t\rho$ approximation.

The results discussed above constitute a microscopic input for transport simulations implementing the production of strange hadrons in HICs. Such a study has been performed, for instance, within the HSD and IQMD approaches at SIS energies [4, 5], the former one featuring full off-shell strange meson dynamics [5]. Further work along these lines is in progress in order to account for a full catalogue of in-medium cross sections for the relevant processes involving

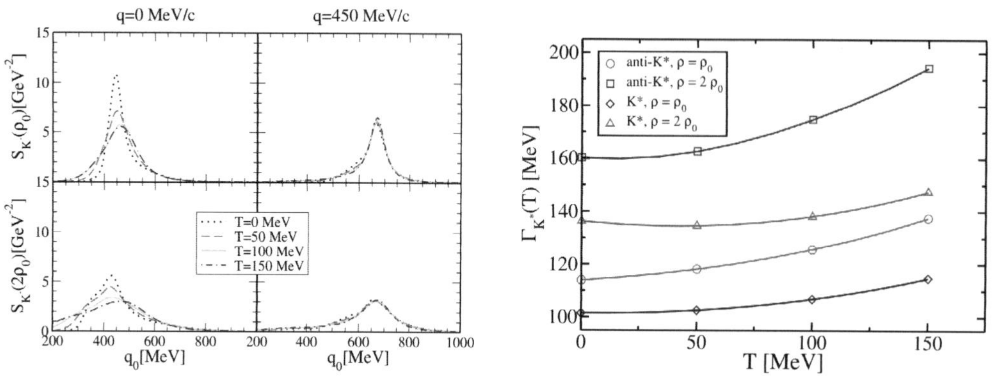

Figure 1. Left: $\bar{K}$ spectral function in hot nuclear matter at different temperatures [14]. Right: Temperature evolution of the $\bar{K}^*(K^*)$ width due to $\bar{K}(K)\pi$ in-medium decay [20].

strange hadrons. We refer to [4] for a report on strangeness production close to threshold in HICs, including predictions for recently proposed angular observables.

2.2. Strange vector mesons in hot and dense matter

Strange vector meson resonances ($\bar{K}^*, K^*$) have triggered substantial attention recently, both theoretically and experimentally. Unlike their unflavoured partners, they do not decay in the dilepton channel, making their experimental detection less clean. Still, both the RHIC low energy scan programme and the HADES experiment in GSI are currently performing measurements in order to extract the in-medium properties of hadronic resonances and, in particular, of strange vector mesons [16, 17]. On the theory side, the properties of $\bar{K}^*$ have been studied in cold nuclear matter [18] starting from a model of the $\bar{K}^*N$ interaction within the hidden gauge formalism (for a recent review see [19]). The $\bar{K}^*$ selfenergy receives two contributions associated to its dominant decay mode $\bar{K}^* \to \bar{K}\pi$ and to the quasi-elastic reaction $\bar{K}^*N \to \bar{K}^*N$ and related absorption channels. The latter mechanism induces a strong broadening of the $\bar{K}^*$ spectral function as a result of the mixing with two $J^P = 1/2^-$ states, the $\Lambda(1783)$ and $\Sigma(1830)$, which are dynamically generated in a parallel way to the $\bar{K}N$ interaction. At $\rho = \rho_0$, the $\bar{K}^*$ width is enlarged beyond 200 MeV, the in-medium $\bar{K}\pi$ cloud contributing about 100 MeV.

The situation for the K^* meson is similar to that of the K and one expects milder changes in the presence of a nuclear medium. The lack of resonant states close to threshold advises to use a $t\rho$ approximation for the K^* selfenergy. The calculation within the hidden gauge formalism is straightforward and reads [20]

$$\Pi_{K^*N} = \frac{1}{2}(t_{K^*p} + t_{K^*n})\rho_0 \frac{\rho}{\rho_0} \simeq 0.22 \frac{m_K}{m_{K^*}} m_{K^*}^2 \frac{\rho}{\rho_0} \ . \tag{1}$$

The latter equality is obtained with the lowest order tree-level amplitudes from the K^*N Lagrangian evaluated at threshold, leading to a positive K^* mass shift of about 50 MeV at $\rho = \rho_0$. When full unitarized scattering amplitudes are implemented, a 20% reduction is obtained over the tree-level result, $\Delta m_{K^*}(\rho_0) \simeq 40$ MeV.

Work is being pursued to extend the former studies to a scenario of nuclear matter at finite temperature. We briefly discuss some results on the two-meson cloud contribution to the in-medium width of strange vector mesons, whereas the full calculation including also the collisional part of the selfenergy will be reported elsewhere. The in-medium width can be obtained straightforwardly from the imaginary part of the selfenergy ($\Gamma_{\bar{K}^* \to \bar{K}\pi} = -\mathrm{Im}\,\Pi_{\bar{K}^*\bar{K}\pi}/P^0$), which for a $\bar{K}^*$ at rest reads

$$\mathrm{Im}\,\Pi_{\bar{K}^*\bar{K}\pi}(P^0,\vec{0}) = -2\pi\, g^2 \int \frac{\mathrm{d}^3q\,\vec{q}^{\,2}}{(2\pi)^3} \int_0^{P^0} \mathrm{d}\omega\, S_{\bar{K}}(\omega,\vec{q})\, S_\pi(P^0-\omega,\vec{q})\,[1+f_\omega+f_{P^0-\omega}] + \mathrm{diff.} \quad (2)$$

The coupling g is fixed to reproduce the $\bar{K}^*$ vacuum decay width and f_ω stands for the Bose distribution function (similar expressions are obtained for K^* and ϕ mesons). The omitted terms, labelled as *diff*, stand for diffusive contributions in which the $\bar{K}^*$ can be absorbed by a thermally excited meson, namely $\bar{K}^*\pi \to \bar{K}$ and $\bar{K}^*K \to \pi$. The Bose enhancement factor becomes important when the light pseudoscalars develop low-energy modes as a consequence of their in-medium dressing. The evolution of the $\bar{K}^*$ and K^* in-medium decay widths with temperature is depicted in Fig. 1 (right). It is worth mentioning that the HADES collaboration have recently reported a reduction of the K^{*0} yield in Ar+KCl collisions at 1.76 A GeV as compared to estimations within the UrQMD transport approach [17].

3. Heavy-flavour relaxation in a hadronic medium

Given the large masses of D and B with respect to other relevant scales, the transport coefficients of these mesons are usually evaluated within a Fokker-Planck equation approach. Two distinct scenarios can be addressed regarding the nature of heavy meson interactions with the hadronic ambient: $\mu_B \simeq 0$ (vanishing baryochemical potential), where the medium is predominantly composed of light mesons; and $\mu_B \neq 0$, where the heavy-meson nucleon interaction sets in and most likely drives the relaxation mechanism [21]. The latter situation is expected to be encountered at low and intermediate energy HICs and in nuclear production reactions (FAIR). We focus here on the first scenario, whereas the second one has been discussed in [22].

An important piece of information is the fact that the spectrum of excitations of the stable open-charm mesons contains two broad resonances with dominant pionic decays in the s-wave: $D_0(2400) \to D\pi$ and $D_1(2427) \to D^*\pi$ (a similar situation is found in the bottom sector). This means that the cross section of the interaction of a $D^{(*)}$ meson with the most abundant constituent of the hadronic medium is dominated by a resonance at low energies. Since a resonant interaction is bound to produce shorter thermalization times, accounting for this feature is important to produce a realistic estimation of the transport coefficients.

The authors of Ref. [6] have evaluated the drag coefficient of D mesons by constructing the scattering amplitudes off π, K, η, ρ, ω, K^* mesons and N, Δ baryons in a phenomenological way, using either Breit-Wigner parameterizations or empirical data. In Refs. [7, 8, 9, 10] an effective field theory constrained by chiral and heavy-quark symmetries has been used to determine the low-energy interactions of D's and B's with the $SU(3)$ octet of light pseudoscalar mesons (see also [23]). An important aspect was introduced in [8, 10] by demanding that the scattering amplitudes satisfy exact unitarity in partial waves, an essential requirement in order to extend the applicability of the low-energy theory up to temperatures $T \simeq m_\pi$. With a minimal set of parameters the unitarized theory dynamically generates the heavy-light meson s-wave resonances in good agreement with the available experimental data both in the charm and bottom sectors.

The (inverse of the) drag coefficient F and the longitudinal diffusion coefficient Γ_0 at $p = 100$ MeV/c for bottomed mesons are shown in Fig. 2 (left and middle panels, respectively), as obtained in [10] . Both drag and diffusion are considerably strengthened in the hotter stages of a HIC, with significant interactions between heavy mesons and the thermal medium (for results in the charm sector see [22]). As expected, the most relevant contribution comes from the pion gas, followed by kaons and (mostly) anti-kaons. At $\mu_B = 0$, the effect of baryons (nucleons and Δ's) is negligible even at the highest considered temperatures. The relaxation time of bottomed mesons travelling with 1 GeV/c momentum is found $\lambda_B \simeq 80$ fm at $T = 150$ MeV, still much greater than the typical lifetime of the hadron gas, thus rendering bottomed mesons as optimal carriers of information of the phase transition upon exiting the interacting region. Comparatively, charmed mesons are more affected by the interactions with the medium ($\lambda_D \simeq 40$ fm at the same conditions), particularly at finite baryochemical potential [21]. The role of unitarization within the heavy-meson chiral effective approach is illustrated in the right panel of Fig. 2.

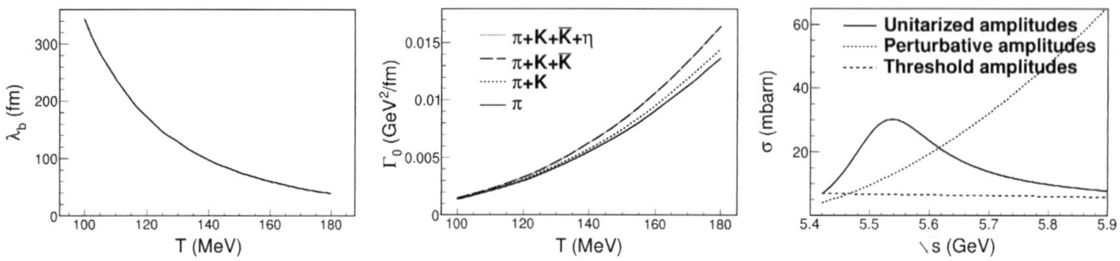

Figure 2. Relaxation time (left) and longitudinal diffusion coefficient (middle) of bottomed mesons at $p = 0.1$ GeV/c as a function of temperature [10]. The total elastic $B\pi$ cross section is shown in the right panel (the solid curve exhibits a B_0 resonance at $\sqrt{s} \simeq 5.53$ GeV).

4. Summary

We have presented a selection of results and discussed work in progress concerning the properties of strange and heavy mesons in hot and dense strongly interacting matter. We have emphasised the benefits of unitarized coupled-channel approaches based on the chiral Lagrangian, which provide the essential features of the interaction of these mesons with the hadronic constituents of the medium in a highly predictive way. The use of realistic hadronic interactions accounting for medium effects in conjunction with microscopic transport approaches qualifies as a powerful device to investigate the complicated reactions that take place in heavy-ion collisions.

Acknowledgments

DC thanks the organizers of FAIRNESS 2013 for a stimulating conference. Work supported by grants BMBF (Germany) no. 05P12RFFCQ and MINECO (Spain) no. FPA-2011-27853-C02-02.

References

[1] Friedman E and Gal A 2007 *Phys. Rept.* **452** 89
[2] Fuchs C 2006 *Prog. Part. Nucl. Phys.* **56** 1
[3] Forster A *et al* 2007, *Phys. Rev.* C **75** 024906
[4] Hartnack C, Oeschler H, Leifels Y, Bratkovskaya E L and Aichelin J 2012 *Phys. Rept.* **510** 119
[5] Cassing W, Tolos L, Bratkovskaya E L and Ramos A 2003 *Nucl. Phys.* A **727** 59
[6] He M, Fries R J and Rapp R 2011 *Phys. Lett.* B **701** 445
[7] Ghosh S, Das S K, Sarkar S and Alam J -e 2011 *Phys. Rev.* D **84** 011503
[8] Abreu L M, Cabrera D, Llanes-Estrada F J and Torres-Rincon J M 2011 *Annals Phys.* **326** 2737
[9] Das S K, Ghosh S, Sarkar S and Alam J -e 2012 *Phys. Rev.* D **85** 074017
[10] Abreu L M, Cabrera D and Torres-Rincon J M 2013 *Phys. Rev.* D **87** 034019
[11] Ramos A and Oset E 2000 *Nucl. Phys.* A **671** 481
[12] Tolos L, Ramos A, Polls A and Kuo T T S 2001 *Nucl. Phys.* A **690** 547
[13] Schaffner J, Bondorf J and Mishustin I N 1997 *Nucl. Phys.* A **625** 325
[14] Tolos L, Cabrera D and Ramos A 2008 *Phys. Rev.* C **78** 045205
[15] Lutz M F M, Korpa C L and Moller M 2008 *Nucl. Phys.* A **808** 124
[16] Blume C and Markert C 2011 *Prog. Part. Nucl. Phys.* **66** 834
[17] Agakishiev G *et al* [HADES Collaboration] 2013 *Eur. Phys. J.* A **49** 34
[18] Tolos L, Molina R, Oset E and Ramos A 2010 *Phys. Rev.* C **82** 045210
[19] Oset E *et al* 2012 *Int. J. Mod. Phys.* E **21** 1230011
[20] Ilner A, Bratkovskaya E L and Cabrera D *in preparation*
[21] Tolos L and Torres-Rincon J M 2013 *Phys. Rev.* D **88** 074019
[22] Torres-Rincon J M *these proceedings*
[23] Fuchs C, Martemyanov B V, Faessler A and Krivoruchenko M I 2006 *Phys. Rev.* C **73** 035204

Journal of Physics: Conference Series **503** (2014) 012018 doi:10.1088/1742-6596/503/1/012018

Recent results on dilepton and strangeness production with HADES

Kirill Lapidus for the HADES collaboration

Excellence Cluster "Universe", Boltzmannstr. 2, 85748 Garching, Deutschland

E-mail: `kirill.lapidus@ph.tum.de`

Abstract.
Selected topics of the HADES research program are discussed with an emphasis on the results obtained by studying proton-proton and proton-niobium collisions at a beam energy of 3.5 GeV. Future directions of the measurements at HADES are outlined.

1. Introduction

HADES is a versatile detector located at GSI Helmholtzzentrum für Schwerionenforschung and currently operating at the SIS18 accelerator in the range of beam energies of 1-2 GeV/u (for nucleus-nucleus collisions), up to 3.5 GeV in proton-induced reactions. A short description of the experimental setup is given in the contribution [1] to these proceedings and a detailed review can be found in [2]. The ultimate goal of the HADES searches is to reveal signals of hadron modification expected to occur at finite net baryonic densities as a manifestation of a partial restoration of the spontaneously broken chiral symmetry. In this respect, primary objects of interest are light unflavoured vector mesons (ρ, ω, ϕ), accessible via di-electron measurements, and (anti)kaons. The strategy is to perform systematical measurements ranging from elementary nucleon-nucleon reactions to collisions of heavy nuclei.

In the following we concentrate mostly on the measurements performed in proton-proton and proton-niobium collisions at a beam kinetic energy of 3.5 GeV. For the results on dilepton and strangeness production in light (C+C) and medium-size (Ar+KCl) systems we refer to [3] and references therein.

2. Measurements in p+p and p+Nb reactions

Studies of dilepton and strangeness production in proton-proton collisions at a beam energy of 3.5 GeV set a reference for the interpretation of the data obtained in proton-niobium collisions at the same energy. The nucleus acts here as a well controlled strongly interacting environment with a known density profile, so the possible in-medium effects at saturation density might be probed.

2.1. Vector mesons: $\omega \to e^+e^-$

A measurement of the electron-positron pair production in proton-proton collisions at a beam kinetic energy of 3.5 GeV was performed by the HADES collaboration [4]. Fig. 1 shows the invariant mass spectrum of signal e^+e^--pairs together with a "cocktail" of individual contributions simulated with the PYTHIA and Pluto event generators (employed for the treatment of the production and decay processes, respectively). The low-mass region $M_{ee} < 0.15$ GeV/c^2 is saturated by the Dalitz-decays of the π^0-meson; the intermediate mass range $0.15 < M_{ee} < 0.5$ GeV/c^2 is dominated by the Dalitz-decays of the η-meson with certain contributions of the channels $\Delta \to Ne^+e^-$ and $\omega \to \pi^0 e^+e^-$. Finally, a distinct peak corresponding to the direct decays of the ω-meson on top of the broad ρ-meson contribution is visible. The description of the experimental data with the PYTHIA cocktail is not satisfactory: there is a large gap below the ω pole.

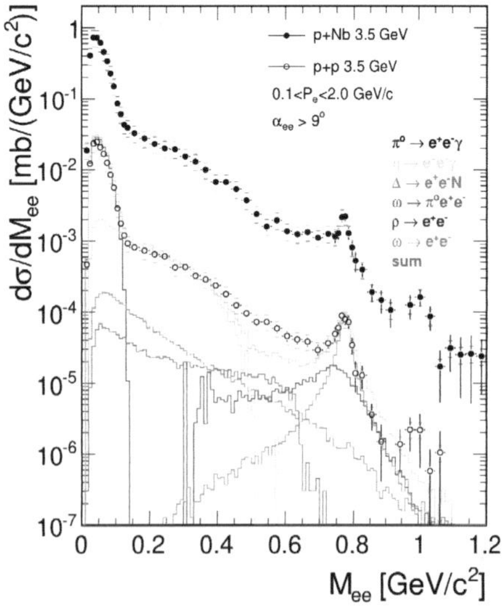

Figure 1. Invariant mass spectra of electron-positron pairs measured in p+p (open circles) and p+Nb (full circles) collisions. For the p+p measurement a dilepton cocktail simulated with PYTHIA is shown.

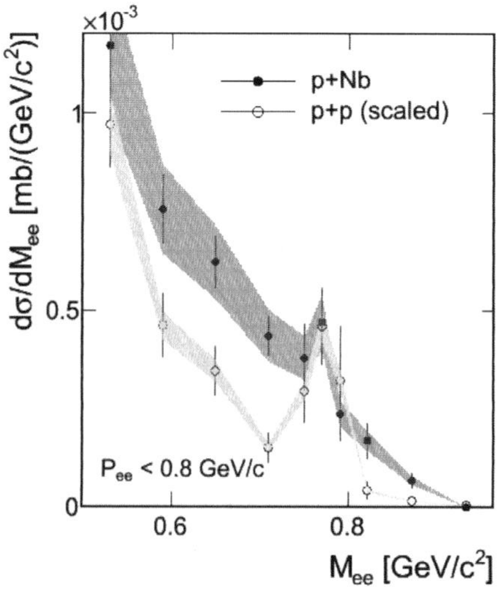

Figure 2. Comparison of the invariant mass spectra measured in p+p and p+Nb reactions for slow electron-positron pairs ($P_{ee} < 0.8$ GeV/c).

A possible remedy for this discrepancy is discussed in [5]. Here a contribution of a number of N^* and Δ^* resonances to the final e^+e^--pair spectrum is included. A strict vector meson dominance model is considered, forcing all electromagnetic transitions to proceed via an intermediate ρ-meson: $N^* \to N\rho \to Ne^+e^-$ and introducing thus a duality between the contribution of the baryonic resonances and the contribution of the off-shell ρ-meson. Within this approach, the missing yield in the invariant mass spectrum below the ω-pole is effectively filled by the contribution of the baryonic resonances. Whereas the specific shares of individual contributions are not known precisely, this interpretation strongly points to the importance of the resonances' contribution in this energy range.

The invariant mass spectrum of pairs reconstructed in proton-niobium collisions [6] is shown on Fig. 1 as well. A comparison to the reference measurement in proton-proton collisions has been made for slow electron-positron pairs ($P_{ee} < 0.8$ GeV/c) in the vicinity of the ω-peak (Fig. 2). To facilitate the comparison, the p+p spectrum has been scaled up according to a Glauber model, see [6] for details. The observed excess located left to the ω-peak might be interpreted as a contribution from baryonic resonances formed in secondary processes, such as πN-reactions. Since these resonances strongly couple, as discussed above, to the off-shell ρ-meson, the presented data contains information about the in-medium ρ-meson spectral function.

2.2. Neutral kaons: $K^0_S \to \pi^+\pi^-$

Analysis of the inclusive K^0 production in proton-proton collisions (preliminary results of which were reported in [7]) yields important constraints for the models describing the production of strangeness in proton-nucleus and nucleus-nucleus reactions. Comparison of the data with a resonance model of kaon production [8] showed that the model gives unsatisfactory description, overestimating the total production cross section and failing to describe the shape of transverse momentum spectra. This disagreement is not surprising, since it is known that this model strongly overestimates the production cross section in a number of channels, such as for example $pp \to p\pi^+\Lambda K^0$ [9].

Analysis of the inclusive neutral kaon production has been continued in p+Nb collisions with a goal to probe the sensitivity of the kaon spectra to the in-medium interaction of kaons in terms of the repulsive in-medium potential. Interpretation of the data with a help of the GiBUU transport model [10], which allows to simulate the propagation of kaons in a mean-field potential generated by the baryonic density, is in its final stage. Model calculations show that the experimental data are sensitive to the effect of the repulsive potential.

Figure 3 shows the region of the K^0 rapidities experimentally accessible in p+Nb reactions.

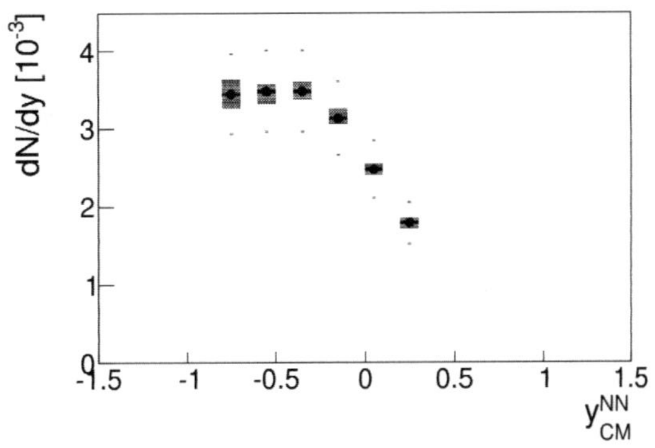

Figure 3. K^0_S rapidity distribution in p+Nb collisions represented in the nucleon-nucleon center-of-mass reference frame.

2.2.1. $\Lambda(1405)$ The strange baryons $\Sigma(1385)$ and $\Lambda(1405)$ were reconstructed at HADES in proton-proton collisions ([11] and [12], respectively). The $\Lambda(1405)$ enjoys particular theoretical attention due to its key role in the understanding of the low-energy antikaon-nucleon interaction. The presence of the $\Lambda(1405)$ just below the $\bar{K}N$ threshold makes the problem of the antikaon-nucleon interaction complex and requires a treatment within a coupled channel approach. Within this approach the $\Lambda(1405)$ is understood as a dynamically generated state with two main contributions: an antikaon-nucleon bound state and a $\Sigma\pi$ resonance.

The reaction $p + p \to p + \Lambda(1405) + K^+$ was analysed and the missing mass spectrum $MM(p, K^+)$, showing a signal of the $\Lambda(1405)$ (the gray band) is presented on Fig. 4. Most interesting is the location of the $\Lambda(1405)$ mass peak — well below 1405 MeV/c^2. This is a first reconstruction of the $\Lambda(1405)$ in proton-proton collisions in the charged decay mode and these data give a new input to models describing the production of $\Lambda(1405)$.

Since an antikaon-nucleon bound state is realised in nature as a contribution to $\Lambda(1405)$, one might ask whether a system of two nucleons bound by an antikaon, a ppK^- state, might exist. Indeed, a hypothesis of an existence of such a system has been put forward in [13, 14].

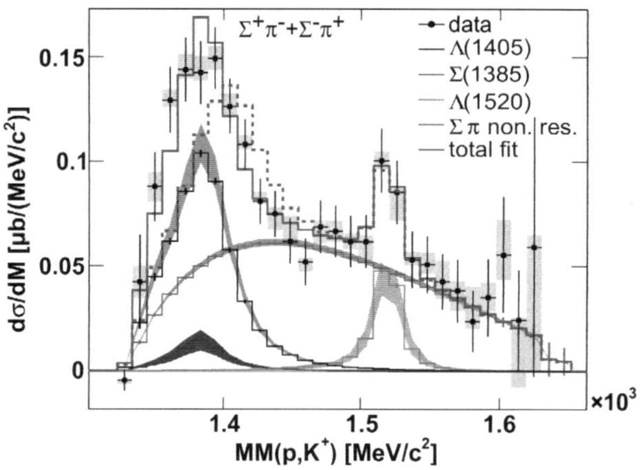

Figure 4. Sum of missing mass $MM(p, K^+)$ distributions reconstructed with the $\Sigma^+\pi^-$ and $\Sigma^-\pi^+$ pairs. The gray band shows the contribution of the $\Lambda(1405)$. Solid gray curve corresponds to the incoherent sum of all relevant channels (see the legend).

The HADES collaboration is currently performing a search for the simplest kaonic cluster in a reaction $pp \to pK^+\Lambda$, more details can be found in [15].

3. Outlook

After a profound upgrade of the experimental apparatus, the HADES detector successfully measured collisions of heavy ions Au+Au at a beam energy of 1.23 GeV/u. A discussion of the dilepton measurement can be found in a dedicated contribution to these proceedings [16].

A number of strange hadrons (charged kaons and antikaons, K^0_S, Λ and ϕ), all produced sub-threshold with respect to free nucleon-nucleon reactions, were reconstructed. A novelty in this experimental campaign was the possibility to reconstruct the reaction plane looking at the spectator protons and light fragments with a dedicated Forward Wall hodoscope, which covers low polar angles. This allows to analyse the anisotropy of the particle emission relative to the reaction plane on the event-by-event basis, which is sensitive to the in-medium interaction of various hadron species with the bulk nuclear matter.

An experiment with a pion beam offers a natural continuation of the HADES measurements of dilepton and strangeness production in proton-proton and proton-nucleus collisions. The usage of a light projectile allows to produce hadrons with low momenta in the target reference frame, ensuring optimal conditions for the observation of the in-medium effects. Of particular interest is to measure production off several nuclear targets (light and heavy) with an aim to perform a comparative analysis.

In future, HADES will continue to explore dense baryonic matter at the FAIR facility and perform measurements at beam energies of up to 8 GeV/u provided by the SIS100 heavy-ion synchrotron. This energy region lies above the vector meson production threshold and has not been yet explored by means of dilepton spectroscopy. The HADES detector will be installed in a common cave with the CBM experiment and the latter will continue measurements at even higher energies. The fitness of the experimental apparatus for the demanding conditions of a high charged-particle occupancy has been already checked with a recent measurement of Au+Au collisions.

Transition to higher beam energies will open a possibility to probe baryonic densities comparable to those achieved in the interior of the neutron stars. Of special interest in this respect is the study of multistrange baryons (Ξ^-, Ω^-) and hypernuclei. They serve as a probe of the hyperon interaction at high baryonic densities — a topic relevant for the question of the strangeness appearance in the interior of neutron stars [17].

4. Conclusions

As discussed in the present contribution, study of elementary nucleon-nucleon reactions delivers a basis for the interpretation of proton-nucleus and nucleus-nucleus collisions. Besides, already the vacuum properties of certain particles such as the ρ-meson (due to its strong coupling to baryonic resonances) or the $\Lambda(1405)$ baryon (due to its complex, molecular-like nature) are of interest by themselves.

Proton-nucleus reactions offer a next level of complexity. Here, the in-medium behaviour of various meson species propagating inside a nucleus, including ω-mesons and neutral kaons, was studied by HADES. A truly heavy colliding system (Au+Au collisions at 1.23 GeV/u) was measured by HADES in 2012 and a number of analyses is on-going with first promising results both in the dilepton and in the strangeness sector.

In the near future, HADES has a unique opportunity to perform measurements with pion-induced reactions characterized by favourable conditions for an observation of in-medium effects. After completion of the experimental program at SIS18, HADES will move to FAIR and will continue measurements at SIS100 at beam energies of up to 8 GeV/u.

Acknowledgments

The collaboration gratefully acknowledges the following funding: INFN-LNS Catania (Italy); LIP Coimbra (Portugal): PTDC/FIS/113339/2009; SIP JUC Cracow (Poland): NN202198639; GSI Darmstadt (Germany): Helmholtz Alliance HA216/EMMI; TU Darmstadt (Germany): VH-NG-823, Helmholtz Alliance HA216/EMMI; HZDR, Dresden (Germany): 283286, 05P12CRGHE; Goethe- University, Frankfurt (Germany): Helmholtz Alliance HA216/EMMI, HIC for FAIR (LOEWE), GSI F&E, BMBF 06FY9100I, HGS-Hire, H-QM research school TU München, Garching (Germany): BMBF 06MT7180; JLU Giessen (Germany): BMBF:05P12RGGHM; University Cyprus, Nicosia (Cyprus): UCY/3411-23100; IPN Orsay, Orsay Cedex (France): CNRS/IN2P3; NPI AS CR, Rez, (Czech Republic): MSMT LG 12007, GACR 13-06759S.

References

[1] Behnke C *et al.* (HADES Collaboration) *These proceedings*
[2] Agakishiev G *et al.* (HADES Collaboration) 2009 *Eur.Phys.J.* **A41** 243–277 (*Preprint* 0902.3478)
[3] Salabura P *et al.* (HADES Collaboration) 2013 *J.Phys.Conf.Ser.* **420** 012013
[4] Agakishiev G *et al.* (HADES Collaboration) 2012 *Eur.Phys.J.* **A48** 64 (*Preprint* 1112.3607)
[5] Weil J, van Hees H and Mosel U 2012 *Eur.Phys.J.* **A48** 111 (*Preprint* 1203.3557)
[6] G. Agakishiev *et al.* [HADES Collaboration], Phys. Lett. B **715**, 304 (2012) [arXiv:1205.1918 [nucl-ex]].
[7] Berger-Chen J C, Fabbietti L and Lapidus K (HADES Collaboration) 2012 *Hyperfine Interact.* **210** 111–114
[8] Tsushima K, Sibirtsev A and Thomas A W 1999 *Phys.Rev.* **C59** 369–387 (*Preprint* nucl-th/9801063)
[9] Nekipelov M, Buscher M, Hartmann M, Keshelashvili I, Kleber V *et al.* 2007 *J.Phys.* **G34** 627 (*Preprint* nucl-ex/0610026)
[10] Buss O, Gaitanos T, Gallmeister K, van Hees H, Kaskulov M *et al.* 2012 *Phys.Rept.* **512** 1–124 (*Preprint* 1106.1344)
[11] Agakishiev G *et al.* (HADES Collaboration) 2012 *Phys.Rev.* **C85** 035203 (*Preprint* 1109.6806)
[12] Agakishiev G *et al.* (HADES Collaboration) 2013 *Phys. Rev. C* **87**(2) 025201
[13] Akaishi Y and Yamazaki T 2002 *Phys.Rev.* **C65** 044005
[14] Yamazaki T and Akaishi Y 2002 *Phys.Lett.* **B535** 70–76
[15] Fabbietti L *et al.* (HADES Collaboration) 2013 *Nucl.Phys.* **A914** 60–68 (*Preprint* 1305.2631)
[16] Harabasz S *et al.* (HADES Collaboration) *These proceedings*
[17] Providncia C, Avancini S S, Cavagnoli R, Chiacchiera S, Ducoin C *et al.* 2013 (*Preprint* 1307.1436)

Journal of Physics: Conference Series **503** (2014) 012019 doi:10.1088/1742-6596/503/1/012019

Overview of the CBM detector system

Tomáš Balog

GSI Helmholtzzentrum für Schwerionenforschung GmbH, Planckstrasse 1, 64291 Darmstadt, Germany

E-mail: t.balog@gsi.de

Abstract. The Compressed Baryonic Matter (CBM) experiment at the future Facility for Antiproton and Ion Research (FAIR) is a fixed target experiment designed to explore the QCD phase diagram in the region of high net-baryon densities. The CBM detector system will access beams directly from the superconducting synchrotrons SIS100 and SIS300. It is designed for interaction rates up to 10^7 Hz to enable measurements of rare observables and diagnostic probes created in the early and dense phase of the fireball evolution. The layout of the CBM detector system is adapted to the experimental requirements concerning the acceptance in the laboratory frame (mid and forward rapidities), reaction rates, radiation tolerance, determination of the vertices with accuracy of 50 μm, particle densities (up to 700 particles passing through the active area of the detector in single central Au+Au collision at 25 GeV/nucleon) and selectivity [1, 2].

The technical challenge for the detector system (Fig. 1) is to identify both hadrons and leptons and to filter the rare probes. The measurements will be performed via nucleus-nucleus, proton-nucleus and proton-proton interactions at different beam energies. Proton-proton interactions are required for the baseline determination ("normal nuclear matter"). For the particle identification, especially multi-strange hyperons, hypernuclei, vector mesons decaying into lepton pairs and particles with charm quarks the background suppression is required even at the highest interaction rates [2]. Thus the track reconstruction of each individual particle passing through the detector system has to be made online and filtered with respect to the physics requirements.

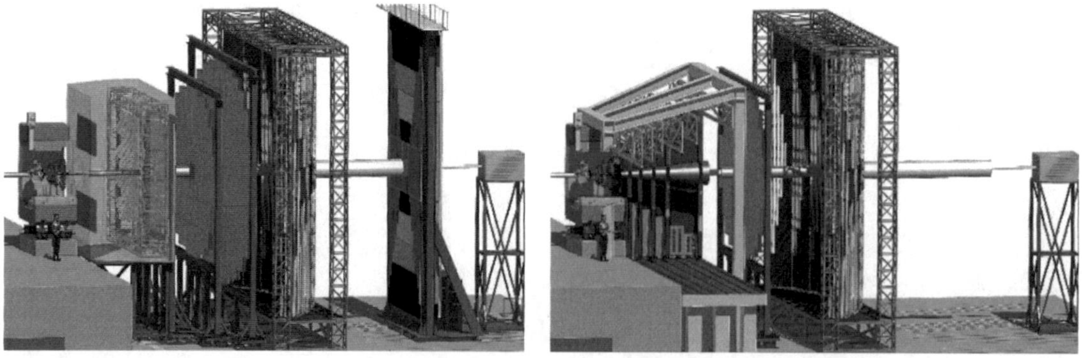

Figure 1. Left - electron configuration and right - muon configuration of the CBM experiment [1].

In such case the hierarchical trigger system can not be applied and self-triggered read-out electronics will be used instead. In addition high speed data processing, transition and analysis using fast algorithms are prerequisites for such system. The essential factor is the data transport and computational throughput rather than the decision latency.

The CBM detector system will provide both electron and muon detection systems in order to take advantage of both measurement methods. Muon and electron pairs created during particle decays will be measured during separate runs in order to control the systematic approach of the measurements. The acceptance of the detector systems as coming from the physics requirements is full 2π for φ (azimuth) and from 2.5° to 25° for Θ (polar angle). In electron configuration following detectors will be used: Micro-vertex Detector (MVD), Silicon Tracking System (STS), both placed in gap of 1 T superconducting magnet, then Ring Imaging Cherenkov Detector (RICH), Transition Radiation Detectors (TRD), Resistive Plate Chambers for time-of-flight measurements (TOF), Electromagnetic Calorimeter (ECAL) and Projectile Spectator Detector (PSD) as hadronic calorimeter. In muon configuration the RICH detector will be replaced by the Muon Detection System (MUCH) and ECAL will be removed [1].

Dipole magnet: The dipole magnet of CBM will be superconducting with a large aperture (Fig. 2). The coils (each with 1749 turns and cooled by liquid helium) will provide a magnetic field with total bending power 1 Tm. The total weight of the magnet will be 160 tons. In the aperture the MVD and STS detector system will be placed [2].

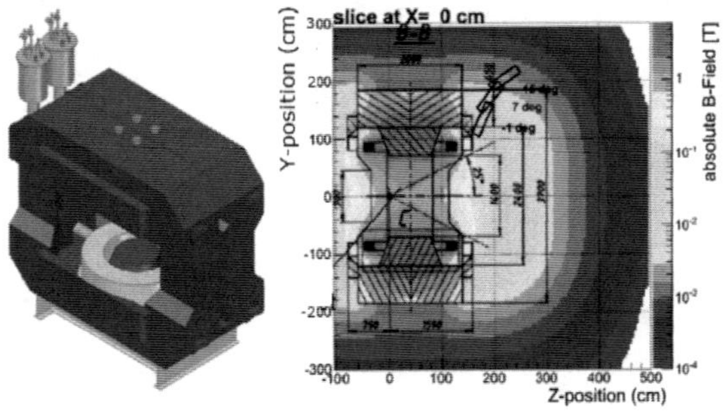

Figure 2. Left - engineering view of the CBM dipole magnet and right - magnetic field map [3, 4].

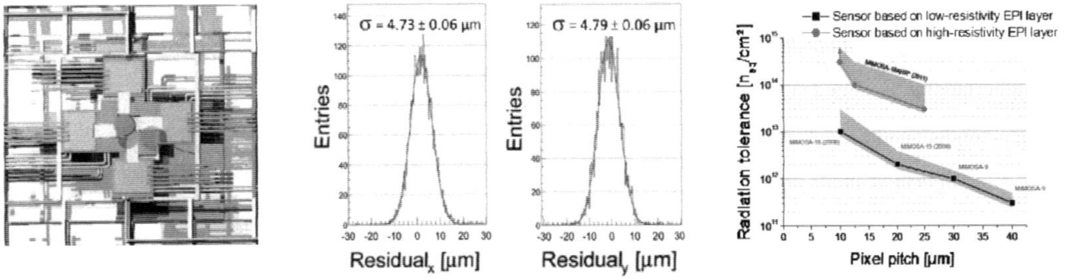

Figure 3. Left - engineering view of the MVD, middle - examples of the hit position resolution in MIMOSA-26 sensors and right - radiation hardness measurements for different sensors tested for the development of the MVD.

Micro-Vertex Detector (MVD): Micro-Vertex Detector will be built from of Monolithic Active Pixel Sensors (MAPS) with pixel sizes $18\times18\ \mu m^2$ and $20\times40\ \mu m^2$. Depending on the pixel size the hit position resolution varies from 3.5 μm to 6 μm resulting in secondary vertex resolution of 50 μm to 100 μm along the beam axis (Fig. 3). The detector will consist of 3 stations placed 5, 10 and 15 cm downstream of the target in vacuum box. The material budget of sensors in each of the stations together with cooling and support structures is kept between 300 μm and 500 μm of silicon equivalent [2]. For a SIS300 running scenario the hardness against nonionizing radiation is required to be above 10^{13} neutron equivalent per cm^2 (n_{eq}/cm^2) and for ionizing radiation to be above 3 Mrad. Time resolution of the sensors is expected to be below 30 μs. Currently tested chip, called MIMOSA-26, fulfils most of the requirements giving the hit position resolution of about 4 μm, and having radiation hardness for non-ionizing radiation well above 10^{13} n_{eq}/cm^2 [3].

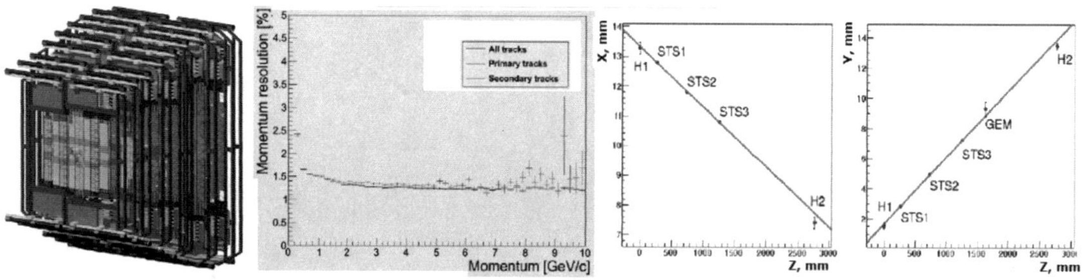

Figure 4. Left - engineering view of the STS, middle - simulated momentum resolution ($\Delta p/p = 1.3\%$ for high momenta) with realistic material budget of all STS stations and two most right - tracks obtained with a test detector system including three STS stations in a 2.4 GeV/c proton beam [2].

Silicon Tracking System (STS): Following the MVD downstream of the target is the Silicon Tracking System (STS), also placed inside the aperture of the dipole magnet. It will provide the track reconstruction and momentum determination of the charged particles. It will be built out of 8 tracking layers populated with double-sided micro-strip silicon sensors placed from 30 to 100 cm distance from the target. To obtain required momentum resolution $\Delta p/p \approx 1\%$ the stations have to have an ultra low material budget (Fig. 4). Thus the front-end electronics will be placed outside the active area of the STS and the sensors will be interconnected to the read-out chip via low-mass cables [2]. The stations are arranged in four duplets. The strips of the sensors are 2, 4, 6 cm and 12 cm when two sensors with 6 cm strip length are daisy-chained. Each of the sensors comprises 1024 channels with strip pitch 58 μm. The stereo angle is 7.5°. The thickness of the used sensors will be 300±15 μm [2].

Ring Imaging Cherenkov Detector (RICH): When CBM will operate in electron configuration the Ring Imaging Cherenkov Detector follows the STS. It is foreseen to be used for the identification and suppression of pions in the momentum range below 10 GeV/c [1]. The gas radiator (length 1.7 m) consists of CO_2 where the pion threshold for Cherenkov is 4.65 GeV/c. The Cherenkov radiation will be reflected by mirrors built from 72 mirror tiles with a curvature of 3 m radius and Al+MgF_2 reflective coating (Fig. 5). The photo-detector plane where photons will be reflected will be built from Multi-Anode Photo Multiplier Tubes (MAPMT) and will be shielded from the magnetic field [2].

Muon Chamber System (MUCH): In the muon configuration of the CBM experiment the STS is followed by the Muon Chamber System. The main role of MUCH is to detect the muons from J/Ψ and light vector meson decays in an environment of high particle densities [1]. MUCH

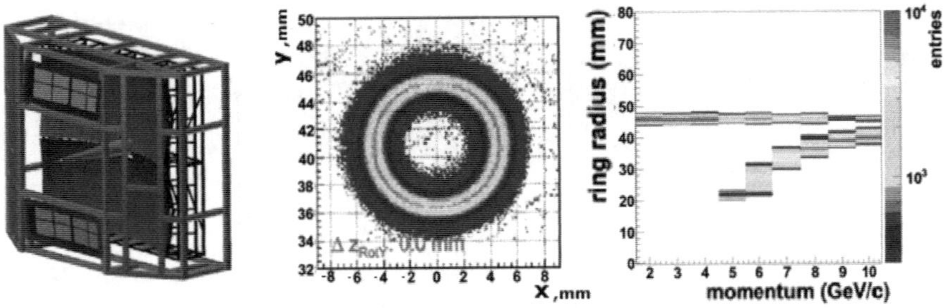

Figure 5. Left - engineering view of the RICH detector, middle -Cherenkov light photons as seen in RICH and right - electron-pion identification as measured during beam-time [4].

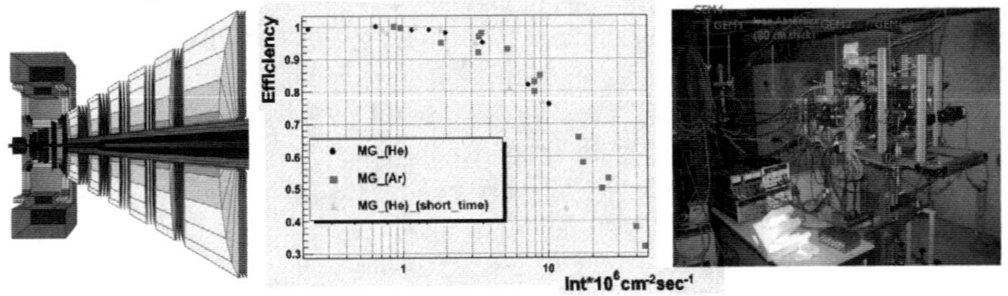

Figure 6. Left - engineering view of the MUCH detector, middle - efficiency as a function of hit rate showing efficiency above 90% up to 5 MHz/cm^2. Right - beam-test setup [4].

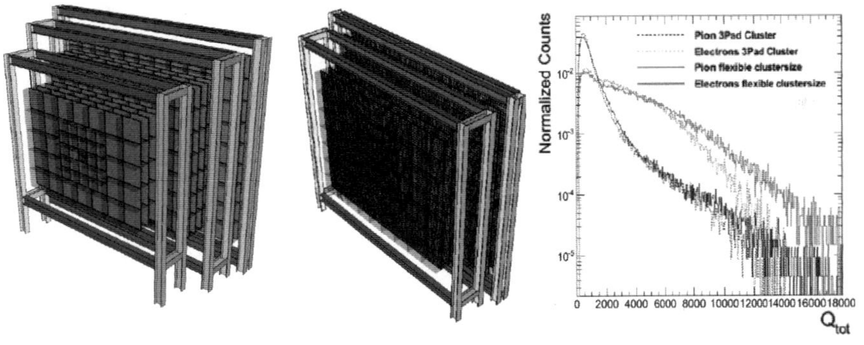

Figure 7. Left - engineering view of the TRD detector for electron setup, middle - engineering view of the TRD detector for muon setup and right - pion (black) and electron (red) identification as measured during an in-beam test [4].

will provide particle tracking and perform momentum dependent muon identification using and instrumented hadron absorber. The absorber will be segmented into several layers with triplets of tracking Gas Electron Multiplier (GEM) detectors in the gaps [2]. The design consists from 6 absorber layers made of iron and 18 gaseous tracking chambers (Fig. 6). The expected hit rate reaches 3 MHz/cm^2 in the first detector system. At SIS100 a MUCH start version will be used which will consist from three chamber triplets with three absorbers [2].

Transition Radiation Detector (TRD): The Transition Radiation Detector will be used for identification of electrons and pions with p $>$ GeV/c ($\gamma \geq 1000$) [1]. It will consist out of three transition radiation detector stations. The TRD readout will be realized in rectangular pads providing resolution of 300 μm - 500 μm across and 3 mm - 30 mm along the pad (Fig. 7) [2]. Every second transition radiation layer is rotated by 90°. The pion suppression factor obtained will be well above 100 at electron identification efficiency 90%. The expected hit rate in TRD is up to 100 kHz/cm^2. In the SIS100 setup only one station of TRD will be used as an intermediate tracker between the STS and the Time-of-Flight wall [2].

Time-of-Flight Wall (TOF): An array of Timing Multi-gap Resistive Plate Chambers will be used for hadron identification via time-of-flight measurements [1]. It will cover an area of 120 m^2 using pad structures for the inner areas and strip structures for the outer zones. The hit rates in the inner zones are expected to reach 25 kHz/cm^2 and 10 kHz/cm^2 in the outer areas. The required time resolution of the chambers is 80 ps (Fig. 8) [2].

CBM calorimeters: In the CBM detector the Electromagnetic Calorimeter (ECAL) will be used to measure photons and neutral mesons (π^0, η) decaying into photons (Fig. 9) [1]. The Projectile Spectator Detector (PSD) will provide measurements of centrality and reaction plane (Fig. 9). It is designed to determine the number of non-interacting nucleons from a projectile nucleus in nucleus-nucleus collisions. Both calorimeters have a "shashlik" structure build from lead/scintillator layers.

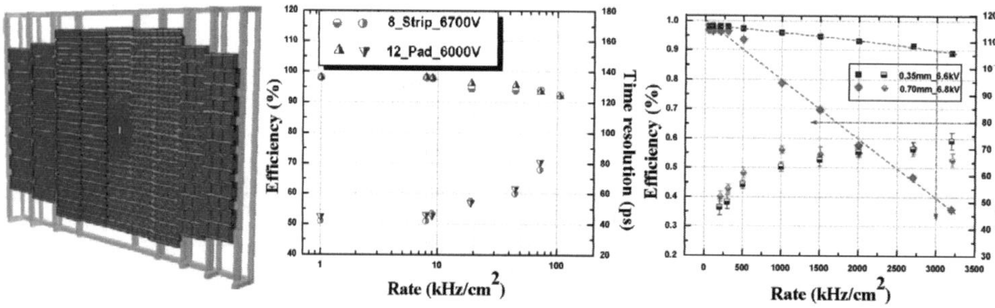

Figure 8. Left - engineering view of the TOF detector, middle - efficiency and time resolution as a function of the hit rate for pads and right - efficiency and time resolution as a function of the hit rate for strips [3, 4].

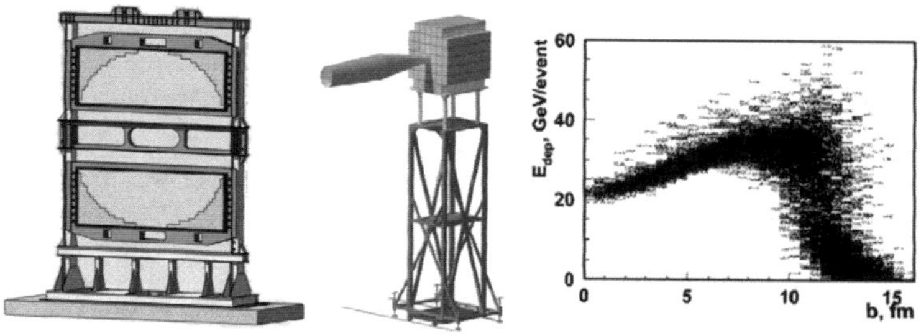

Figure 9. Left - engineering view of the ECAL, middle - engineering view of the PSD and right - centrality determination using PSD as simulated for Au+Au collisions at 10 AGeV.

References

[1] B. Friman et al. (Eds.), The CBM Physics Book, Lect. Notes Phys. 814, Springer-Verlag, 2011
[2] GSI Report 2013-4, Technical Design Report for the CBM Silicon Tracking System (STS), October 2013, http://repository.gsi.de/record/54798
[3] GSI Report 2013-1, May 2013, GSI Scientific Report 2012, http://repository.gsi.de/record/52876
[4] CBM Progress Report 2012, April 2013, https://www-alt.gsi.de/documents/DOC-2013-Mar-49.html

Journal of Physics: Conference Series **503** (2014) 012020 doi:10.1088/1742-6596/503/1/012020

D-meson diffusion in hadronic matter

Juan M Torres-Rincon[1,0], Luciano M Abreu[2], Daniel Cabrera[3], Felipe J Llanes-Estrada[4], Laura Tolos[1,3]

[1] Institut de Ciències de l'Espai (IEEC/CSIC), Campus Universitat Autònoma de Barcelona, Facultat de Ciències, Torre C5, E-08193 Bellaterra, Spain
[2] Instituto de Física, Universidade Federal da Bahia, 40210-340, Salvador, BA, Brazil
[3] Frankfurt Institute for Advances Studies. Johann Wolfgang Goethe University, Ruth-Moufang-Strasse 1, 60438 Frankfurt am Main, Germany
[4] Departamento de Física Teórica I, Universidad Complutense, 28040 Madrid, Spain

Abstract. We present effective-field-theory results with unitarized interactions on the D-meson transport coefficients in a gas populated by light mesons and baryons at finite temperature and baryochemical potential. The Fokker-Planck equation is used to compute the drag force, the relaxation time and the diffusion coefficients of D mesons for collisions at FAIR. At finite baryochemical potential, the combined effect of net baryonic density and sizable meson-baryon interaction makes the D mesons to relax more efficiently than in the $\mu_B = 0$ case. We also describe the connection with the quark-gluon plasma phase in adiabatic trajectories on the phase diagram at both zero and finite baryochemical potential.

1. Introduction

Some of the most useful tools to study the quark-gluon plasma (QGP) phase –formed in the early stages of a relativistic heavy-ion collision– are heavy quarks. Due to their large mass ($m_c, m_b \gg \Lambda_{QCD}, T$) they can be treated as Brownian particles interacting with lighter partons. Not only the average momentum of the heavy quarks is modified by the collisions, but also the momentum distribution is broadened. The efficiency of these two processes is quantified by the *drag force* and the *momentum-space diffusion coefficient*, respectively.

The previous picture also holds after hadronization. The role of the Brownian particle is played by a D or B meson, whereas the thermal bath is now composed of pions, kaons, nucleons and other light hadrons with whom the heavy meson may interact. Thus, the transport coefficients of D mesons [1, 2] and B mesons [3] become of key importance in the equilibration of these hadrons. The presence of these coefficients will amount on the modification of the final heavy meson spectra and affect the flow harmonics v_n and the nuclear suppression factor R_{AA}. To study these transport coefficients and quantify the energy loss of heavy quarks in the hot and dense medium we use quantum kinetic theory.

We focus on the heavy-ion physics to be produced at the Facility for Antiproton and Ion Research (FAIR) [4], where the collision energy will be lower than that for collisions at the Large Hadron Collider, for instance. Lower energies typically produce a net baryon density in the central rapidity region. This fact will imply physics at non-zero baryochemical potential in the QCD phase diagram, maybe close to the conjectured critical point. For these collisions,

[0] Speaker

the available energy density makes difficult the production of B mesons. Therefore, in this manuscript we focus our attention to the calculation of D-meson transport coefficients [2].

In Sec. 2 we introduce the Fokker-Planck kinetic equation and detail the physical interpretation of the D-meson transport coefficients. In Sec. 3 we briefly describe the interaction of D mesons with lighter hadrons, both mesons and baryons. Finally, in Sec. 4 we present our main results and conclusions.

2. Fokker-Planck equation and transport coefficients

The collective description of D mesons is encoded in their distribution function $f(t,p)$. The temporal evolution of $f(t,p)$ follows a kinetic equation defined in the phase space, which in the case of a Brownian particle in a thermal bath composed of lighter particles is the *Fokker-Planck equation*:

$$\frac{\partial f(t,\mathbf{p})}{\partial t} = \frac{\partial}{\partial p_i}\left\{F_i(\mathbf{p})f(t,\mathbf{p}) + \frac{\partial}{\partial p_j}\left[\Gamma_{ij}(\mathbf{p})f(t,\mathbf{p})\right]\right\} , \tag{1}$$

where $i = 1,2,3$ is the spatial index, F_i is the drag force or friction term accounting for the average momentum change, and Γ_{ij} is the momentum-space diffusion matrix which accounts for the broadening of the momentum distribution. The computation of these coefficients is performed through their microscopic expressions in terms of momentum averages,

$$F_i(\mathbf{p}) = \int d\mathbf{k}\; w(\mathbf{p},\mathbf{k})\; k_i \ , \qquad \Gamma_{ij}(\mathbf{p}) = \frac{1}{2}\int d\mathbf{k}\; w(\mathbf{p},\mathbf{k})\; k_i k_j \ , \tag{2}$$

where the interaction measure $w(\mathbf{p},\mathbf{k})$ represents the probability of a D meson with momentum $\mathbf{p}$ to suffer a collision with a light particle, loosing a momentum $\mathbf{k}$:

$$\begin{aligned} w(\mathbf{p},\mathbf{k}) &= g_l \int \frac{d^3q}{(2\pi)^9}\, n_{F,B}(E_l(q),T)\, \left[1 \pm n_{F,B}(E_l(q+k),T)\right] \frac{1}{2E_D(p)}\frac{1}{2E_l(q)}\frac{1}{2E_D(p-k)} \\ &\times \frac{1}{2E_l(q+k)}(2\pi)^4\delta(E_D(p)+E_l(q)-E_D(p-k)-E_l(q+k))\ \overline{|\mathcal{M}|^2} \ , \end{aligned} \tag{3}$$

with g_l being the spin-isospin degeneration of the light particle, $n_{F,B}$ its equilibrium distribution function following Fermi-Dirac or Bose-Einstein statistics, respectively; $E_D(p)$ is the energy of the D meson with momentum $\mathbf{p}$, $E_l(q)$ the energy of the light particle with momentum $\mathbf{q}$ and $\overline{|\mathcal{M}^2|}$ is the average scattering matrix element squared of the binary collision.

If we consider an isotropic gas and focus on the *static limit* (where $p \to 0$) there is only one independent transport coefficient. For definiteness, we take it to be the drag force $F(p) = F_i p^i/p^2$, while the momentum-space diffusion coefficient Γ is related to F through the Einstein relation $\Gamma = F m_D T$ (where m_D is the D-meson mass and T the temperature of the thermal medium).

In several talks at this workshop some emphasis was put on the concept of *relaxation time*. This coefficient describes how efficient the relaxation of the non-equilibrium distribution function is, while it relaxes towards the equilibrium state (Boltzmann's H theorem). Consider for simplicity the one dimensional case of Eq. (1) and assume constant transport coefficients F and Γ. The analytical solution of the kinetic equation is

$$f(t,p) = \sqrt{\frac{F}{2\pi\Gamma(1-e^{-2Ft})}}\exp\left[-\frac{F}{2\Gamma}\frac{(p-p_0e^{-Ft})^2}{1-e^{-2Ft}}\right] , \tag{4}$$

where p_0 is the initial D-meson momentum. It is immediate to see that F^{-1} should play the role of an intrinsic time for the relaxation process. In particular, the average momentum reads

$$\langle p\rangle \equiv \int_{-\infty}^{\infty} dp\; pf(t,p) \Big/ \int_{-\infty}^{\infty} dp\; f(t,p) = p_0\, e^{-\frac{t}{\tau_R}} , \tag{5}$$

where we have defined the relaxation time $\tau_R \equiv 1/F$, which provides a characteristic time accounting for the exponential decay of the average momentum.

We can also introduce the *diffusion coefficient in position space* D_x. The standard result for the average square displacement

$$\langle (x(t) - x_0)^2 \rangle = 2D_x t \ , \tag{6}$$

shows that D_x is a measure of the "speed" of the diffusion in space. A large D_x implies that the D meson can diffusively travel larger distances for a given time. In the static limit this coefficient is related to the drag force by the relation $D_x = T/(m_D F)$.

All these coefficients depend on the temperature and baryochemical potential. In a heavy-ion collision both variables change with time along the fireball's trajectory in the phase diagram. For simplicity, we assume adiabatic trajectories –with a constant entropy per baryon consistent with the energies expected at FAIR collisions– which give us the implicit relation $\mu_B = \mu_B(T)$.

3. Interaction and effective field theories

The interaction between the D mesons and the light hadrons –as it lies in the nonperturbative and confined regime of QCD– can be described by effective models. The global symmetries of the QCD Lagrangian should be implemented into the effective models allowing a reliable description of the hadronic interaction. We take the *chiral and heavy-quark-spin symmetries* to be the fundamental guides in order to construct our effective Lagrangian (therefore, it results in a double expansion). In addition to these symmetries, our third principle is the exact satisfaction of the *unitarity condition* of the scattering matrix.

To account for the D meson–light meson scattering we introduce pions, kaons, antikaons and η mesons. Details on the effective theory based on chiral and heavy-quark-spin expansions can be found in Refs. [1, 2, 5]. Using Feynman rules we construct the perturbative amplitude (or potential) V for each scattering channel. In the meson-baryon sector we also construct the effective Lagrangian based on heavy-quark-spin symmetry which accounts for the interaction of D mesons with nucleons and Δ baryons. Details on the effective theory and the computation of the potential V can be found in [2, 6, 7, 8, 9].

In both sectors we apply a unitarization method [10] (based on a Bethe-Salpeter (BS) equation) to extend the interaction description up to temperatures $T \simeq 140$ MeV without loosing physical sense. In addition, we are able to dynamically generate resonant states like the $D_0(2400)$ in the meson–meson sector and the $\Sigma_c(2800)$ and $\Lambda_c(2595)$ in the meson–baryon sector, which are automatically taken into account in the interaction. The scattering matrix element in Eq. (3) is related to the perturbative amplitude as:

$$\mathcal{M}(s) \propto \frac{V(s)}{1 - G(s)V(s)} \ , \tag{7}$$

where G represent the diagonal matrix of the two intermediate propagators in the BS equation.

4. Results and Conclusions

We present our main results taken from Ref. [2]. In Fig. 1 we show the drag coefficient $F(p = 100 \text{ MeV})$ for a D meson as a function of the temperature when the baryochemical potential is set to zero. The largest contribution comes from the pions, as these mesons populate the gas at temperatures below the hadronization one. The kaons, antikaons and η mesons contribute appreciably only at high temperatures $T > 100$ MeV, being the baryon contribution $(N + \Delta)$ negligible at $\mu_B = 0$.

The position-space diffusion coefficient D_x is shown at $\mu_B = 0$ in Fig. 2, when all hadrons are included in the calculation. We normalize this coefficient by the thermal wavelength $2\pi T$

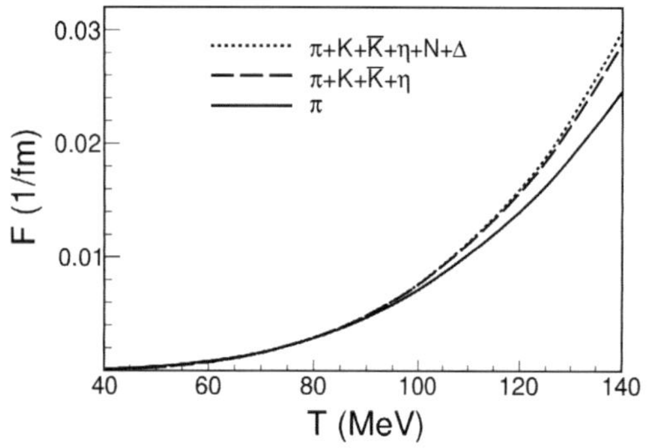

Figure 1. Drag force $F(p)$ for a D meson with momentum $p = 100$ MeV as a function of temperature for several species in the thermal bath at $\mu_B = 0$ (LHC collisions). Clearly the dominant contribution comes from pions as they are the most abundant in the bath. The contribution of baryons is practically negligible. The diffusion coefficient in momentum space is related to this coefficient via the Einstein relation $\Gamma = F m_D T$.

to construct an adimensional number. The calculation for the QGP (Rapp and Hees) is a T-matrix calculation from Ref. [11] and the lattice-QCD computation (Banerjee, Datta, Gavai and Majumdar) is taken from Ref. [12].

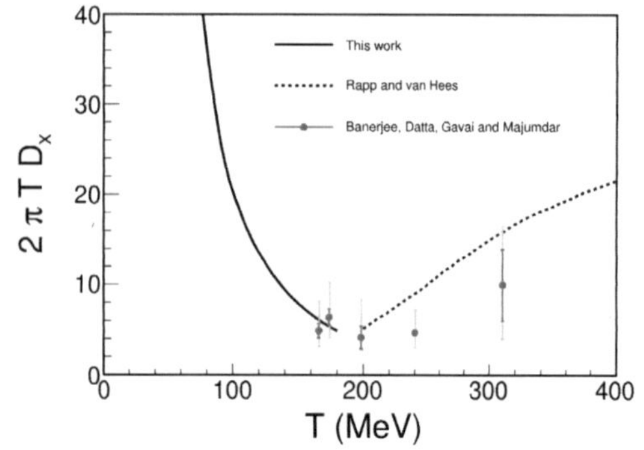

Figure 2. Diffusion coefficient multiplied by the thermal wavelength $2\pi T$ as a function of temperature, around the crossover temperature at $\mu_B = 0$. The solid line represents our results for hadronic matter with all species included. The dashed line shows the result of Ref. [11] for a quark-gluon plasma. Dots represent the result from the lattice-QCD calculation extracted from Ref. [12].

Turning to the $\mu_B \neq 0$ case –relevant for the physics at FAIR– we consider three characteristic adiabatic trajectories in the phase diagram for different initial collision energies ($\sqrt{s} \approx 5 - 40A$ GeV [4]). In the left panel of Fig. 3 we sketch our trajectories on the phase diagram. They correspond to fixed entropy per baryon of $10 - 30$ [13]. In the right panel we plot our results in the hadron gas (at low temperature) and the results of a perturbative QGP at finite quark chemical potential based on the computation of Ref. [14].

Finally, we show the relaxation time for each adiabatic trajectory plus the result at $\mu_B = 0$ as a function of temperature in Fig. 4. Notice that the presence of baryons at finite μ_B enhances the D-meson stopping and the relaxation time is accordingly reduced. Therefore collisions at FAIR put the D mesons closer to the equilibrium state than at the LHC. These relaxation times should be compared to the fireball's lifetime which is of the order of $t_f \sim 10$ fm. In any case, the D mesons cannot totally relax through collision as $\tau_R > t_f$, confirming that D mesons should be sensible to some properties of the initial QGP.

In summary, we have presented our latest results on the transport coefficients of D mesons that describe the energy loss and diffusion of these states in a hadronic medium at finite temperature and baryochemical potential. Focusing on FAIR physics, we have obtained that the collisions with baryons make the D mesons more thermally relaxed than the case at vanishing baryochemical potential. We have obtained a nice connection with the results of QGP at high temperature, providing an indication for a possible minimum of $2\pi T D_x$ at both the crossover temperature (at $\mu_B = 0$) and at the first-order transition temperature (at FAIR collisions).

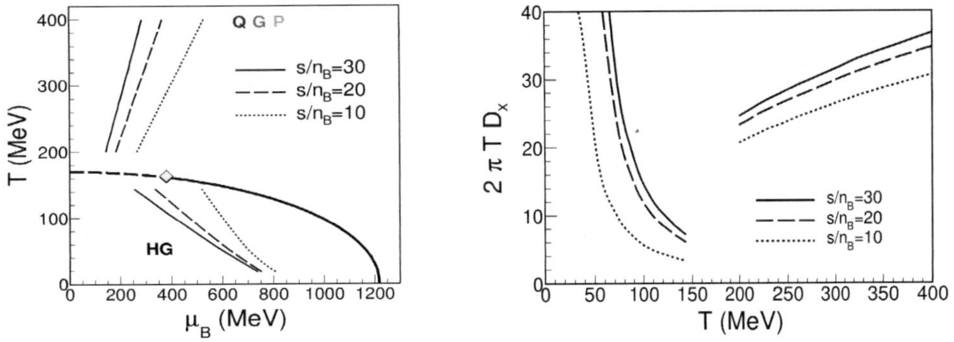

Figure 3. Left panel: Adiabatic trajectories for different initial conditions at FAIR collisions. Right panel: D_x as a function of temperature for the three adiabatic trajectories.

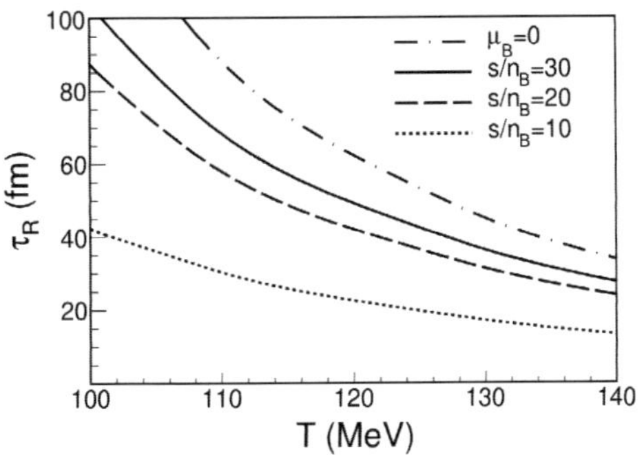

Figure 4. Relaxation time for the adiabatic trajectories as a function of temperature. The trajectories with larger baryonic density (those with smaller entropy per baryon) make the D meson to have a lower relaxation time and hence, to follow a more efficient relaxation via baryonic collisions.

Acknowledgments

JMTR acknowledges the organizers of FAIRNESS 2013 for the hospitality received and the opportunity to present these results. This work has been financed by grants FPA2010-16963, FPA2011-27853-C02-01, FPA2011-27853-C02-02 (Spain), BMBF (Germany) under project no. 05P12RFFCQ, EU Integrated Infrastructure Initiative Hadron Physics Project under Grant Agreement n. 227431 and Grant No. FP7-PEOPLE-2011-CIG under Contract No. PCIG09-GA-2011-291679. LMA thanks CAPES (Brazil) for partial financial support. DC acknowledges support from Centro Nacional de Fisica de Partículas, Astropartículas y Nuclear (Consolider - Ingenio 2010) and LT acknowledges support from the Ramón y Cajal Research Programme.

References

[1] Abreu L M, Cabrera D, Llanes-Estrada F J, Torres-Rincon J M 2011 *Annals Phys.* **326** 2737
[2] Tolos L and Torres-Rincon J M 2013 *Phys. Rev.* D **88** 074019
[3] Abreu L M, Cabrera D and Torres-Rincon J M 2013 *Phys. Rev.* D **87** 034019
[4] Friman B *et al* 2001 *Lect. Notes Phys.* **814**, 1
[5] Guo F K, Hanhart C and Meissner U G, 2009 *Eur. Phys. J.* A **40** 171
[6] Mizutani T and Ramos 2006 *Phys. Rev.* C **74** 065201
[7] Garcia-Recio C *et al* 2009 *Phys. Rev.* D **79** 054004
[8] Garcia-Recio C, Nieves J, Romanets O, Salcedo L L and Tolos L 2013 *Phys. Rev.* D **87** 074034
[9] Romanets O *et al* 2012 *Phys. Rev.* D **85** 114032
[10] Oller J A and Oset E, 1997 *Nucl. Phys.* A **620** 438
[11] Rapp R and van Hees H 2010 *Quark Gluon Plasma 4* (World Scientific) 111
[12] Banerjee D, Datta S, Gavai R and Majumdar P 2012 *Phys. Rev.* D **85** 014510
[13] Bravina L V *et al.* 2008 *Phys. Rev.* C **78** 014907
[14] Moore G D and Teaney D 2005 *Phys. Rev.* C **71** 064904

FAIRNESS 2013: FAIR NExt generation of ScientistS 2013 IOP Publishing
Journal of Physics: Conference Series **503** (2014) 012021 doi:10.1088/1742-6596/503/1/012021

Radiation quality of cosmic ray nuclei studied with Geant4-based simulations

Lucas N. Burigo[1,2], **Igor A. Pshenichnov**[1,3], **Igor N. Mishustin**[1,4] and **Marcus Bleicher**[1,2]

[1] Frankfurt Institute for Advanced Studies, Johann Wolfgang Goethe University, 60438 Frankfurt am Main, Germany
[2] Institut für Theoretische Physik, Johann Wolfgang Goethe University, 60438 Frankfurt am Main, Germany
[3] Institute for Nuclear Research, Russian Academy of Sciences, 117312 Moscow, Russia
[4] Kurchatov Institute, Russian Research Center, 123182 Moscow, Russia

E-mail: burigo@fias.uni-frankfurt.de

Abstract. In future missions in deep space a space craft will be exposed to a non-negligible flux of high charge and energy (HZE) particles present in the galactic cosmic rays (GCR). One of the major concerns of manned missions is the impact on humans of complex radiation fields which result from the interactions of HZE particles with the spacecraft materials. The radiation quality of several ions representing GCR is investigated by calculating microdosimetry spectra. A Geant4-based Monte Carlo model for Heavy Ion Therapy (MCHIT) is used to simulate microdosimetry data for HZE particles in extended media where fragmentation reactions play a certain role. Our model is able to reproduce measured microdosimetry spectra for H, He, Li, C and Si in the energy range of 150–490 MeV/u. The effect of nuclear fragmentation on the relative biological effectiveness (RBE) of He, Li and C is estimated and found to be below 10%.

1. Introduction

Future space travels shall take humans to the Moon for extended periods and even beyond in interplanetary missions [1, 2]. Not only further technological developments will be necessary, but also a better understanding of the health risks to crew members exposed to galactic cosmic rays (GCR). The GCR spectrum is composed mainly by protons and helium nuclei with only a fraction of about 1% due to heavier ions showing a broad energy spectra peaked between 100–1000 MeV/u. Although the flux of high energy and charge (HZE) particles is low, a high biological effectiveness of such particles increases their contribution to the received biological dose. Since HZE particles can not be effectively shielded during the journey following chronic radiation effects are expected to be significant [3]

Shielding of crew members is indispensable for their protection from protons of solar particle events and from the proton component of GCR. However, the linear energy transfer (LET) of a HZE particle increases due to the energy loss in spacecraft materials, and a complex radiation field is created due to nuclear fragmentation reactions. Dose monitors for such radiation fields shall be able to cope with a broad LET spectrum. In particular, a Tissue Equivalent Proportional Counter (TEPC) has been applied for microdosimetry measurements aboard Space Shuttle [4]. TEPC emulates a micrometer volume of human tissue and measures lineal energy, y, of stochastic events. Lineal energy is defined as the energy deposited to a sensitive volume of TEPC divided by its mean chord length [5]. The probability distributions of y (microdosimetry spectra) measured by TEPC can be used in estimating LET and RBE of complex radiation fields [6, 7].

Our previous investigations of the response of TEPC to radiation fields relevant to space research [8] are extended to detailed simulations of microdosimetry spectra. In the present work such spectra resulting from irradiation of extended media with protons, helium and HZE particles in the energy range of 150–490 MeV/u are compared to experimental data. The contributions of secondary fragments created in fragmentation of beam nuclei are calculated. The FAIR facility will allow microdosimetry measurements and radiobiological experiments with HZE particles at higher energies. This will help to reduce the uncertainties in estimating the health risks of astronauts due to expose to GCR.

2. Materials and Methods

The radiation environment inside a spacecraft due to GCR irradiation can be studied by Monte Carlo method. The interaction of GCR with various materials can be simulated with the Geant4 toolkit [9, 10]. Originally being developed for experiments in high energy physics, the code is now widely used for modeling in space research. It comprises software libraries with a broad set of functionalities for the simulation and analysis of particle propagation in various media. A user should implement its own software application tailored to the particular research task.

The Monte Carlo model for Heavy Ion Therapy (MCHIT) is a Geant4-based application intended to study physics processes relevant to ion-beam cancer therapy [11, 12]. MCHIT has been used to describe a wide set of experimental data including depth-dose profiles for protons and carbon nuclei in tissue-like materials, yields of secondary nuclear fragments, energy spectra and angular distributions of secondary neutrons. Microdosimetry spectra for neutrons, protons and light ions measured with TEPCs were also calculated [13, 14]. MCHIT is currently built with Geant4 of version 9.5 with patch 02. The electromagnetic processes are simulated by means of G4EmPenelope including models for ionization process of gas media by ions. The fast stage of nucleus-nucleus collision is modeled by Light Ion Binary Cascade model for helium and lithium projectiles and by Quantum Molecular Dynamics model for carbon and silicon projectiles. Further details on the physics list and respective models are given elsewhere [14].

In this study the MCHIT model is used for investigating the effect of nuclear fragmentation reactions on microdosimetry spectra and RBE for protons, helium and HZE particles. Microdosimetry spectra measured at NIRS, Japan for 160 MeV ^{1}H, 150 MeV/u ^{4}He and 490 MeV/u ^{28}Si [15] as well as at GSI, Germany for 185 MeV/u ^{7}Li and 300 MeV/u ^{12}C [16] were taken for benchmarking. The measurements at NIRS were performed with a wall-less TEPC emulating a cylindrical tissue volume of 0.72 μm in diameter while at GSI a walled TEPC was applied emulating a spherical volume of 2.7 μm in diameter. At NIRS the wall-less TEPC was placed behind range shifters made of PMMA, while the walled TEPC was placed inside a water phantom at GSI.

RBE is estimated by MCHIT coupled with the Microdosimetric-Kinetic (MK) model developed by Hawkins [6] and extended by Kase *et. al* [7]. The impact of nuclear fragmentation reactions on the relative biological effectiveness (RBE) of radiation fields by helium, lithium and carbon at different water-equivalent depths is evaluated. Details of the calculational procedure can be found elsewhere [14].

3. Results and Discussions

Microdosimetry spectra were simulated for various ions showing a general agreement with experimental data as presented in figure 1. Panel (a) shows the spectrum for homogeneous proton irradiation of a wall-less TEPC behind a 163 mm-we (water-equivalent) range shifter. The spectrum is peaked at lineal energy $y = 1$ keV/μm extending up to 20 keV/μm. In panel

(b) the spectrum for helium beam with same wall-less TEPC but behind a range shifter of 157.1 mm-we is presented. In this case the spectrum is peaked at ~15 keV/μm due to primary ions while secondary protons give the main contribution for events below 2 keV/μm. Lineal energy events due to helium projectiles extend up to 200 keV/μm. Panels (c) and (d) show the spectra measured with a walled TEPC irradiated by a pencil-like lithium beam inside a water phantom at the plateau and Bragg peak of the depth-dose distribution, respectively. The disagreement between experimental data and simulation results in panel (c) is likely related to pile-up of events during data acquisition in experimental set-up as discussed elsewhere [14]. The change of radiation quality with depth in water is clearly seen. Not only the contribution of lithium changes due to the reduction of kinetic energy, but also the role played by secondary particles evolves. At the plateau position, the contribution of projectile-like helium fragments is seen between 0.2–18 keV/μm, and of target-like fragments between 18–100 keV/μm. This is explained by the fact that projectile-like fragments have velocities similar to the primary ion but smaller stopping power due to a smaller nuclear charge. The contribution of primary ions is peaked at 5 keV/μm and extends up to 22 keV/μm. When TEPC is moved deeper into phantom, the peak is shifted to ~60 keV/μm and a shoulder in the spectrum at lower y values is clearly visible due to the contribution of helium and hydrogen fragments. Panels (e) and (f) show similar measurements with a pencil-like carbon beam at plateau and peak positions, respectively. Also in this case a clear change of radiation quality is observed with depth where a variety of secondary particles along with the primary carbon ions contribute to the spectra. The maxima are observed at 16 keV/μm and 120 keV/μm at plateau and peak, respectively. The underestimation of the satellite peak observed in the spectrum at the peak position indicates that the Geant4 models for nuclear fragmentation may need further improvements. The panels (g) and (h) present the spectra measured with the wall-less TEPC behind two range shifters of different thickness irradiated by silicon ions. As seen, the shape of the calculated spectra is defined by light ($Z < 7$) and heavy ($Z > 6$) nuclei, which contribute to low ($y < 40$ keV/μm) and high ($y > 4$ keV/μm) linear energy events. Behind a range shifter of 135 mm-we the spectrum is peaked at 100 keV/μm, while the peak is shifted to 300 keV/μm when the thickness of the range shifter is increased to 159.6 mm-we.

Microdosimetry spectra for 152 MeV/u ^{4}He, 176 MeV/u ^{7}Li and 290 MeV/u ^{12}C ions corresponding to the same range in water were calculated with MCHIT at various depths inside a water phantom. The resulting spectra were used as an input to MK model in order to estimate the RBE_{10} corresponding to 10% survival of Human Salivary Gland (HSG) cells after irradiation. The obtained RBE_{10} as a function of depth is shown in figure 2. Microdosimetry spectra were also calculated at same positions for simulations when nuclear fragmentation reactions are neglected. The corresponding RBE_{10} is also presented in figure 2 for comparison. One can see that nuclear fragmentation causes a reduction of RBE_{10} in case the radiation field would be composed only by primary ions. The RBE_{10} for helium and lithium is only significantly decreased at the far end of the ion range while for carbon ions the RBE_{10} values at all investigated positions are decreased by less than 10%. This shows that the loss of a primary carbon ion and yield of secondary fragments shall result in less biological effect for this particular end-point.

4. Conclusions

Microdosimetry spectra for hydrogen, helium, lithium, carbon and silicon ions in the energy range relevant to GCR can be successfully calculated with Geant4/MCHIT. The measurements of microdosimetry spectra behind shielding open another possibility to validate Monte Carlo transport codes. Nuclear fragmentation reactions that happen in the shielding of a spacecraft decrease the RBE_{10} for survival of HSG cells with respect to primary nuclei. Therefore, not only the physical dose, but also biological effects of radiation can be reduced by such shielding.

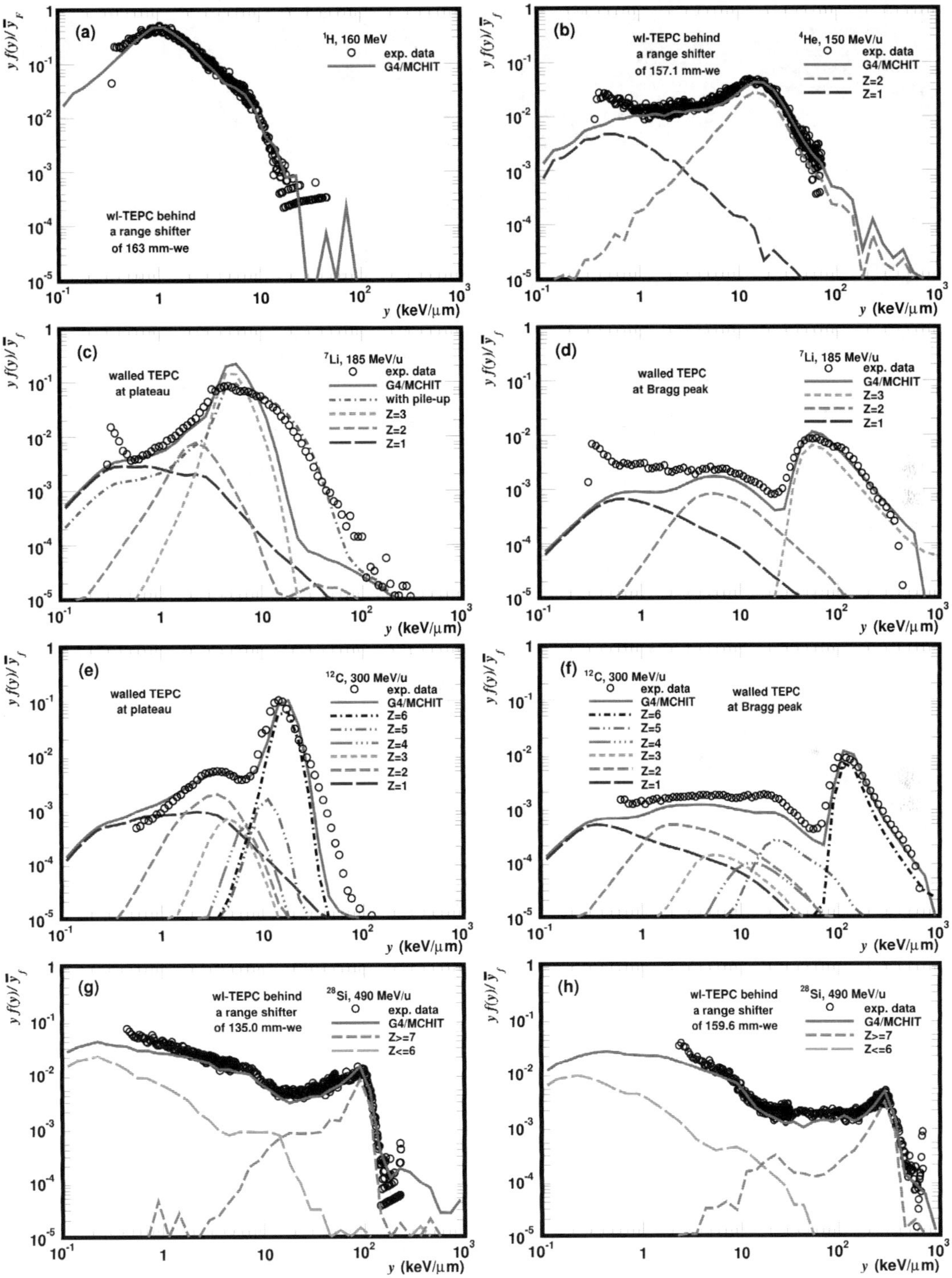

Figure 1. Microdosimetry spectra for (a) ^{1}H, (b) ^{4}He, (c-d) ^{7}Li, (e-f) ^{12}C and (g-h) ^{28}Si at different water-equivalent depths. Experimental data from [15, 16] are shown.

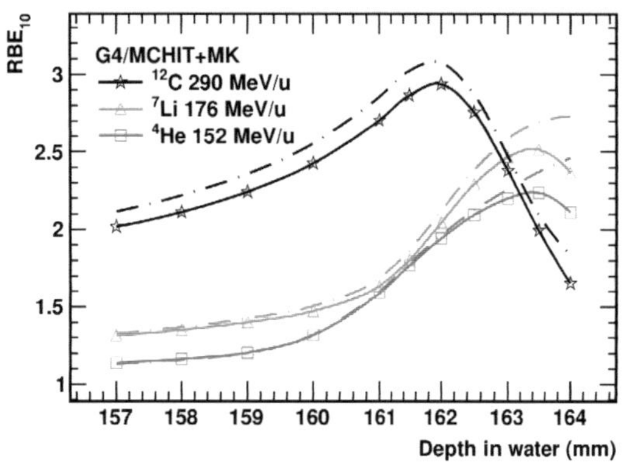

Figure 2. RBE_{10} for HSG cells as a function of depth in water after irradiation by light nuclei. Solid line present expected results calculated by MCHIT+MK models when all interaction processes are taken into account in simulations. Dashed lines present hypothetical RBE_{10} in the absence of nuclear fragmentation of the considered beams.

Acknowledgments

The presented results were obtained in the framework of NanoBIC-NanoL project. L.N.B. is grateful to the Beilstein Institute for support. This work was also partially supported by HIC for FAIR within the Hessian LOEWE-Initiative. Our calculations were performed at the Center for Scientific Computing (CSC) of the Goethe University Frankfurt. We are grateful to the staff of the Center for support.

References

[1] NASA The NASA Vision for Space exploration URL http://www.nasa.gov/exploration/home/index.html
[2] ESA Aurora's roadmap to Mars URL http://www.esa.int/Our_Activities/Human_Spaceflight/Exploration/Aurora_s_roadmap_to_Mars
[3] Durante M and Cucinotta F A 2008 *Nat Rev Cancer* **8** 465–472
[4] Badhwar G D, Cucinotta F A, Braby L A and Konradi A 1994 *Radiat. Res.* **139** 344–351
[5] ICRU 1983 *Report No. 36, Microdosimetry* (Bethesda, MD: ICRU)
[6] Hawkins R B 2003 *Radiation Research* **160** 61–69
[7] Kase Y *et al* 2006 *Radiat. Res.* **166** 629–638
[8] Burigo L, Pshenichnov I, Mishustin I and Bleicher M 2013 *Journal of Physics: Conference Series* **426** 012006
[9] Agostinelli S *et al* 2003 *Nucl. Instrum. Methods A* **506** 250–303
[10] Allison J *et al* 2006 *IEEE T. Nucl. Sci.* **53** 270–278
[11] Mishustin I, Pshenichnov I and Greiner W 2010 *Eur. Phys. J. D* **60** 109–114
[12] Pshenichnov I, Botvina A, Mishustin I and Greiner W 2010 *Nucl. Instrum. Methods B* **268** 604–615
[13] Burigo L, Pshenichnov I, Mishustin I and Bleicher M 2013 *Nucl. Instrum. Methods B* **310** 37–53
[14] Burigo L, Pshenichnov I, Mishustin I and Bleicher M 2014 *Nucl. Instrum. Methods B* **320** 89–99
[15] Tsuda S *et al* 2012 *J. Radiat. Res.* **53** 264–271
[16] Martino G, Durante M and Schardt D 2010 *Phys. Med. Biol.* **55** 3441–3449

Understanding the origin of the elements: experiments at the dripline

Christoph Langer

National Superconducting Cyclotron Laboratory and Joint Institute for Nuclear Astrophysics, Michigan State University, East Lansing, MI 48824, USA

E-mail: `langer@nscl.msu.edu`

Abstract. The field of Nuclear Astrophysics deals with the question of the origin of the elements. Nuclear physics plays a major role in the understanding and modeling of different processes of heavy element production and therefore, experimental and theoretical studies on nuclear systems are needed.
By means of two important explosive astrophysical scenarios, current selected experimental progress, achieved at the GSI facility in Darmstadt, Germany, and at the NSCL at MSU, USA, will be reviewed. A short outlook on future developments will be given at the end.

1. Introduction

In an impressive publication by M. Burbidge, G. Burbidge, W. Fowler and F. Hoyle (B^2FH) from 1957 [1] the authors suggested an answer to the long-standing question: what is the origin of the heavy elements in the universe? Their remarkable argument is based on the very promising idea, that nucleosynthesis is linked with (ongoing) processes inside stars and in stellar explosions. Nuclear physics plays a key role in the understanding and modeling of the complex processes taking place in stars. Besides astrophysical processes involving stable isotopes or isotopes close to stability (which will not be subject of this review), short-lived radioactive nuclei with large proton-to-neutron ratios (and vice versa) are considered to be important in the different nucleosynthesis processes proposed in B^2FH.
With the advent of radioactive beam facilities it became possible to gain experimental access to nuclei far away from stability. This has enabled a variety of studies of basic properties of the complex many-body quantum system, like *e.g.* the mass and level structure, which in turn stimulates the development and improvement of the underlying nuclear theory. Huge progress has been made over the last decades and the astrophysical models have benefited from reliable input provided from nuclear physics. Recent experimental efforts and results motivated by two different nucleosynthesis processes will be discussed in Sec. 2 and 3. A short outlook will be given in Sec. 4.

2. The rapid proton-capture process

The rp process is an extended series of fast (p, γ) reactions along isotonic chains, interspersed with slower β^+-decays and electron-capture reactions. Within typically 10-100 s, the rp process spans a nuclear reaction network of up to the $A \approx 100$ region [2, 3]. The high densities and high temperatures involved indicate an explosive astrophysical site like a type I x-ray burst.

A type I x-ray burst is a thermonuclear explosion ignited in the outer envelope of an accreting neutron star, situated in a binary system with a relatively unevolved companion star. Mainly H/He-rich material is transferred onto the accretion disk formed around the neutron star and in most cases, the accreted material is burned unstably on very regular timescales. Around a hundred of such x-ray bursters have been identified in the Galaxy, thus being the most frequently observed explosion, see *e.g.* Refs. [4, 5].
Accurate modeling of type I x-ray bursts still exhibits large uncertainties mainly owing to many unconstrained proton-capture reaction rates. In order to eliminate nuclear physics uncertainties induced into the astrophysical models, reaction studies involving exotic proton-rich nuclei are necessary. Since most of the involved nuclei are very short-lived, and, thus, can not be fabricated into targets, the reaction studies are performed in inverse kinematics (unstable nuclei are used as projectiles, not as targets). Moreover, direct (p, γ) and (α, γ) reactions usually have extremely small cross sections making the direct measurement extremely difficult if not impossible with current radioactive beam facilities. To overcome this problem, indirect studies have been used in many cases to constrain experimentally important reaction rates identified in sensitivity studies [6]. As an example: the astrophysical temperature-dependent capture-reaction rate, when proceeding through a narrow isolated resonance, can be approximated by

$$\langle \sigma v \rangle \propto \Sigma_r (\omega\gamma)_r \cdot e^{-\frac{E_r}{k_B T}}, \tag{1}$$

with σ being the cross section, v the relative velocity of the particles, $(\omega\gamma)_r$ the resonance strength (depending on the spins and partial widths of the resonance and the particle), E_r the energy of the resonance, and $k_B T$ the Boltzmann constant multiplied with the temperature T. As can be seen from this equation, experimental extraction of spectroscopic properties (like *e.g.* the resonance energy E_r) can be used to indirectly obtain stellar reaction rates.
A powerful indirect reaction method is Coulomb breakup. It can be used to extract properties like *e.g.* spectroscopic information and radiative-capture cross sections via detailed balance (*i.e.* the time-reversed reaction is measured), see *e.g.* Refs. [7, 8]. Versatile setups are used to measure the reaction residues and γ-rays after decay of the excited states, and the extracted kinematical information can be combined to reconstruct the excitation energy spectrum or the momentum distributions.
In a recent experiment at the R^{3}B-LAND setup at GSI, the excitation energies of low-lying levels in proton-rich ^{31}Cl have been extracted by employing the Coulomb-breakup method and reconstruction of the excitation energy spectrum. The ^{30}S$(p, \gamma)^{31}$Cl reaction is considered to be an important reaction during the rp process due to its low Q-value of ~300 keV. The rate had been determined by theoretical calculations [9], and an unambiguous experimental confirmation, especially of the important first low-lying state, was still missing. The left part of Fig. 1 shows the energy-differential excitation spectrum of ^{31}Cl [10]. The two-lying peaks can be clearly identified and are in good agreement with the theoretical calculations in [9].
In-flight γ-ray spectroscopy is another powerful tool to gain spectroscopic information on exotic nuclei. In this experimental approach, mainly γ-decaying states are populated via a variety of different possible reaction mechanisms, like *e.g.* nucleon transfer, knock-out, and Coulomb excitation, at corresponding beam energies. The emitted γ-rays are subsequently detected in dedicated detector systems (usually in coincidence with the reaction residue) and detailed spectroscopic information of excited states can be deduced. Especially new instrumental tools, like *e.g.* the GRETINA array in the US [12] and AGATA in Europe [13], enable studies with unprecedented sensitivity involving nuclei so far out of reach.
The right part of Fig. 1 shows a preliminary spectrum of Doppler-corrected γ-ray transitions in proton-rich ^{58}Zn, populated in a ^{57}Cu(d, n) transfer reaction at 75 MeV/u, measured with GRETINA in conjunction with the S800 spectrometer [14] at the NSCL. Similar to ^{31}Cl, ^{58}Zn is considered to be an important nucleus during rp-process nucleosynthesis [6], and X-ray burst

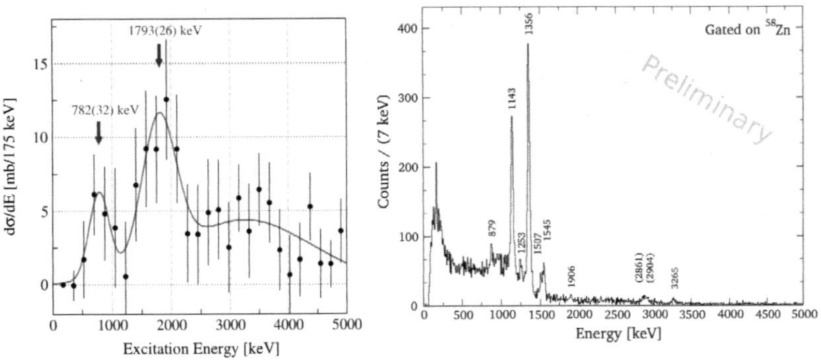

Figure 1. Left: Results of a Coulomb-breakup experiment carried out at the R^3B-LAND setup at GSI using ^{31}Cl as an incoming beam. The excitation spectrum reveals two important resonances (energies marked with arrows), which play a crucial role for the $^{30}S(p,\gamma)^{31}Cl$ reaction during rp-process nucleosynthesis [10]. Right: A Doppler-corrected γ-ray spectrum of ^{58}Zn, measured with GRETINA at the NSCL [11]. The reaction $^{57}Cu(p,\gamma)^{58}Zn$ is of major interest for x-ray burst models.

models predicting the light curve and the composition of the ashes rely on accurate input for this particular reaction. So far it was not possible to extract experimentally spectroscopic information for ^{58}Zn, which mainly determines the $^{57}Cu(p,\gamma)^{58}Zn$ reaction rate. Using the improved sensitivity of the GRETINA array, it was subsequently possible to extract the information needed to improve the reaction rate by orders of magnitude [11].
Measurements studying the capture reactions directly are in most cases favorable in comparison to indirect methods since any dependence on theory is minimized. In a recent proof-of-principle experiment performed at the storage ring ESR at GSI the $^{96}Ru(p,\gamma)$ reaction cross section was successfully measured [15] at the relevant stellar (low) energies. In that experiment a ^{96}Ru beam was injected into the ring, where a liquid hydrogen target was installed. Because of the many circulations of the ions in the ring, a high target luminosity of 10^{25} cm^{-2}s^{-1} was achieved overcoming the low cross sections. First results are in good agreement with theory and the whole concept is, thus, very promising.

3. The rapid neutron-capture process

About half of the elements heavier than iron are created during the rapid neutron-capture process (r process). Up to now, the astrophysical site of the r process remains unclear, partly because the involved nuclear physics is still uncertain and relies on theoretical input parameters, such as masses, β-decay half-lives, and neutron-capture reaction rates far away from stability [16, 17].
Mass measurements employing new techniques [18] provide a powerful tool to experimentally constrain atomic masses, which are crucial input parameters for r-process models. In a recent experiment utilizing the Canadian Penning Trap (CPT) at the ATLAS facility at Argonne National Lab, the masses of 33 r-process nuclides were measured, ranging from ^{130}In to very neutron-rich ^{146}Cs [19]. The authors used different r-process simulations implementing the newly measured masses for the involved nuclei. According to their models, a significant increase of processing time through that region is obtained when comparing the results of the simulations using state-of-the-art mass models. This points towards insufficiently constrained mass models for accurate r-process simulations. New facilities and techniques are needed to further improve the existing mass models.

The r-process reaction flow and the overall processing time also depends on the β-decay half-lives of certain important isotopes. Huge experimental effort is therefore put into lifetime measurements, and recent progress has been achieved to constrain global models and to accurately predict the reaction flow, see *e.g.* Refs. [20, 21].
In order to calculate neutron-capture reaction rates, precise knowledge of the evolution of nuclear shell structure towards very neutron-rich systems is required. Similar to the rp-process reaction studies, different experimental techniques can be employed to study the structure of most exotic neutron-rich nuclei.
One of the major changes in the evolution of shell structure are the observed dramatic changes in the well-known magic numbers $(2, 8, 20, 28, ..)$. This involves *e.g.* the disappearance of the $N = 20$ magic number, and the appearance of a new magic number with $N = 16$, resulting in a new doubly magic nucleus ^{24}O (see Ref. [22] and references therein for a good overview). A key role in the theoretical understanding of these phenomena is played by the tensor short-range interaction of the nuclear force, and here especially by the monopole interaction [23].
Moreover, inclusion of 3N-forces seems to be able to resolve some still standing open questions, like the abrupt change of the dripline behavior going from oxygen isotopes to the fluorine chain. In a recent experiment at the R^{3}B-LAND setup at GSI, the authors of [24] measured the excitation energy spectrum of unbound ^{26}O. In order to understand the data, they explicitly included 3N-forces into their model, with which they are able to quantitatively explain the abrupt change of the dripline behavior and the structure of ^{26}O. Although ^{26}O is not situated on the standard r-process reaction path, the result improved the structural description and understanding of very neutron-rich nuclei, especially in the light region. This can be modified and used for nuclei which are relevant for modern r-process calculations, and shows nicely the interplay between Nuclear Structure Physics and Nuclear Astrophysics. Experiments like these are needed in the future to get a more complete picture of the nuclear force, and thus, being able to predict neutron-capture rates with high accuracy, making astrophysical model predictions eventually more reliable.

4. Outlook

Radioactive beam facilities play a crucial role in the understanding of the origin of the elements. For certain studies, however, the present available facilities do not have enough beam intensity and major upgrades are needed in order to pursue studies involving nuclei so far unreachable.
Most of the neutron-rich isotopes situated on the assumed r-process path can not be produced with the current facilities. To overcome this problem, many astrophysical nuclear physics input parameters need to be extrapolated, which, in turn, carries large uncertainties. With future facilities currently under construction or already in the commissioning phase, a new era will be possible using most exotic beams with high intensities and beam quality.
Direct reaction rate measurements can also benefit from new facilities such as the proposed storage ring at the future NuSTAR experiment (see *e.g.* Refs. [25, 26] and references therein). Based on the high target luminosity achieved through the circulation of a cooled radioactive beam in the storage ring, direct reactions like (p, γ) or (α, γ) involving the most exotic nuclei can be studied.
Complementary, the ReA3 experiment at the NSCL (right now in the commissioning phase) provides high-intensity stopped and re-accelerated radioactive beams at around 3 MeV/u and, thus, enables reaction studies at highest intensities. In a later driver accelerator upgrade (*FRIB*, see [27]) coupled to ReA3, even the most exotic nuclei can be produced, stopped and re-accelerated, and subsequently used for reaction studies. In this facility the *Separator for Capture Reactions* (SECAR) will be used for direct reaction rate studies related to Nuclear Astrophysics. The recoil separator, the *SuN* (Summing NaI) detector, ANASEN, and the gas-jet target JENSA will be extremely powerful for further, so far unreachable, studies (see [28] for more information).

The author is therefore very much looking forward into the bright and exciting future in the field of Nuclear Physics and Nuclear Astrophysics.

Acknowledgments

The author wants to thank the R^3B-LAND collaboration and the collaboration at NSCL and GRETINA. The author acknowledges support from HIC for FAIR. This work was supported by NSF grants PHY11-02511 and PHY08-22648 (Joint Institute for Nuclear Astrophysics), and by the HGF Young Investigators Project VH-NG-327.

References

[1] Burbidge E M, Burbidge G R, Fowler W A and Hoyle F 1957 *Rev. Mod. Phys.* **29**(4) 547–650
[2] Wallace R and Woosley S 1981 *The Astrophysical Journal Supplement Series* **45** 389–420
[3] Schatz H, Aprahamian A, Görres J, Wiescher M, Rauscher T, Rembges J, Thielemann F K, Pfeiffer B, Möller P, Kratz K L *et al.* 1998 *Physics Reports* **294** 167–263
[4] Parikh A, José J, Sala G and Iliadis C 2012 *Progress in Particle and Nuclear Physics*
[5] Galloway D K, Muno M P, Hartman J M, Psaltis D and Chakrabarty D 2008 *The Astrophysical Journal Supplement Series* **179** 360
[6] Parikh A, José J, Moreno F and Iliadis C 2008 *The Astrophysical Journal Supplement Series* **178** 110
[7] Bertulani C A and Baur G 1988 *Physics Reports* **163** 299 – 408
[8] Baur G, Bertulani C and Rebel H 1986 *Nuclear Physics A* **458** 188 – 204
[9] Wrede C, Caggiano J, Clark J, Deibel C, Parikh A and Parker P 2009 *Physical Review C* **79** 045808
[10] Langer C *et al.* 2014 *accepted for publication in Physical Review C*
[11] Langer C *et al.* 2014 *(in preparation)*
[12] Paschalis S *et al.* 2013 *Nuclear Instruments and Methods in Physics Research Section A: Accelerators, Spectrometers, Detectors and Associated Equipment* **709** 44 – 55
[13] Akkoyun S *et al.* 2012 *Nuclear Instruments and Methods in Physics Research Section A: Accelerators, Spectrometers, Detectors and Associated Equipment* **668** 26 – 58
[14] Bazin D, Caggiano J, Sherrill B, Yurkon J and Zeller A 2003 *Nuclear Instruments and Methods in Physics Research Section B: Beam Interactions with Materials and Atoms* **204** 629–633
[15] Zhong Q, Aumann T, Bishop S, Blaum K *et al.* 2010 *Journal of Physics: Conference Series* vol 202 p 012011
[16] José J and Iliadis C 2011 *Reports on Progress in Physics* **74** 096901
[17] Arcones A and Montes F 2011 *The Astrophysical Journal* **731** 5
[18] Meisel Z and George S 2013 *International Journal of Mass Spectrometry* **349-350** 145 – 150
[19] Van Schelt J, Lascar D, Savard G *et al.* 2013 *Phys. Rev. Lett.* **111**(6) 061102
[20] Hosmer P, Schatz H, Aprahamian A, Arndt O, Clement R, Estrade A, Kratz K L, Liddick S, Mantica P, Mueller W *et al.* 2005 *Physical Review Letters* **94** 112501
[21] Montes F, Estrade A, Hosmer P, Liddick S, Mantica P, Morton A, Mueller W, Ouellette M, Pellegrini E, Santi P *et al.* 2006 *Physical Review C* **73** 035801
[22] Krücken R 2010 *ArXiv e-prints* (*Preprint* 1006.2520)
[23] Otsuka T, Suzuki T, Fujimoto R, Grawe H and Akaishi Y 2005 *Physical Review Letters* **95** 232502
[24] Caesar C, Simonis J, Adachi T, Aksyutina Y *et al.* (R3B collaboration) 2013 *Phys. Rev. C* **88**(3) 034313
[25] Aumann T 2007 *Progress in Particle and Nuclear Physics* **59** 3–21
[26] Reifarth R *et al.* 2013 *ArXiv e-prints* (*Preprint* 1310.1632)
[27] URL http://www.frib.msu.edu/
[28] URL http://www.nscl.msu.edu/exp/sr

Search for mesic nuclei in the photoproduction of $\eta\pi^0$ mesons off light nuclei

Irakli Keshelashvili, Alexander Käser

Department of Physics, University of Basel, CH-4056 Basel, Switzerland

E-mail: i.keshelashvili@unibas.ch

Abstract. For the understanding of the strong nuclear force, the study of interaction between mesons and nuclei is important. In case of long lived charged mesons, like π and K, secondary beams can be used for experiments. But the situation is different for the short lived neutral mesons, like π^0, η and η'. The only access is indirect, making use of final-state interaction (FSI). The mesons are produced with some initial reaction in the nucleus and then their interaction with the same nucleus is studied. It is much discussed whether it is possible to form, via the strong interaction, quasi-bound states of mesons and nuclei, which would be the ideal tool for such studies. We will present preliminary results from the Crystal Ball/TAPS experiment at the Mainz MAMI accelerator for double neutral meson photoproduction, showing interesting potential to study meson nucleon interaction and discuss their relevance. The very first attempt has been made to measure the quasi-free behavior of the photoproduction of $\eta\pi^0$ mesons from the ^2H.

1. Introduction

In the past three decades, many attempts to study meson nucleon interaction has been made. Using charged pion beams off different light targets has been investigated to find possible meson nucleon bound state [1, 2]. Furthermore, the TAPS experimet at MAMI accelerator has tried to investigate photon induced production of π^0 and η for the different light targets, like deuteron, helion (^3He) and helium (^4He) [3, 4, 5, 6, 7]. All attempts were inconclusive towards finding meson nucleon bound state. In the most cases, the single meson production was in focus of interest, it was caused mainly due to limited capabilities of old detector systems. But using modern, fast and more dedicated/customisable trigger systems, combined with fast data acquisition systems, allow us to investigate multiple meson production with high statistics. The η production at the ANKE spectrometer in the reaction $pd \rightarrow^3 He\eta$ [8] and the recent measurement by our group using Crystal Ball/TAPS detector of the reaction $\gamma^3He \rightarrow^3 He\eta$ [9] shows a very rapid rise of the total cross section close to threshold implying a very large η^3He scattering length and hence the existence of a quasibound state very close to threshold.

Unfortunately, coherent photoproduction of η-mesons is forbidden for the spin, isospin $J = I = 0$ nucleus since η-photoproduction in the threshold region is dominated by an isovector, spin-flip transition [10]). To the contrary, coherent threshold photoproduction of $\eta\pi^0$-pairs will be not suppressed, since it does not require a nucleon spin-flip. Due to the specific decay chain and the emission of the η-mesons in s-wave, one may expect strong contributions where the available kinetic energy is taken by the pion and the relative momentum of nucleus and η-meson is small, which are ideal conditions for the formation of a quasi-bound state. Very preliminary

Journal of Physics: Conference Series **503** (2014) 012023 doi:10.1088/1742-6596/503/1/012023

results from the coherent photoproduction of $\eta\pi^0$ pairs off the deuteron [11] seem to support this expectation.

2. Experiment

A schematic view of the Crystal Ball/TAPS experimental setup is shown in figure 1. The beam for this experiment is produced at the so called continuous wave accelerator facility MAMI [12, 13] in Mainz. Using an electron beam of fixed 1.5 GeV energy on a copper radiator a bremsstrahlung photon beam is produced. The energy of the photons was determined by a momentum analysis of the scattered electron in a magnetic spectrometer (Glasgow photon tagger [14, 15, 16]). After collimation the beam impinged on the target (liquid deuterium or ^{3}He, respectively). The target is surrounded by a thin plastic scintillator barrel [17], which was used for charged particle identification, and the spherical electromagnetic calorimeter Crystal Ball [18]. This detector consists of 672 NaI crystals and covers 94% of 4π steradians. The acceptance in the forward direction of the Crystal Ball is covered by the TAPS detector [19, 20] which for the part of the data taking was made of 384 BaF_2 crystals and for the second part of measurement, its 6 crystals where exchanged by 24 $PbWO_4$ crystals in order to meet higher countrate. During all measurements, in front of every crystal a 5mm plastic scintillators are installed as a charged particle identification/veto detector. As trigger condition a deposited energy sum of 300 MeV in the Crystal Ball and a total multiplicity of two or more hits in whole calorimeter was requested.

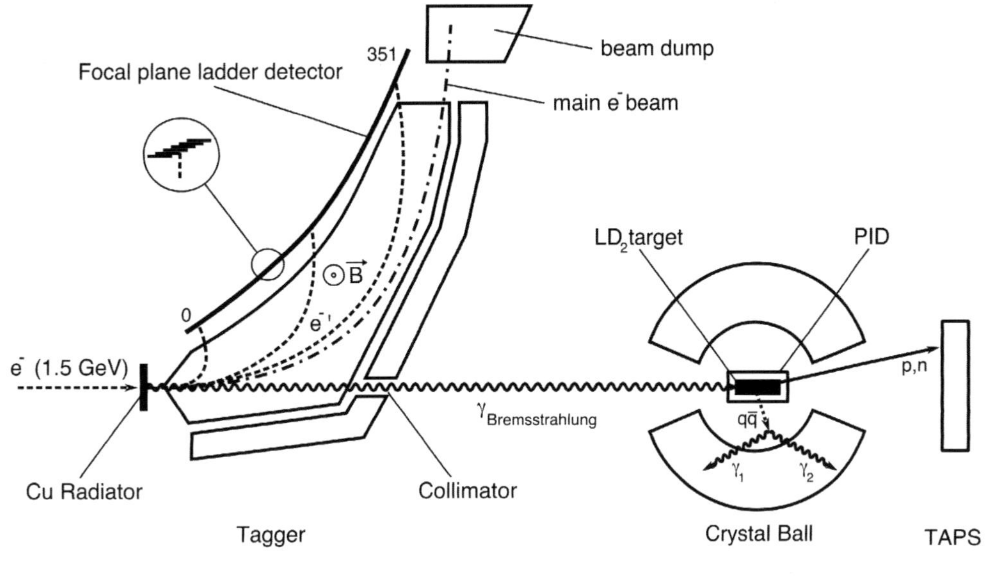

Figure 1. Scheme of the experimental setup.

3. Data Analysis

The analyzed reactions were $\gamma d \to \eta\pi^0 p(n)$ (photoproduction off quasi-free protons bound in the deuteron), $\gamma d \to \eta\pi^0 n(p)$ (photoproduction off quasi-free neutrons bound in the deuteron), $\gamma d \to \eta\pi^+ n(p)$ (photoproduction off quasi-free protons bound in the deuteron) and $\gamma d \to \eta\pi^- p(n)$ (photoproduction off quasi-free neutrons bound in the deuteron). As a first step charged particles were distinguished from neutral particles. This was done, using the veto detectors and the PID and in TAPS with time-of-flight vs. energy spectra. The η and π^0 were detected via their 2γ decay channels and after applying a χ^2 test, identified via a invariant mass

Journal of Physics: Conference Series **503** (2014) 012023 doi:10.1088/1742-6596/503/1/012023

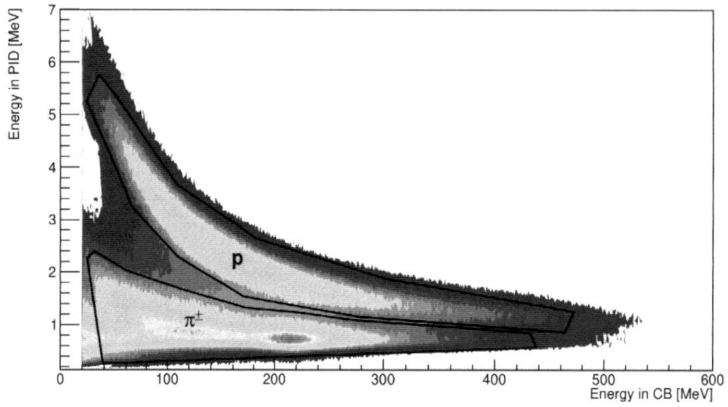

Figure 2. Energy deposited in the PID plotted against energy plotted in the CB. The poligons depicture the regions in which a charged particle is considered to be a proton or a charged pion

analysis. The remaining neutral particles of the χ^2 test (in case of 3 or 5 neutral hits) were treated as neutron candidates. Charged pions and protons in the crystal ball were identified with dE-E cuts as shown in figure 2. To get rid of background from competing reactions a precise missing mass analysis was performed using a Monte Carlo simulation of the detector systems for each respective reaction.

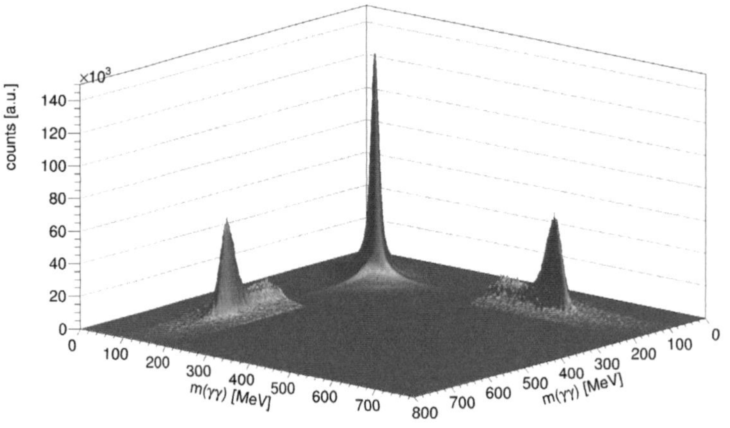

Figure 3. Invariant mass of the best $\gamma\gamma$ pairs after the χ^2 test

4. Results

The total cross sections of the reactions $\gamma d \to \eta\pi^0 p(n)$, $\gamma d \to \eta\pi^0 n(p)$, $\gamma d \to \eta\pi^- p(p)$ and $\gamma d \to \eta\pi^+ n(n)$ were extracted from experimental data. Figure 5 shows the ratios of the total cross sections as function of the final state center of mass Energy. It is clearly visible, that the cross sections fulfill the following equations:

$$\begin{aligned} \sigma(\gamma p \to \eta\pi^0 p) &= \sigma(\gamma n \to \eta\pi^0 n) \\ 2\sigma(\gamma p \to \eta\pi^+ n) &= 2\sigma(\gamma n \to \eta\pi^- p) \end{aligned} \quad (1)$$

Together with a isospin analysis considering the proper Clebsch-Gordon coefficients, this confirmes, that the reaction has a dominant contribution from photoexcitation of a $\Delta^\star$-resonance decaying via either a $\eta\Delta$ or $\pi N^\star$ intermediate state. The specific cascade can be identified via the analysis of the invariant mass distribution of the nucleon-meson pairs. A preliminary analysis considering invariant mass distributions as well as angular distributions suggests, that quasifree $\eta\pi$ photoproduction is dominated by the following cascade decay:

$$\gamma N \to D_{33}(1700) \to \eta\Delta(1232)\eta\pi N$$

Final results will also include an analysis of the coherent photoproduction of $\eta\pi$ pairs off the deuteron. This is a well suited tool to investigate the formation of η-mesic nuclei. Sigificant contributions are thought to come from events where the pion takes most of the available kinetic energy and the relative moment of η and nucleus is small, which increases the probability of forming a bound state.

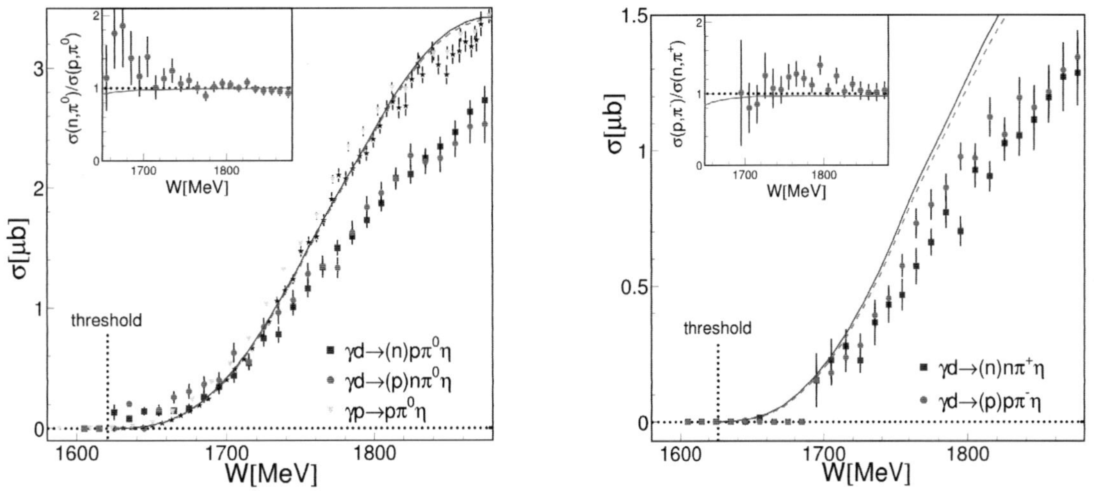

Figure 4. Very preliminary total cross sections for the $\pi^0\eta N$ left and for the $\pi^\pm\eta N$ final states as function of reconstructed final state invariant mass $W(N_p, \pi^0, \eta)$

References

[1] R.S.Bhalerao and L.C.Liu, Phys. Rev. Lett. 54 (1985) 865.
[2] L.C.Liu and Q.Haider, Phys. Rev. C 34 (1986) 1845.
[3] B. Krusche et al., Phys. Rev. Lett. 74 (1995) 3736.
[4] B. Krusche et al., Phys. Lett. B 397 (1997) 171.
[5] V. Hejny et al., Eur. Phys. J. A 6 (1999) 83.
[6] V. Hejny et al., Eur. Phys. J. A 13 (2002) 493.
[7] M. Pfeiffer et al., Phys. Rev. Lett. **92**, 252001 (2004).
[8] T. Mersmann et al., Phys.Rev.Lett. **98** 242301 (2007)
[9] F. Pheron et al., Phys. Lett. B **709**, 21 (2012).
[10] B. Krusche and S. Schadmand, Prog. Part. Nucl. Phys. **51**, 399 (2003).

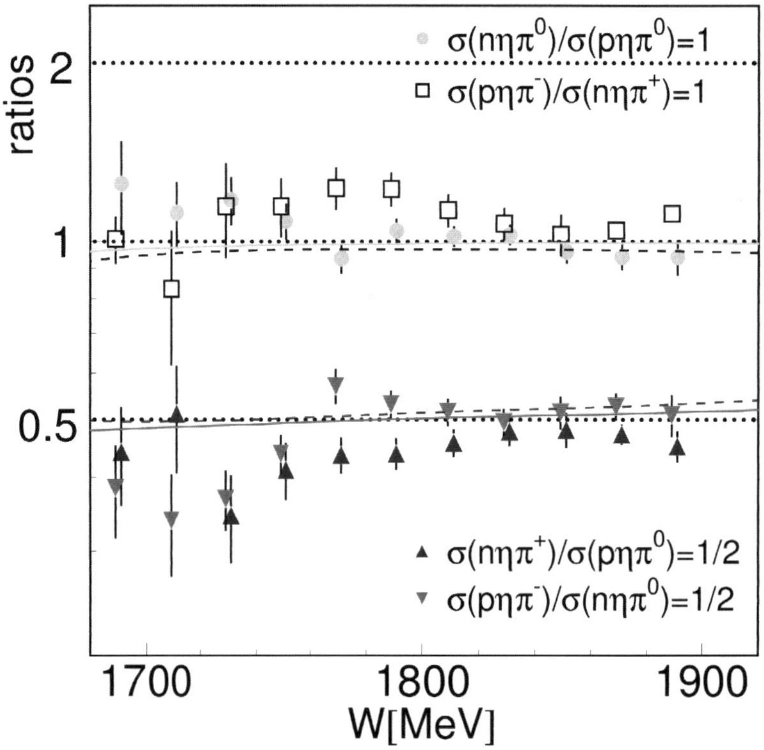

Figure 5. Very preliminary ratios of total cross sections as function of final state invariant mass $W(N_p, \pi, \eta)$.

[11] B. Krusche, Eur. Phys. J. Special Topics **198**, 199 (2011).
[12] H. Herminghaus et al., *IEEE Trans. Nucl. Sci.*, **30**, 3274, (1983).
[13] K.-H. Kaiser et al., *Nucl. Instrum. Methods A*, **593**, 159, (2008).
[14] I. Anthony et al., *Nucl. Instrum. Methods A*, **301**, 230, (1991).
[15] S. J. Hall et al., *Nucl. Instrum. Methods A*, **368**, 698, (1996).
[16] J. C. McGeorge et al., *Eur. Phys. J. A*, **37**, 129, (2008).
[17] D. Watts., *Proc. of the 11th International Conference on Calorimetry in Particle Physics, Perugia, Italy, 2004, World Scientific, Singapore, 2005*, (2005).
[18] A. Starostin et al., *Phys. Rev. C*, **64**, 055205, (2001).
[19] R. Novotny., *IEEE Trans. Nucl. Sci.*, **38**, 379, (1991).
[20] A. R. Gabler et al., *Nucl. Instrum. Methods A*, **346**, 168, (1994).

Journal of Physics: Conference Series **503** (2014) 012024 doi:10.1088/1742-6596/503/1/012024

Simulations on the measurement of the D_s meson semileptonic form factor with the PANDA detector

Lu Cao, James Ritman

Institut für Kernphysik, Forschungszentrum Jülich, D-52425 Jülich, Germany

E-mail: `l.cao@fz-juelich.de`

Abstract. We report on software development to evaluate the expected performance of the PANDA detector to measure the semileptonic form factor $D_s \to e + \nu + \pi, \eta, \eta'$. The related decay models in this chain are checked via a Dalitz plot analysis; the present version of EvtGen in PandaRoot has been enhanced by a new model describing the $D_s \to KK\pi$ decay. With the help of theoretical predictions of the cross section, the expected count rate is estimated.

1. Introduction

The PANDA experiment [1] is one of the major projects of the FAIR facility in Darmstadt and will study the interactions between an intense, phase space cooled beam of antiprotons provided by the High Energy Storage Ring (HESR) and hydrogen or heavier nuclear targets in the momentum range of 1.5-15 GeV/c. In order to serve the wide physics potential with antiprotons, PANDA is designed as a general purpose detector covering nearly the complete solid angle for both neutral and charged particles with good momentum and particle identification (PID) resolution as well as excellent vertex determination. By performing resonance and threshold scans with the high precision anti-proton beams, PANDA will achieve more than an order of magnitude higher mass resolution compared to existing facilities, e.g. the B-factories.

The semileptonic D_s decays are governed by both the weak and strong forces; extraction of the weak CKM parameters requires knowledge of strong interaction. These can be parameterized by the form factor encapsulating the QCD bound state effects. Techniques such as lattice quantum chromodynamics (LQCD) offer increasingly precise calculations of these form factors, but as the uncertainties in the predictions shrink, experimental validation of the results becomes increasingly important.

Recent data from CLEO on semileptonic $D \to \pi, K$ decays [2, 3] have been used for more accurate determinations of the CKM matrix elements $|V_{cd}|$ and $|V_{cs}|$ [4], and some of those authors are currently extending that work to the η and η' in the final state. The principal difference is that now gluonic contributions couple to the singlet component of the η and η', and the mass corrections are more important due to the larger strange quark mass. The theoretical uncertainty to the ratio of the branching ratios $\mathcal{B}(D_s^+ \to e + \nu + \eta')/\mathcal{B}(D_s^+ \to e + \nu + \eta)$ is dominated by uncertainty to the gluonic contribution, thus an improvement of the relative branching ratio can help to pinpoint the gluonic contribution [5].

In the semileptonic decay of D_s meson, there is one neutrino in the event. The achievable performance of the PANDA detector for these types of reactions has not yet been studied in detail; however, this is expected to work very well based upon the design performance and

Journal of Physics: Conference Series **503** (2014) 012024 doi:10.1088/1742-6596/503/1/012024

experience with other detector systems due to the kinematic constraints which allow these events to be completely reconstructed despite one particle not being measured. The goal of this study is to evaluate and optimize the physics performance of the PANDA detector to measure the form factor of semileptonic D_s meson decays. In the following sections, we will introduce the new DS_DALITZ decay model and the details of the reconstruction procedure. The output histograms are shown with the reconstruction results. Finally, the expected rate to measure the events is estimated by using the reconstruction efficiency presented here and a theoretical estimate of the production cross section.

2. Simulation and decay model

Antiproton annihilation on a proton target will be used to generate D_s pair at PANDA with the beam momentum of 8 GeV/c. One of the D_s mesons will be used as a "trigger", and can be reconstructed via a decay branch that is relatively common and has a simple final state, e.g. hadronic mode $D_s \to KK\pi$. For a reasonable simulation, we import a new decay model DS_-DALITZ in the PandaRoot [6] software, which is the official framework of the Collaboration for both simulation and analysis, and is mainly based on the object oriented data analysis framework ROOT [7]. DS_DALITZ includes contributions from the following resonances: $K^*(892)K^+$, $K_0^*(1430)K^+$, $f_0(980)\pi^+$, $\Phi(1020)\pi^+$, $f_0(1370)\pi^+$ and $f_0(1710)\pi^+$. The original code is partly from the EvtGen package developed jointly for the BaBar and CLEO collaboration, where the $D_s \to KK\pi$ decay is implemented as a branch of the existing D_DALITZ model. We tested its performance in Monte Carlo simulation via a Dalitz plot analysis and the result is shown in Figure 1, where the substructure of resonances is clear and comparable with the experimental measurements [8–10]. Figure 2 shows the Dalitz plot obtained by the CLEO experiment with 14,400 events [9]. In the semileptonic channel, we adopt the ISGW2 model to simulate the decay $D_s^+ \to e + \nu + \eta'$ and the ETA_DALITZ model for the $\eta \to \pi^+\pi^-\pi^0$ decay.

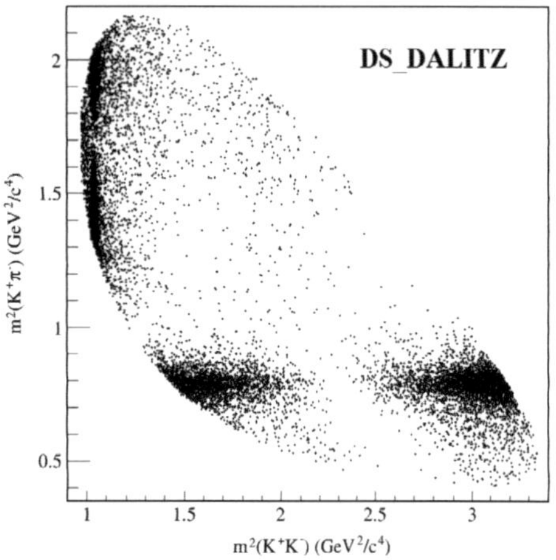

Figure 1. Output of Dalitz plot in MC simulation with DS_DALITZ model.

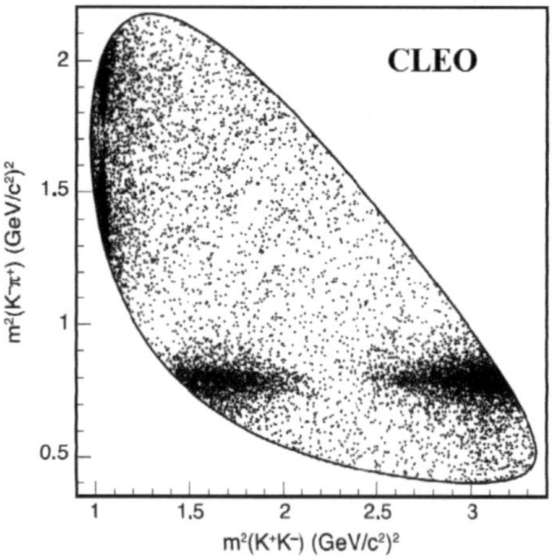

Figure 2. Measured Dalitz plot for the decay $D_s^+ \to K^+K^-\pi^+$ [9].

3. Reconstruction

In the reconstruction procedure, we focus on developing the software and evaluating the expected precision of these measurements using Monte Carlo simulation of the PANDA detector. In order to extract the invariant mass squared of the lepton-neutrino system, we have to reconstruct the semileptonic decay product η and the other charged D_s meson from the initial $p\bar{p}$ system.

We have reconstructed the hadronic channel $D_s^- \to K^+K^-\pi^-$ with 100,000 events, where the mass window for D_s^- is 0.5 GeV/c^2. Using the present software, we perform the vertex fitting and mass constraint fitting for all candidates. The reconstruction efficiency of D_s^- is about 18%. As shown in Figure 3, the mass resolution is 17.7 MeV/c^2. Figure 4 summarizes the expected momentum and vertex resolutions of the reconstructed D_s meson.

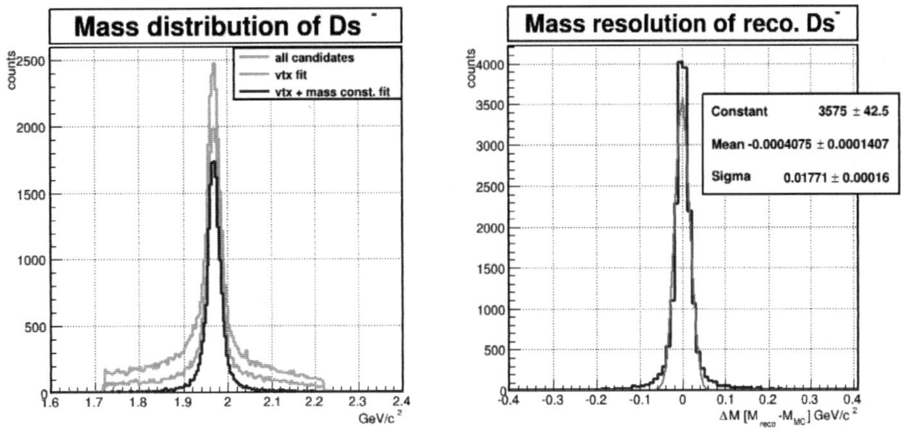

Figure 3. Mass distribution and resolution of reconstructed D_s^-. $KK\pi$ invariant mass distributions (left) for all candidates (orange), after the vertex fit (green), and after the additional mass constraint fit (blue). Invariant mass distribution of the reconstructed D_s^- (right) and a Gaussian fit to determine the resolution.

The reconstruction strategy for this semileptonic decay chain has several steps: $D_s^+ \to \nu_e e^+\eta$; $\eta \to \pi^+\pi^-\pi^0$, and $\pi^0 \to \gamma\gamma$. We started from the combination of two photons which can be detected by the electromagnetic calorimeter (EMC) equipped in the PANDA target spectrometer. The pre-selection has been applied for the reconstruction according to the subdetector specific properties of the EMC [11], e.g. the minimal laboratory angle between two photons should be larger than the EMC spatial resolutions σ_θ: on the backward end-cap $\sigma_\theta = 0.5^o$, on the barrel $\sigma_\theta = 0.3^o$ and on the forward end-cap $\sigma_\theta = 0.1^o$. The mass constraint fitting is performed on the $\gamma\gamma$ system to select the "best" fitted π^0 for reconstructing the mother particle η (see Figure 5). The charged tracks π^+ and π^- determine the vertex reconstruction of $\eta \to \pi^+\pi^-\pi^0$, and in this step the performance of the fitter plays an essential role on the total efficiency of the reconstructed η, since the neutral daughter π^0 has achieved a high efficiency with the mass resolution of 3.2 MeV/c^2. After applying these fit procedures, 26% of the events have been reconstructed (see Figure 6). The preliminary results of reconstruction efficiency ϵ and resolutions σ of the D_s and η are listed in Table 1.

4. Summary

The hadronic decay $D_s \to KK\pi$ and the decay $\eta \to \pi^+\pi^-\pi^0$ emanating from the semileptonic decay $D_s^+ \to \nu_e e^+\eta$ have been investigated within the PandaRoot software in order to estimate the resolution and efficiency to measure these channels. Table 1 summarizes the obtained efficiency and resolutions of the reconstructed D_s and η. The next step will be to expand the kinematic fit to include the unmeasured neutrino in the semileptonic decay branch.

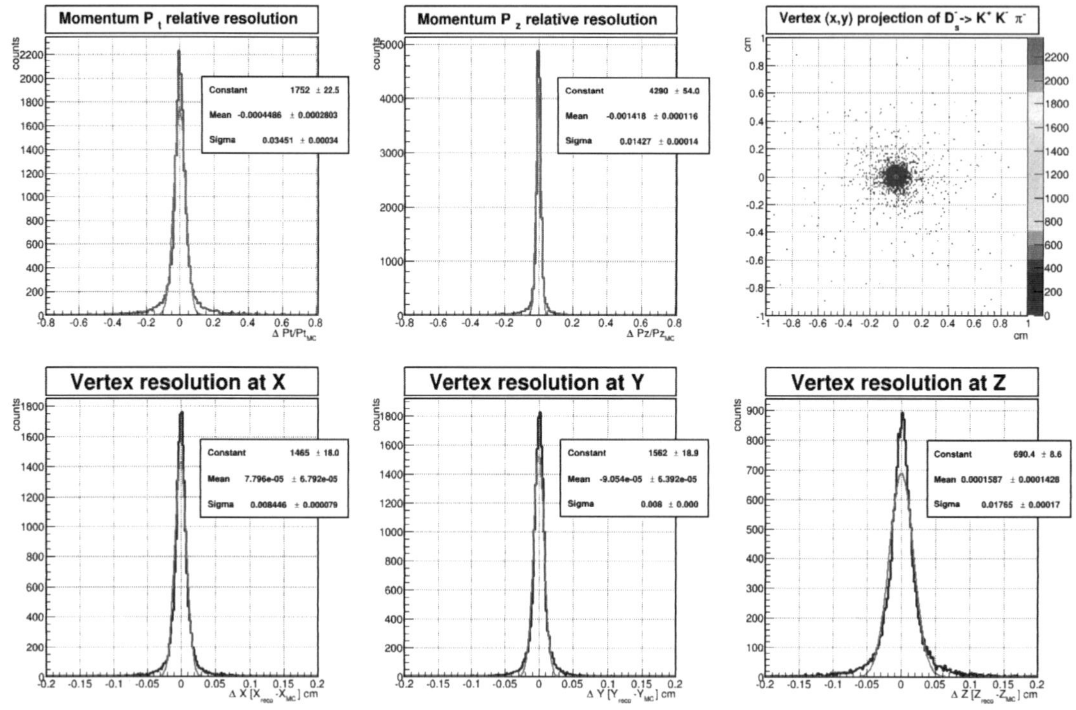

Figure 4. Resolutions of kinematic properties of reconstructed D_s^-. The top left and center frames indicate the relative resolution for the transverse and longitudinal momentum, respectively. The top right frame indicates the reconstructed vertex distribution for the D_s^- decay branch, and the lower line indicates the vertex resolution in the X,Y,Z projections.

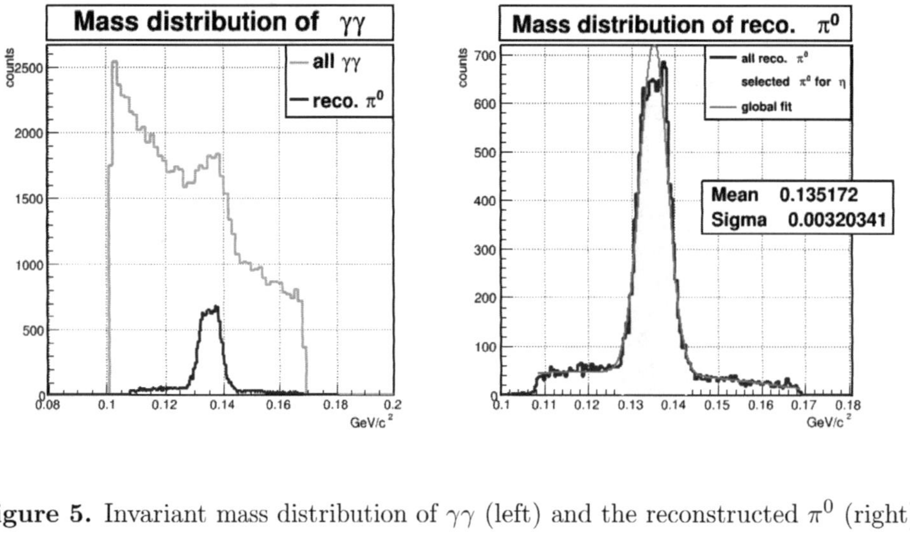

Figure 5. Invariant mass distribution of $\gamma\gamma$ (left) and the reconstructed π^0 (right).

The efficiency of the $\nu_e e^+$ system should be the product of the efficiencies of D_s^- and η (listed in Table 1). At this moment, the value we obtained is about 4.7%. The cross section of D_s

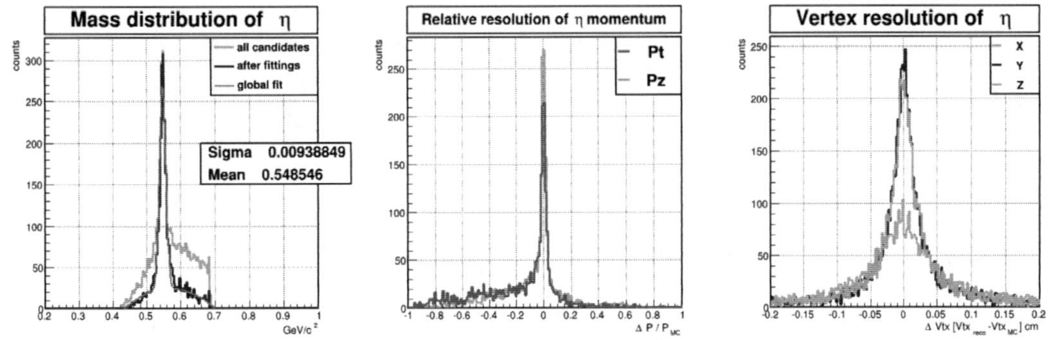

Figure 6. Resolution of mass and kinematic properties of reconstructed η. Three pion invariant mass distribution from eta decays (left), transverse (blue) and longitudinal (green) momentum resolution of the reconstructed η, and the vertex resolution in all three Cartesian projections.

pair at PANDA is roughly estimated to be 10 nb with the beam momentum of 8 GeV/c [12]. With the average branching ratios $\mathcal{B}(D_s^+ \to \eta e^+ \nu_e) = 2.67\%$ and $\mathcal{B}(D_s^- \to K^+ K^- \pi^-) = 5.49\%$, the production rate is estimated to be approximately 200 events per month with the luminosity of 10^{32} cm^{-2}s^{-1}. This result is not the final estimate, since the present PandaRoot software is under development. The reconstruction efficiency including the resolutions can be improved in the near future. On the other hand, an explicit prediction of cross section from the theory is urgently needed. This issue will be also interesting for the theorists working on the different models.

Table 1. Reconstruction results of D_s^- and η.

Particle	eff. [%]	Mass reso. [MeV/c^2]	Vertex reso. [μm]			Momentum reso. [%]	
			x	y	z	p_t	p_z
D_s^-	18	17	78	80	177	3.5	1.4
η	26	9	318	287	675	3.8	4.3

References

[1] Erni W *et al.* (PANDA Collaboration) 2009 (*Preprint* hep-ex/0903.3905)
[2] Besson D *et al.* (CLEO Collaboration) 2009 *Phys.Rev.D* **80** 032005 (*Preprint* hep-ex/0906.2983)
[3] Ge J Y *et al.* (CLEO Collaboration) 2009 *Phys.Rev.D* **79** 052010 (*Preprint* hep-ex/0810.3878)
[4] Khodjamirian A, Klein C, Mannel T and Offen N 2009 *Phys.Rev.D* **80** 114005 (*Preprint* hep-ph/0907.2842)
[5] Yelton J *et al.* (CLEO Collaboration) 2009 *Phys.Rev.D* **80** 052007 (*Preprint* hep-ex/0903.0601)
[6] Spataro S (PANDA Collaboration) 2008 *Journal of Physics: Conference Series* **119** 032035 , and references therein
[7] http://root.cern.ch
[8] Frabetti P *et al.* 1995 *Physics Letters B* **351** 591 – 600
[9] Mitchell R E *et al.* (CLEO Collaboration) 2009 *Phys. Rev. D* **79**(7) 072008 (*Preprint* hep-ex/0903.1301)
[10] del Amo Sanchez P *et al.* (BABAR Collaboration) 2011 *Phys. Rev. D* **83**(5) 052001
[11] Erni W *et al.* (PANDA Collaboration) 2008 (*Preprint* physics.ins-det/0903.3905)
[12] Haidenbauer J 2012 Priv. comm.

Journal of Physics: Conference Series **503** (2014) 012025 doi:10.1088/1742-6596/503/1/012025

Evolution of elliptic and triangular flow as a function of beam energy in a hybrid model

J. Auvinen

Frankfurt Institute for Advanced Studies (FIAS), Ruth-Moufang-Strasse 1, D-60438 Frankfurt am Main, Germany

E-mail: auvinen@fias.uni-frankfurt.de

H. Petersen

Frankfurt Institute for Advanced Studies (FIAS), Ruth-Moufang-Strasse 1, D-60438 Frankfurt am Main, Germany

Institut für Theoretische Physik, Goethe Universität, Max-von-Laue-Strasse 1, D-60438 Frankfurt am Main, Germany

E-mail: petersen@fias.uni-frankfurt.de

Abstract. Elliptic flow has been one of the key observables for establishing the finding of the quark-gluon plasma (QGP) at the highest energies of Relativistic Heavy Ion Collider (RHIC) and the Large Hadron Collider (LHC). As a sign of collectively behaving matter, one would expect the elliptic flow to decrease at lower beam energies, where the QGP is not produced. However, in the recent RHIC beam energy scan, it has been found that the inclusive charged hadron elliptic flow changes relatively little in magnitude in the energies between 7.7 and 39 GeV per nucleon-nucleon collision. We study the collision energy dependence of the elliptic and triangular flow utilizing a Boltzmann + hydrodynamics hybrid model. Such a hybrid model provides a natural framework for the transition from high collision energies, where the hydrodynamical description is essential, to smaller energies, where the hadron transport dominates. This approach is thus suitable to investigate the relative importance of these two mechanisms for the production of the collective flow at different values of beam energy. Extending the examined range down to 5 GeV per nucleon-nucleon collision allows also making predictions for the CBM experiment at FAIR.

1. Introduction

In 2010, the RHIC beam energy scan program was launched to study the features of the QCD phase diagram and to search for signs of the possible first-order phase transition between the confined and deconfined matter [1]. The existence of a critical point marking the boundary of cross-over and the aforementioned first-order phase transition in the plane of baryochemical potential μ_B and temperature T was predicted by lattice calculations [2, 3, 4]; it has, however, been put to question by the continuum extrapolated results [5, 6] which suggest the phase transition remaining cross-over also at large values of μ_B.

Elliptic flow v_2 is one of the key observables that supports the formation of a strongly coupled quark-gluon plasma at the highest energies of RHIC and the Large Hadron Collider (LHC). Thus the naive expectation for v_2 in the beam energy scan would be a decrease at lower beam energies where the hydrodynamic phase is short or the QGP is not created at all. However, the measured inclusive charged hadron elliptic flow v_2 demonstrates relatively little dependence on the collision energy $\sqrt{s_{NN}}$ between 7.7 and 39 GeV [7].

One possible method for investigating the importance of the hydrodynamical evolution for the flow production is the hybrid approach, where one uses a transport model for the non-equilibrium phases at the beginning and in the end of a heavy-ion collision event, and hydrodynamics for the intermediate hot and dense stage and the phase transition between the quark-gluon plasma and hadronic matter. This approach should be applicable for a wide range of heavy ion collision energies and thus optimal for studying the beam energy dependence of the flow observables down to $\sqrt{s_{NN}} = 5$ GeV, an energy reachable also at the future heavy ion collisions at FAIR.

2. Hybrid model

In this study, a transport + hydrodynamics hybrid model by Petersen *et al.* [8] is utilized. The initial state in this model is produced by the Ultrarelativistic Quantum Molecular Dynamics (UrQMD) string / hadronic cascade [9, 10]. The transition to hydrodynamics is done when the two colliding nuclei have passed through each other: $t_{\text{start}} = \max\{2R(\gamma_{CM}^2 - 1)^{-1/2}, 0.5 \text{ fm}\}$, where R represents the nuclear radius and $\gamma_{CM} = (1 - v_{CM}^2)^{-1/2}$ is the Lorentz factor. The minimum time of 0.5 fm has been determined by the model results at the collision energy $\sqrt{s_{NN}} = 200$ GeV [11]. At t_{start}, the energy-, momentum- and baryon number densities of the particles, represented by Lorentz-contracted 3D Gaussian distributions with the width $\sigma = 1.0$ fm, are mapped onto the hydro grid. Spectator particles are excluded from this procedure and propagated separately in the cascade.

The SHASTA algorithm [12, 13] is used to solve the (3+1)-D ideal hydrodynamics evolution equations. The equation of state (EoS) is based on a hadronic chiral parity doublet model with quark degrees of freedom, coupled to Polyakov loop to include the deconfinement phase transition [14]. It possesses the important feature of being applicable also at finite baryon densities. At the end of the hydrodynamical evolution, the active EoS is changed to the hadron gas EoS, so the active degrees of freedom on both sides of the transition hypersurface match exactly [15].

The particlization, i.e. the transition from hydro to transport, is done when the energy density ϵ reaches the critical value $2\epsilon_0$, where $\epsilon_0 = 146$ MeV/fm^3 is the nuclear ground state energy density. The particle distributions are generated according to the Cooper-Frye formula from the iso-energy density hypersurface, which is constructed using the Cornelius hypersurface finder [16]. The rescatterings and final decays of these particles are then computed in the UrQMD. The final distribution of particles can then be directly compared against the experimental data. It has been tested that the hybrid model has a reasonable agreement with the experimental data for particle m_T spectra at midrapidity $|y| < 0.5$ for energies ranging from $E_{\text{lab}} = 40$ AGeV to $\sqrt{s_{NN}} = 200$ GeV [17, 18].

3. Results

3.1. Elliptic flow

In this study, the flow coefficients v_n are computed from the particle momentum distributions using the event plane method [19, 20]. This, together with the new implementation of the Cooper-Frye hypersurface finder and particlization, forms the core difference compared to previous studies of elliptic flow in the same hybrid approach [21, 22]. The primary interest in the following is to see, if the experimentally observed weak sensitivity of the elliptic flow on the collision energy is manifested also in the hybrid model results.

Figures 1a and 1b show the p_T-integrated elliptic flow v_2 produced in Au+Au -collisions for the p_T range 0.2 - 2 GeV, compared with the STAR data for the (0-5)% and (30-40)% centrality classes. In the model these are respectively represented by the impact parameter intervals $b = 0 - 3.4$ fm and $b = 8.2 - 9.4$ fm. Figure 1c shows the differential $v_2(p_T)$ for $b = 6.7 - 8.2$ fm,

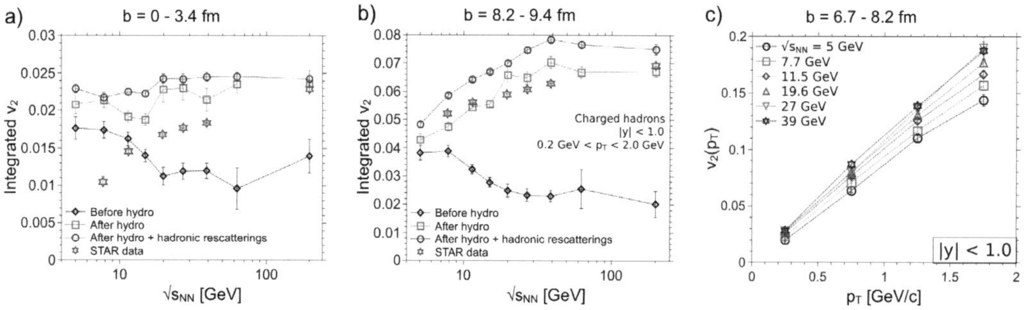

Figure 1. Integrated v_2 for $\sqrt{s_{NN}} = 5-200$ GeV, at the beginning of hydrodynamical evolution (diamonds), immediately after particlization (squares) and after the full simulation (circles) in a) central collisions and b) midcentral collisions, compared with the STAR data [7, 23] (stars). c) Differential v_2 at midrapidity $|y| < 1.0$ for $\sqrt{s_{NN}} = 5 - 39$ GeV in impact parameter range $b = 6.7 - 8.2$ fm.

which roughly corresponds to (20-30)% centrality class. Figures 1a and 1b also demonstrate the magnitude of v_2 at three different times: just before the hydrodynamics phase begins, right after the hydrodynamics phase has ended and particlization has been done, and after the hadronic rescatterings have been performed in the UrQMD (in other words, after the full evolution).

In the most central collisions the effect of the hadronic rescatterings is negligible; in the impact parameter range $b = 8.2 - 9.4$ fm the rescatterings contribute about 10% on the final result. The hydrodynamics also produce very little elliptic flow at $\sqrt{s_{NN}} \leq 7.7$ GeV; for the mid-central collisions, v_2 is in practice completely produced by the transport dynamics, which include resonance formation and decay and string excitation and fragmentation processes. These initial dynamics, which are often neglected in other hybrid approaches, gain importance at lower energies. On the other hand, above $\sqrt{s_{NN}} = 19.6$ GeV the hydrodynamic phase is clearly the dominant source of v_2.

The simulation results overshoot the experimental data for all collision energies. This suggests that either the viscous corrections should be included, or the energy density value chosen for particlization should be higher. In the most central collisions below $\sqrt{s_{NN}} = 11.5$ GeV, the model appears to produce too much flow already at transport phase; here having agreement with the data would require modifications in how the pre-equilibrium phase is handled. However, for the purposes of this study, the most important feature is the good qualitative agreement in the midcentral collisions, as here the flow effects are at their largest. Also $v_2(p_T)$ (Fig. 1c) has relatively weak dependence on $\sqrt{s_{NN}}$, which is in accordance with the STAR results.

3.2. Triangular flow

Based on the above results, it appears that the hydrodynamically produced elliptic flow indeed vanishes, as was the naive expectation, but this is partially compensated by the increased flow production in the transport phase and so the observed v_2 has only weak collision energy dependence. To study this phenomenon further, we do the same analysis for another flow observable: the triangular flow v_3, which originates purely from the event-by-event variations in the initial spatial configuration of the colliding nucleons, and is thus largely independent of the collision geometry.

As illustrated by Figure 2a, the p_T-integrated v_3 increases from ≈ 0.01 to above 0.015 with increasing collision energy in the most central collisions, whereas in midcentrality $b = 6.7 - 8.2$ fm there is a rapid rise from ≈ 0 at $\sqrt{s_{NN}} = 5$ GeV to the value of ≈ 0.02 for $\sqrt{s_{NN}} \geq 27$ GeV. The collision energy dependence is seen also for midcentral $v_3(p_T)$ in Fig. 2b. The energy dependence of v_3 in midcentral collisions qualitatively resembles the hydrodynamically produced

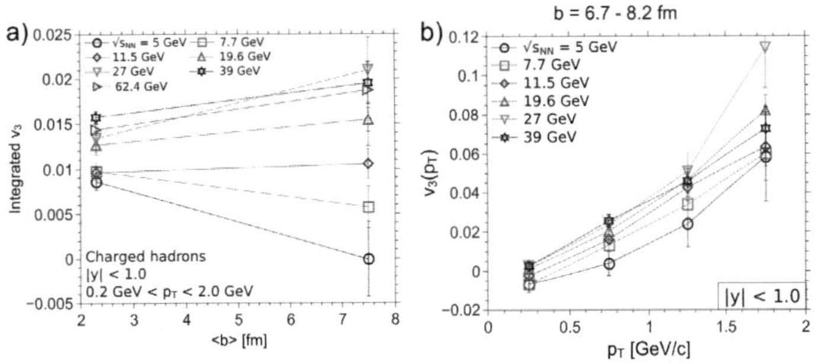

Figure 2. a) Integrated v_3 at midrapidity $|y| < 1.0$ in central ($b = 0 - 3.4$ fm) and midcentral ($b = 6.7 - 8.2$ fm) collisions for collision energies $\sqrt{s_{NN}} = 5 - 62.4$ GeV. b) $v_3(p_T)$ in midcentral collisions for $\sqrt{s_{NN}} = 5 - 39$ GeV.

v_2 in Figure 1b. Thus for the higher flow coefficients, which are more sensitive to viscosity, the transport part of the model is unable to compensate for the diminished hydro phase.

3.3. Effect of initial geometry

Let us then investigate in more detail the effect of initial collision geometry on the flow coefficients. Figure 3a illustrates the collision energy and centrality dependencies of the average initial state spatial eccentricity $\langle \epsilon_2 \rangle$ and triangularity $\langle \epsilon_3 \rangle$, where the eccentricity and triangularity are defined as in [25] and calculated at the beginning of hydrodynamical evolution t_{start}.

The average eccentricity and triangularity are of the same magnitude in the most central collisions, where the nuclear overlap region is nearly isotropic. The situation changes in midcentral collisions, where, due to the collision geometry, $\langle \epsilon_2 \rangle$ is clearly larger than $\langle \epsilon_3 \rangle$. The observed dependence on collision energy is largely explained by t_{start}, which changes rapidly at low energies, from 5.19 fm at $\sqrt{s_{NN}} = 5$ GeV to 1.23 fm at $\sqrt{s_{NN}} = 19.6$ GeV. At low energies there is thus enough time for the pre-equilibrium transport to decrease the initial spatial anisotropies.

Figure 3b shows the coefficients v_2 and v_3 scaled with $\langle \epsilon_2 \rangle$ and $\langle \epsilon_3 \rangle$, respectively. The relation of the elliptic flow to the initial eccentricity remains largely unchanged for the whole collision energy range, while the v_3 response to the triangularity of the initial state saturates only after 19.6 GeV. This suggests that the hadronic medium is too viscous to convert initial state fluctuations into triangular flow, and a sufficiently long-living intermediate phase with a low-viscosity fluid is needed for the v_3 production.

4. Summary

We have demonstrated that the experimentally observed behavior of v_2 as a function of collision energy $\sqrt{s_{NN}}$ can be qualitatively reproduced utilizing a hybrid transport + hydrodynamics approach. The diminished hydrodynamical evolution for v_2 production at lower collision energies is compensated by the pre-equilibrium transport dynamics. This compensation does not apply to triangular flow v_3, which decreases considerably faster, reaching zero in midcentral collisions at $\sqrt{s_{NN}} = 5$ GeV. This makes v_3 the better signal for the formation of quark-gluon plasma in heavy ion collisions. However, according to the preliminary STAR data, v_3 remains constant in central collisions at $\sqrt{s_{NN}} = 7.7 - 27$ GeV [26], which suggests that the low-viscous state of nuclear matter is manifested at the lower collision energies in greater extent than expected. The flow coefficients thus remain interesting observables also for the heavy ion collisions at FAIR energies.

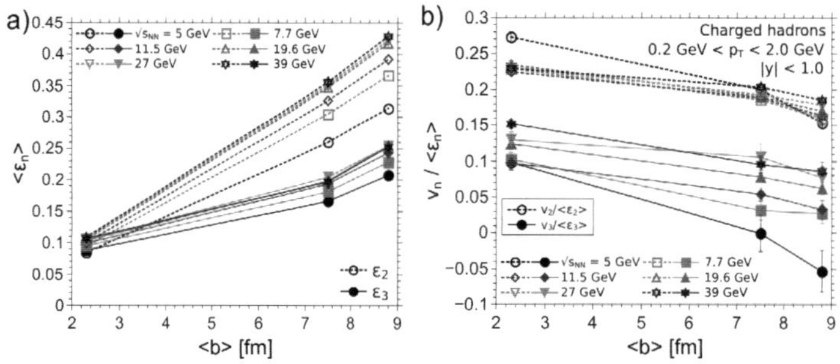

Figure 3. a) Average eccentricity $\langle\epsilon_2\rangle$ (open symbols) and triangularity $\langle\epsilon_3\rangle$ (filled symbols) as a function of average impact parameter $\langle b\rangle$ for the collision energy range $\sqrt{s_{NN}} = 5 - 39$ GeV. b) Scaled flow coefficients $v_2/\langle\epsilon_2\rangle$ and $v_3/\langle\epsilon_3\rangle$ as a function of average impact parameter $\langle b\rangle$ for $\sqrt{s_{NN}} = 5 - 39$ GeV.

5. Acknowledgements

The authors acknowledge funding of the Helmholtz Young Investigator Group VH-NG-822. This work was supported by the Helmholtz International Center for the Facility for Antiproton and Ion Research (HIC for FAIR) within the framework of the Landes-Offensive zur Entwicklung Wissenschaftlich-ökonomischer Exzellenz (LOEWE) program launched by the State of Hesse. Computational resources have been provided by the Center for Scientific Computing (CSC) at the Goethe-University of Frankfurt.

References

[1] Kumar L [STAR Collaboration] 2011 *Nucl. Phys.* A **862-863** 125
[2] Fodor Z and Katz S D 2004 *J. High Energy Phys.* JHEP04(2004)050
[3] Ejiri S 2008 *Phys. Rev.* D **78** 074507
[4] Gavai R V and Gupta S 2008 *Phys. Rev.* D **78** 114503
[5] Aoki Y, Endrodi G, Fodor Z, Katz S D and Szabo K K 2006 *Nature* **443** 675
[6] Endrodi G, Fodor Z, Katz S D and Szabo K K 2011 *J. High Energy Phys.* JHEP04(2011)001
[7] Adamczyk L *et al* [STAR Collaboration] 2012 *Phys. Rev.* C **86** 054908
[8] Petersen H, Steinheimer J, Burau G, Bleicher M and Stocker H 2008 *Phys. Rev.* C **78** 044901
[9] Bass S A, Belkacem M, Bleicher M, Brandstetter M, Bravina L, Ernst C, Gerland L and Hofmann M *et al* 1998 *Prog. Part. Nucl. Phys.* **41** 255
[10] Bleicher M, Zabrodin E, Spieles C, Bass S A, Ernst C, Soff S, Bravina L and Belkacem M *et al* 1999 *J. Phys.* G **25** 1859
[11] Petersen H, Coleman-Smith C, Bass S A and Wolpert R 2011 *J. Phys.* G **38** 045102
[12] Rischke D H, Bernard S and Maruhn J A 1995 *Nucl. Phys.* A **595** 346
[13] Rischke D H, Pursun Y and Maruhn J A 1995 *Nucl. Phys.* A **595** 383 [Erratum 1996 *Nucl. Phys.* A **596** 717]
[14] Steinheimer J, Schramm S and Stöcker H 2011 *Phys. Rev.* C **84** 045208
[15] Steinheimer J, Dexheimer V, Petersen H, Bleicher M, Schramm S and Stoecker H 2010 *Phys. Rev.* C **81** 044913
[16] Huovinen P and Petersen H 2012 *Eur. Phys. J.* A **48** 171
[17] Auvinen J and Petersen H 2013 *PoS* **CPOD2013** 034
[18] Auvinen J and Petersen H 2013 *Preprint* arXiv:1310.1764 [nucl-th]
[19] Poskanzer A M and Voloshin S A 1998 *Phys. Rev.* C **58** 1671
[20] Ollitrault J Y 1997 *Preprint* nucl-ex/9711003
[21] Petersen H and Bleicher M 2009 *Phys. Rev.* C **79** 054904
[22] Petersen H and Bleicher M 2010 *Phys. Rev.* C **81** 044906
[23] Adams J *et al* [STAR Collaboration] 2005 *Phys. Rev.* C **72** 014904
[24] Adamczyk L *et al* [STAR Collaboration] 2013 *Phys. Rev.* C **88** 014904
[25] Petersen H, Qin G-Y, Bass S A and Muller B 2010 *Phys. Rev.* C **82** 041901
[26] Pandit Y [STAR Collaboration] 2012 Talk at Quark Matter 2012

The relation between cross-section, decay width and imaginary potential of heavy quarkonium in a quark-gluon plasma

Miguel Ángel Escobedo

Physik-Department, Technische Universität München, James-Franck-Str. 1, 85748 Garching, Germany

E-mail: miguel.escobedo@ph.tum.de

Abstract. Heavy quarkonium suppression was proposed long ago as a signal of the formation of a deconfined phase in heavy ion collisions. Originally the mechanism responsible for this suppression was thought to be color screening. However perturbative computations and recent lattice studies suggest the existence of an imaginary part of the potential which could have a more important role for suppression than screening. In this work we review some general aspects of effective field theories for heavy quarkonium in the medium and discuss the physical phenomena behind the imaginary part of the potential and the decay width of heavy quarkonium and their corresponding cross sections.

1. Introduction

Heavy quarkonium is a meson formed by two heavy quarks whose mass m_Q is much bigger than Λ_{QCD}. They are quite different to other mesons. The first difference is that a lot of the physics relevant to study this system happens at a energy scale m_Q where perturbation theory is applicable. For example the typical size of the more deeply bounded quarkonium states is of order $1/(m_Q v)$, where v is the velocity of the heavy quarks around the center of mass, while for other mesons the size is of order $1/\Lambda_{QCD}$. A second difference is that heavy quarkonium is a non-relativistic system, meaning that $v \ll 1$. For this reason it is expectable that at some accuracy it can be described by a Schrödinger equation with some potential, hence we can say that it is the QCD analogous to the hydrogen atom. On the other hand this small v can spoil naive perturbation theory in a way that will be discussed later.

The original idea of using quarkonium suppression as a probe of deconfinement in heavy ion collisions is found in [1]. Since then this phenomena has been studied experimentally in many facilities, as for example SPS, RHIC and LHC. In the future the CBM experiment will be able to probe the situation when the quark-gluon plasma has a large chemical potential. Even though quarkonium suppression is a well established experimental fact an understanding of the responsible mechanism and a quantitative description of experimental data is still missing.

The dissociation mechanism that was originally considered in [1] was color screening. In order to illustrate this mechanism we can make use of what happens in the perturbative limit of QCD. There the potential between two infinitely heavy color charges in the vacuum will have the form of a Coulomb potential $V(r) = -\frac{\alpha}{r}$. Then the two heavy charges will feel an attractive force. If instead of the vacuum we are in a medium in which color screening takes place then we will have instead of a Coulomb a Yukawa potential $V(r) = -\frac{\alpha}{r}e^{-m_D r}$ where m_D is a Debye mass that will encode the information about the medium. If the two charges are separated by a distance that is much bigger than $1/m_D$ they are not going to see each other so in an effective way they will behave like individual particles.

However another mechanism for dissociation might also exists. The heavy quarkonium state will have some probability to be hit by some of the particles in the medium and this hit might induce a decay. This means that the quarkonium state might still exist but with a very short mean life. Indeed in a perturbative computation in [2] it was shown that the correlator of the heavy quark part of the electromagnetic current fulfils a Schrödinger equation with a potential that has an imaginary part. This imaginary part was bigger than the real part for the situation in which screening is important and hence it indicates that in the perturbative limit the decay width makes the different quarkonium states dissociate before screening becomes relevant.

In the previous discussion we have been talking about potentials and the Schrödinger equation without talking about how they can be obtained from QCD. The question can be rephrased in the following way. Is there a controlled expansion in which we can obtain a Schrödinger equation as a leading order approximation and we can estimate the error done by making such approximation? A way to answer this question that has been very successful in the vacuum is by using effective field theories (EFTs). This program for finite temperature T and zero chemical potential μ has been started in the last years [3, 4]. Using these techniques the potential of [2] has been confirmed for $T \gg m_D \sim 1/r$. The case $1/r \gg T \gg E \gg m_D$ was studied in [5] where corrections to decay width that can not be encoded in a potential were found. This result was later compared with lattice computations [6] yielding similar results. The case $T \sim 1/r$ was studied in muonic hydrogen [7]. In this work we are going to focus on the recent investigations that have been done to translate the previous results in terms of cross sections [8, 9]. Opposite to what happens to the decay width these cross sections do not depend on the nature of the medium as long as it is isotropic and homogeneous, so they can be directly applied to mediums with a finite chemical potential as the one that is expected to be formed in CBM.

This work is organized as follows. In section 2 we review non-relativistic EFTs, in section 3 we discuss the process responsible of the decay of quarkonium in the medium at low densities and temperatures that is called gluo-dissociation, in section 4 we discuss the process that dominates at high densities and temperatures, inelastic parton scattering. Finally, section 5 is devoted to the conclusions.

2. Non-relativistic EFTs for heavy quarkonium

The general idea of EFTs was introduced in [10]. An EFT is a field theory which gives exactly the same result as a more general one but on a limited kinematic regime. They are useful for studying problems in which different energy scales are important. In most QCD processes Λ_{QCD} is an important scale but in heavy quarkonium we also have another important scale m_Q that fulfils $m_Q \gg \Lambda_{QCD}$. Moreover, because heavy quarkonium is a non-relativistic system, there appears other energy scales as the inverse of the typical radius $1/r \sim m_Q v$ and $E \sim m_Q v^2$ the binding energy.

Non-relativistic QCD (NRQCD) [11] is an EFT in which modes with a virtuality of order m_Q^2 are integrated out. The fact that heavy quarks are heavy ensures that NRQCD can be derived from QCD using perturbation theory. In this EFT the heavy quark instead of being represented by a bispinor field is represented by two spinor fields, one for the heavy quark and another for the heavy anti-quark. Approximate symmetries as the spin symmetry can already be seen by looking at the NRQCD lagrangian.

Potential NRQCD (pNRQCD) [12] is an EFT in which modes with a virtuality of order $1/r^2$ are integrated out. In this case pNRQCD can be derived from NRQCD using perturbation theory only for states whose size is smaller than $1/\Lambda_{QCD}$. This is expected to happen for the more deeply bound states as $\Upsilon(1S)$ with a Bohr radius of order $1/a_0 \sim 1355\,\texttt{MeV}$ or J/Ψ with $1/a_0 \sim 770\,\texttt{MeV}$ (both values are taken from [13]). In the static limit agreement between high orders computations in pNRQCD and the lattice can be found up to distances of order $0.2\,fm$ and in fact it can be used to make a competitive determination of α_s [14]. Another important issue when doing computations is how Λ_{QCD} is related with the binding energy E. Through this work we are going to assume that $E \gg \Lambda_{QCD}$ when computations at this scale are needed. This might be reasonable for $\Upsilon(1S)$ and J/Ψ where $E \sim 350\,\texttt{MeV}$ [13].

In pNRQCD when $1/r \gg \Lambda_{QCD}$ there exist two kind of fields that represent heavy quark and heavy antiquarks pairs, a singlet field and an octet field. At leading order they both fulfil a Schrödinger equation with a potential that encodes physics that is integrated out when going from NRQCD to pNRQCD.

The main advantage of using EFTs is that one can define a power counting such that the size of each contribution can be predicted just looking at the Lagrangian. As an example let us look in the computation of the binding energy of a given state in perturbation theory. If this is done directly in QCD we will find an expansion of the following type

$$E = m_Q \alpha_s \sum_{n=0}^{\infty} \alpha_s^n A_n(v) \tag{1}$$

where A_n are coefficients that in general are going to depend on ratios of the different scale that appear in the problem and therefore on v. There is nothing that forbids that a given A_n goes to infinity as $v \to 0$. If this happens we lost control of our perturbative expansion. This is indeed the case of some diagrams that appear in heavy quarkonium physics as for example the box diagram that is enhanced by a $m_Q r$ power. If the same computation is done using pNRQCD the series can be reorganized as

$$E = m_Q \alpha_s v^2 \sum_{n,m} \alpha_s^n v^m B_{n,m} \tag{2}$$

where now we know that $B(n,m)$ will be of order 1.

With the aim of having a way to correctly define the potential and the Schrödinger equation and to have a well defined power counting non-relativistic EFTs were extended to finite temperature. In general the presence of a medium introduces another set of energy scales in the problem and the physics is going to be different depending on the relation with the scales that define the bound state in the vacuum. In a perturbative medium, as the one we are going to consider, the induced scales are the following. There is a scale p_{Hard} which corresponds to the typical energy of the partons in the medium. This energy scale might be integrated out to yield an EFT called Hard Dense Loop (HDL) (Hard Thermal Loop for $\mu = 0$) [15]. Another important scale is the so-called Debye mass m_D that corresponds to the inverse of the distance in which screening effects become important

$$m_D^2 = \frac{N_c g^2 T^2}{6} + \frac{g^2 C_F N_F}{12}\left(T^2 + \frac{3\mu^2}{\pi^2}\right) \tag{3}$$

Depending on the relation between p_{Hard} and the rest of scales the dominant mechanism responsible for the decay width is going to be different. It happens that the key relation is that of m_D with the binding energy. If $E \gg m_D$ the dominant mechanism is gluo-dissociation while if $m_D \gg E$ it is inelastic parton scattering. This conclusion can be easily obtained using pNRQCD power counting [4, 7]. In the next two sections we are going to discuss these two processes in detail.

3. Gluo-dissociation

When $E \gg m_D$ the decay width is dominated by the imaginary part of the leading order singlet self-energy. In the case of zero chemical potential this was computed for $T \gg E$ in [5]

$$\delta\Gamma_n = \frac{1}{3}N_C^2 C_F \alpha_s^3 T - \frac{16}{3m} C_F \alpha_s T E_n + \frac{4}{3} N_C C_F \alpha_s^2 T \frac{2}{mn^2 a_0} \quad (4)$$

and for $T \sim E$ in [3]

$$\delta\Gamma_n = \frac{4}{3}\alpha_s C_F T \langle n| r_i \frac{|E_n - h_o|^3}{e^{\beta|E_n - h_o|} - 1} r_i |n\rangle \quad (5)$$

Another way to compute this decay width that has been used in the past is to compute the cross section for this process at $T = 0$ and later convolute it with a partonic distribution function. There are two important questions to discuss. If this convolution is justified by perturbative thermal field theory and how the cross section obtained in pNRQCD relates with the one commonly used in phenomenological analyses. The cross section for gluo-dissociation was computed in [16]. They used an OPE based on the multipole expansion. This expansion is also present in pNRQCD Lagrangian so in this sense the result has to be similar. The difference comes from the fact that they use the large N_c limit while this approximation was not needed to compute the decay width in pNRQCD.

In order to extract the cross section from our previous computation of the decay width we use the cutting rules that were generalized to a thermal medium in [17]. For the case of gluo-dissociation in fact the naively expected convolution is obtained

$$\Gamma = \int \frac{d^3k}{(2\pi)^3} f(k)\sigma(k) \quad (6)$$

with a cross section for the $1S$ state

$$\sigma_{1S}(k) = \frac{\alpha_s C_F}{3} 2^{10}\pi^2 \rho(\rho+2)^2 \frac{E_1^4}{mk^5}\left(t(k)^2 + \rho^2\right) \frac{\exp\left(\frac{4\rho}{t(k)}\arctan\left(t(k)\right)\right)}{e^{\frac{2\pi\rho}{t(k)}} - 1}, \quad (7)$$

where $t(k) \equiv \sqrt{k/|E_1| - 1}$ and $\rho = \frac{1}{N_c^2 - 1}$. This result was also found in [18] using different techniques. More details about gluo-dissociation in pNRQCD can be found in [8].

4. Inelastic parton scattering

This process dominates the decay width for $m_D \gg E$. Here we are going to focus in the regime $p_{Hard} \gg 1/r \sim m_D$ that is the more relevant one for dissociation. The process we are studying corresponds to $HQ + p \to Q + \bar{Q} + p$ which means that a heavy quarkonium HQ is hit by a parton p and goes into two heavy quarks Q plus a parton. An approximation that is used in phenomenological analyses [19] is $\sigma(HQ + p \to Q + \bar{Q} + p) = 2\sigma(Q + p \to Q + p)$ (quasi-free approximation). In the right hand side the cross section computed in [20] is used. This approach neglects the interference terms that arise in the interaction of the heavy quark and the heavy antiquark. In the case in which the thermal corrections to the potential can be considered a perturbation one can use the cutting rules to extract a cross section that will not make this approximation. Doing this we notice two things. First that the naive convolution formula can not be applied in this case because we have a parton in the final state

$$\Gamma = \int \frac{d^3k}{(2\pi)^3} f(k)(1 + f(k))\sigma(k, m_D) \quad (8)$$

and second that $\sigma(k, m_D)$ depends now on the medium through m_D that acts as a thermal mass. For $1S$ states we find

$$\sigma_{1S}(k, m_D) = 8\pi C_F \alpha_s^2 N_F a_0^2 \left(-\frac{3}{2} + 2\log\left(\frac{2}{x}\right) + \log\left(\frac{y^2}{1+y^2}\right) - \frac{1}{y^2}\log(1+y^2)\right) \quad (9)$$

where a_0 is the Coulombic Bohr radius, $x = m_D a_0$ and $y = k a_0$. In conclusion, the EFT computation takes into account correctly the effects of Pauli blocking and Bose enhancement and improves on the quasi-free approximation. More details about inelastic parton scattering in EFT can be found in [9]

5. Conclusions

EFTs have been used to compute a wide range of temperature regimes in the case of small chemical potential. This research has shown the importance of the imaginary part of the potential and the decay width for dissociation. In this work we discussed the physical process behind them in different regimes. For gluo-dissociation we found that the pNRQCD improved the previously known cross-section by going beyond the large N_c approximation. For inelastic parton scattering it can be shown that the imaginary part of the potential improves over the quasi-free approximation in taking into account the interference terms.

Acknowledgements

I acknowledge Nora Brambilla, Jacopo Ghiglieri and Antonio Vairo for collaboration in the two papers in which this work is based. I also acknowledge financial support from the DFG project BR4058/1-1.

[1] T. Matsui and H. Satz, Phys. Lett. B **178** (1986) 416.
[2] M. Laine, O. Philipsen, P. Romatschke and M. Tassler, JHEP **0703** (2007) 054 [hep-ph/0611300].
[3] M. A. Escobedo and J. Soto, Phys. Rev. A **78** (2008) 032520 [arXiv:0804.0691 [hep-ph]].
[4] N. Brambilla, J. Ghiglieri, A. Vairo and P. Petreczky, Phys. Rev. D **78** (2008) 014017 [arXiv:0804.0993 [hep-ph]].
[5] N. Brambilla, M. A. Escobedo, J. Ghiglieri, J. Soto and A. Vairo, JHEP **1009** (2010) 038 [arXiv:1007.4156 [hep-ph]].
[6] G. Aarts, C. Allton, S. Kim, M. P. Lombardo, M. B. Oktay, S. M. Ryan, D. K. Sinclair and J. I. Skullerud, JHEP **1111** (2011) 103 [arXiv:1109.4496 [hep-lat]].
[7] M. A. Escobedo and J. Soto, Phys. Rev. A **82** (2010) 042506 [arXiv:1008.0254 [hep-ph]].
[8] N. Brambilla, M. A. Escobedo, J. Ghiglieri and A. Vairo, JHEP **1112** (2011) 116 [arXiv:1109.5826 [hep-ph]].
[9] N. Brambilla, M. A. Escobedo, J. Ghiglieri and A. Vairo, JHEP **1305** (2013) 130 [arXiv:1303.6097 [hep-ph]].
[10] S. Weinberg, Physica A **96** (1979) 327.
[11] W. E. Caswell and G. P. Lepage, Phys. Lett. B **167** (1986) 437; G. T. Bodwin, E. Braaten and G. P. Lepage, Phys. Rev. D **51** (1995) 1125 [Erratum-ibid. D **55** (1997) 5853] [hep-ph/9407339].
[12] A. Pineda and J. Soto, Nucl. Phys. Proc. Suppl. **64**, 428 (1998) [hep-ph/9707481]; N. Brambilla, A. Pineda, J. Soto and A. Vairo, Nucl. Phys. B **566** (2000) 275 [hep-ph/9907240]. N. Brambilla, A. Pineda, J. Soto and A. Vairo, Rev. Mod. Phys. **77** (2005) 1423 [hep-ph/0410047].
[13] A. Pineda, Heavy quarkonium and nonrelativistic effective field theories. Ph. D thesis. University of Barcelona.
[14] A. Bazavov, N. Brambilla, X. Garcia i Tormo, P. Petreczky, J. Soto and A. Vairo, Phys. Rev. D **86** (2012) 114031 [arXiv:1205.6155 [hep-ph]].
[15] E. Braaten and R. D. Pisarski, Nucl. Phys. B **337** (1990) 569.
[16] G. Bhanot and M. E. Peskin, Nucl. Phys. B **156** (1979) 391.
[17] R. L. Kobes and G. W. Semenoff, Nucl. Phys. B **272** (1986) 329.
[18] F. Brezinski and G. Wolschin, Phys. Lett. B **707** (2012) 534 [arXiv:1109.0211 [hep-ph]].
[19] X. Zhao and R. Rapp, Phys. Rev. C **82** (2010) 064905 [arXiv:1008.5328 [hep-ph]].
[20] B. L. Combridge, Nucl. Phys. B **151** (1979) 429.

Journal of Physics: Conference Series **503** (2014) 012027 doi:10.1088/1742-6596/503/1/012027

The Compressed Baryonic Matter experiment

Sélim SEDDIKI, on behalf of the CBM Collaboration
Gesellschaft für Schwerionenforschung mbH (GSI)
Planckstraße 1, 64291 Darmstadt, Germany
E-mail: S.Seddiki@gsi.de

Abstract. The Compressed Baryonic Matter (CBM) experiment is a next-generation fixed target detector which will operate at the future Facility for Antiproton and Ion Research (FAIR) in Darmstadt. The goal of this experiment is to explore the QCD phase diagram in the region of high net baryon densities using high-energy nucleus-nucleus collisions. Its research program includes the study of the equation-of-state of nuclear matter at high baryon densities, the search for the deconfinement and chiral phase transitions and the search for the QCD critical point. The CBM detector is designed to measure both bulk observables with a large acceptance and rare diagnostic probes such as charm particles, multi-strange hyperons, and low mass vector mesons in their di-leptonic decay. The physics program of CBM will be summarized, followed by an overview of the detector concept, a selection of the expected physics performance, and the status of preparation of the experiment.

1. Introduction

CBM is a fixed target experiment which will operate at the future FAIR facility (GSI, Darmstadt). This experiment seeks to study fundamental properties of the strong interaction and its underlying theory, the Quantum Chromo-Dynamics (QCD). Some of the most important features of the strong interaction are not quantitatively understood, like the phenomenon of color confinement in ordinary nuclear matter and the origin of most of the mass of light hadrons (thus, of most of the visible universe). In practice, CBM will study the properties of QCD matter at very high net baryon densities, in similar conditions as the ones presumably existing in the core of neutron stars.

Above a certain critical energy density, nuclear matter is believed to exist in a deconfined state of quarks and gluons. This is illustrated in Figure 1: such a state can be produced by heating and/or compressing nuclear matter. In accelerator laboratories, this is done by colliding heavy nuclei, thus creating a dense and hot fireball in the overlap region between the two nuclei. At the RHIC and the LHC colliders, collisions at ultra-relativistic energies lead to an extremely hot but almost net-baryon free fireball. According to lattice QCD, the phase transition from confined to deconfined matter at high temperatures and vanishing net baryon densities is a smooth cross-over, while at moderate temperatures but higher baryon densities a first-order phase transition may take place [1]. Exploring the latter region of the QCD phase diagram, poorly known theoretically and experimentally, is the object of a vast physics program at lower beam energies, which is currently carried out at SPS/CERN and RHIC/BNL and which will be continued later at NICA/JINR and FAIR. Possible indications of the onset of deconfinement have already been observed by NA49 (at the SPS), and confirmed recently by STAR (at the RHIC), at a beam energy of around 30 AGeV [2], which lies within the FAIR energy range.

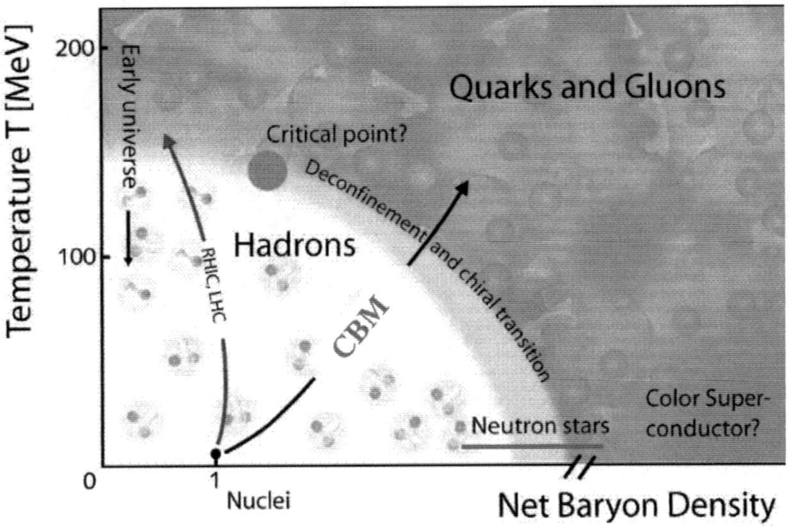

Figure 1. Sketch of the QCD phase diagram of nuclear matter as a function of temperature and net baryon density (see text).

The two different natures of the deconfinement phase transition in the high temperature and high density regions of the phase diagram suggest the existence of a QCD critical point (see Figure 1). The existence and location of this point are till today under debate since lattice QCD calculations have large systematic uncertainties at finite net baryon density [1]. This important feature of the QCD phase diagram thus motivates an experimental research program, which will be undertaken with CBM at FAIR.

The question of the hadron masses is supposedly related to the phenomenon of spontaneous breaking of chiral symmetry, a fundamental property of QCD (in the limit of vanishing quark masses). There are robust theoretical predictions that chiral symmetry is at least partially restored in dense matter [3], and that, among other effects, the in-medium masses of hadrons (containing light quarks) in dense baryonic matter can differ from their corresponding free masses. It is one of the most important goals of CBM to search for such in-medium effects.

An experimental evidence of the first-order phase transition, the QCD critical point and in-medium modifications of hadron masses in dense baryonic matter would be a breakthrough for understanding the properties of the strong interaction.

2. CBM physics program and observables

The FAIR facility will offer unique opportunities for the investigation of the QCD phase diagram at extreme net baryon densities, besides serving a variety of other research fields (hadron physics with anti-proton beams, nuclear structure physics with radioactive beams, and plasma physics with highly pulsed ion beams). For the nuclear collision program, two synchrotrons SIS100 and SIS300 (providing a magnetic rigidity of 100 and 300 Tm, respectively) will deliver fully stripped heavy ion beams up to uranium with extremely high intensities (of up to 2×10^9 ions per second) at beam energies from 2 AGeV to 10 AGeV, and from 15 AGeV to 35 AGeV, respectively. Lighter ions (Z/A = 0.5) will be accelerated up to 45 AGeV, while proton beams will be available up to 30 GeV and 90 GeV at SIS100 and SIS300, respectively.

The broad energy range provided by FAIR will allow producing nuclear matter at the maximal net baryon density achievable with heavy ion collisions [4, 5]. For example, according to various

Figure 2. Layout of the CBM detector setup (see text). Two configurations are foreseen: one dedicated to the measurements of hadrons and electrons (left side), and another one optimized for detecting muons (right side).

transport models [5], the density reaches up to about twelve times the normal nuclear matter density in the core and at the early stage of central Au(20 AGeV)+Au collisions. Another important feature of this accelerator facility is that it will serve several (up to 5) experiments in parallel, allowing CBM operating with heavy ion beams up to 4 months per year. Both the unprecedented beam intensities and the high beam availability will allow CBM measuring rare probes, like charm hadrons and light vector mesons (in their di-leptonic decay), for the first time in the FAIR energy range. These rare probes also include multi-strange particles which are scarcely measured in heavy ion collisions at these energies.

CBM is designed to perform high statistics measurements of hadronic, leptonic and photonic probes with a large acceptance. Its physics program includes a large set of observables [5], among which: the yields and collective flow of strange and charm hadrons is expected to reflect the onset of deconfinement. The collective flow of hadrons (including strange and charm particles) is also expected to be sensitive to the equation-of-state of nuclear matter at the early stage of the reactions. Particle production at threshold energies (charm at SIS300 and possibly multi-strange particles at SIS100) is also expected to deliver important information about the equation-of-state of nuclear matter. In addition, event by event, non statistical, fluctuations of various quantities (particle yields and yield ratios) related to conserved quantum numbers (baryon, charge, and strangeness) would signal a QCD critical point. Finally, in-medium modifications of hadron masses, in particular of open charm hadrons [6] and light vector mesons, may provide valuable insight into the expected restoration of chiral symmetry in dense baryonic matter.

The experimental task of CBM is to measure all these observables in A+A, p+A and p+p collisions, as a function of collision energy and system size, with high precision and statistics, and to search for discontinuities in the aforementioned dependencies which would signal for example a first-order deconfinement phase transition. This vast physics program will be carried out by performing nuclear collisions at extremely high interaction rates.

3. CBM detector, experimental challenge and physics performance

Figure 2 shows a schematic view of the detector concept. Open charm and multi-strange hyperons will be measured through their weak decay into charged hadrons, while charmonium states and light vector mesons will be detected via their decay into both di-electrons and di-muons. The CBM detector setup thus features a flexible arrangement of particle identification detectors: one configuration is dedicated to the identification of hadrons and electrons, while another one is specialized in detecting muons. The second configuration includes absorbers which can efficiently absorb all particles except muons and is not compatible with the first one. In both configurations, the detector comprises a low mass Silicon Tracking System (STS) installed inside a superconducting dipole magnet, a Transition Radiation Detector (TRD) and a Time-Of-Flight (TOF) detector. In addition, a forward hadronic calorimeter, called Projectile Spectator Detector (PSD), will serve for event characterization (collision centrality and event plane determination). The electron and hadron configuration additionally comprises a high resolution Micro-Vertex Detector (MVD), located close to the target and in front of the STS, a Ring Imaging CHerenkov (RICH) and an Electromagnetic CALorimeter (ECAL). As for the muon configuration, these two last sub-detectors are replaced by a MUon CHamber (MUCH), which consists of absorber layers interlaced with tracking stations. The dimensions and positions of all detectors (except the PSD) are chosen such that for central A+A collisions at 25 AGeV and a magnetic dipole field of B = 1 T about 70% of the emitted charged particles are accepted (including mid-rapidity and forward rapidity particles, at low and high transverse momenta).

All detector components have been designed to operate at unprecedentedly high collision rates in order to perform high statistics and systematic measurements. For example, up to 10 MHz and on the order of 100 kHz collision rates are foreseen for J/Ψ and open charm measurements, respectively. This puts very strong constraints on the design of the detector components and readout electronics in terms of rate capability. Also, rare probes in A+A collisions at FAIR energies have multiplicities several order of magnitude lower than that of abundant particles (e.g. pions, protons and electrons). Measuring these rare probes thus calls for very efficient background rejection strategies, which rely on the detector capability for track reconstruction and particle identification.

The STS detector is meant to reconstruct the charged particles' trajectory inside a magnetic dipole field and to determine their momentum. It consists of 8 stations of thin and high resolution double-sided silicon strip sensors, stabilized by a low mass carbon support structure [7]. Detailed detector simulation studies, including in particular a realistic material budget with support structure and cables, demonstrated a high track reconstruction efficiency (of about 95%) and a good momentum resolution ($\Delta p/p$ on the order of 1%), over a large range (p from 0.1 GeV/c to 12 GeV/c). The tracking algorithm has also been proven to be sufficiently fast to deal with the high multiplicity per event in heavy ion reactions and the very high collision rates foreseen [8].

The MVD detector will serve to reconstruct the displaced decay vertex of open charm hadrons. This is crucial to reject the tremendous background which originates dominantly from the collision vertex. The setup will consist of 2 or 3 stations of thin and highly granular commercial CMOS silicon pixel sensors, placed inside a vacuum vessel. The stations are both stabilized and cooled by a low mass (diamond) support structure. Detector simulations, including a realistic material budget and CMOS typical performance, have demonstrated that the decay vertex of open charm hadrons can be reconstructed with a resolution typically below 100 μm [5].

Detailed simulation studies also demonstrated that the combined RICH and TRD detectors can be used to suppress pions mis-identified as electrons (by about a factor of 10^4), while keeping the electron reconstruction efficiency on the order of 50 %. The TOF detector, in combination with the STS providing particle momenta, aims at identifying hadrons. It features a time resolution of about 80 ps, and allows separating well protons and kaons up to about p=6 GeV/c. Pions and kaons can also be distinguished up to p=3.5 GeV/c. Above this range, the RICH detector can be used to further discriminate one another.

The extremely high particle flux expected through the detector (up to several 10^9 particles per second) calls for very high speed, free streaming data read-out, which is an unique feature compared with previous heavy ion experiments. The data will be transported to a large and high performance processor farm with about 60000 computing cores, where partial event

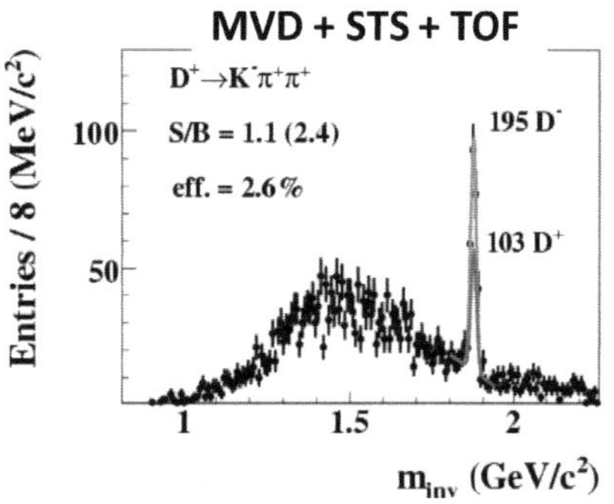

Figure 3. Invariant mass of π^+ π^+ K^- triplets surviving the selection procedure (including D^+ signals and background triplets) in central Au(25 AGeV)+Au collisions.

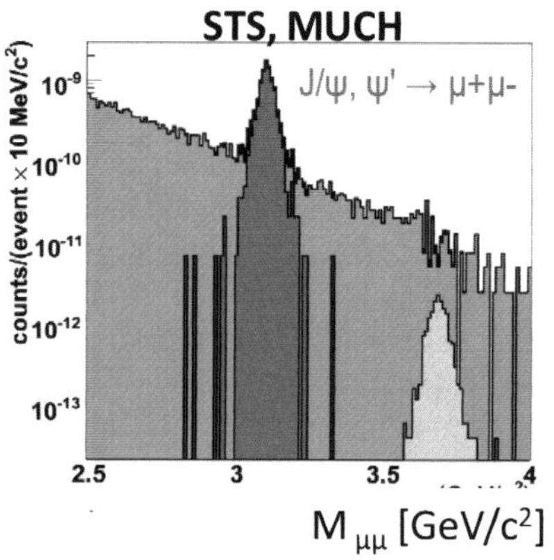

Figure 4. Invariant mass of μ^+ μ^- pairs surviving the selection procedure (including J/Ψ meson signals - red histogram, Ψ' meson signals - green histogram - and background pairs) in central Au(25 AGeV)+Au collisions.

reconstruction (tracking, vertexing, etc) and ultra fast on-line event selection will be performed 'on the fly'. This will allow selecting events which contain signal candidates (e.g. open charm displaced decay vertex, J/Ψ decay muons), making it possible to operate at the foreseen collision rates while keeping the event archiving rate at an acceptable level, on the order of 25 kHz.

Detailed detector simulations have been carried out, including realistic background generation and detector description. They demonstrate the detector performance for reconstructing the various probes foreseen by the experiment, including rare probes (see e.g. [9]). Examples of the expected performance are shown in Figures 3 and 4 in the case of central Au(25 AGeV)+Au collisions. As can be seen, the $D^+ \to \pi^+\ \pi^+\ K^-$ signal can be extracted from the combinatorial background with high signal-to-background ratio and good reconstruction efficiency, thanks to the high vertexing precision of the MVD. Same conclusions are drawn for J/Ψ mesons, for which the 'muon' setup has been used in the simulations.

All detector components are at the stage of prototyping and are being tested on beam, and technical design reports are either submitted (see e.g. [7]) or in the process of submission. The construction phase of the FAIR accelerator is planned to end in 2017, when the experiment apparatus will be installed in the CBM cave. First measurements with nuclear collisions will start in 2019 at the SIS100 synchrotron, which will provide, for the first time, charm data in p+p and p+A collisions at about 30 GeV beam energy and below, and will open the investigation of very interesting strangeness physics in A+A collisions up to 10 AGeV.

References

[1] Fodor Z and Katz S D 2002 *JHEP* **03** 014; de Forcrand P *et al.* 2007 *PoSLAT2007* **178**
[2] Alt C *et al.* 2008 *Phys. Rev.* C **77** 024903; Abelev B I *et al.* 2010 *Phys. Rev.* C **81** 024911
[3] Lutz M *et al.* 1992 *Nucl. Phys.* A **542** 521
[4] Arsene I C *et al.* 2007 *Phys. Rev.* C **75** 034902
[5] Friman B *et al.* 2011 The CBM physics book *Lect. Notes Phys. (Berlin, Springer)* **814** 1-980
[6] Mishra A *et al.* 2004 *Phys. Rev.* C **69** 015202; Cassing W *et al.* 2001 *Nucl. Phys.* A **691** 753
[7] Heuser J *et al.* 2013 Technical design report for the CBM Silicon Tracking System (STS) *GSI Report 2013-4*
[8] Akishina V *et al.* 2013 *CBM progress report 2012* 92-93
[9] Lebedeva E *et al.* 2013 *CBM Progress Report 2012* 106; Galatyuk T *et al.* 2011 *Indian J. Phys.* **85** 81-85

FAIRNESS 2013: FAIR NExt generation of ScientistS 2013 IOP Publishing
Journal of Physics: Conference Series **503** (2014) 012028 doi:10.1088/1742-6596/503/1/012028

Track measurement in the high multiplicity environment at the CBM Experiment

Pradeep Ghosh[1,2], for the CBM Collaboration

[1]Goethe University, Max-von-Laue-Straße 1, D-60438 Frankfurt am Main
[2]GSI Helmholtz Center for Heavy Ion Research GmbH, Planckstraße 1, D-64291 Darmstadt

E-mail: `Pr.Ghosh@gsi.de`

Abstract. In the Compressed Baryonic Matter (CBM) Experiment at FAIR, the Silicon Tracking System (STS) will perform track reconstruction and momentum determination of the charged particles created in interactions of heavy-ion beams with nuclear targets. The STS will consist of 8 tracking layers located at distances between 30 cm and 100 cm downstream of the target inside the 1 Tm magnetic dipole field. An ultra-low material budget is required to achieve momentum resolution of the order of Δp/p = 1%. Therefore, the front-end electronics is placed outside the physics aperture. The active volume of the STS is built from 300 μm thick double-sided silicon microstrip sensors mounted onto lightweight carbon fiber support ladders. The sensors will be read out through ultra-thin micro-cables with fast self-triggering electronics at the periphery of the stations where also other infrastructure such as cooling can be placed. In this paper, the development status of the detector system, highlighting the overview of the STS layout, tracking algorithm and performance simulations are presented.

1. The CBM experiment

The Compressed Baryonic Matter Experiment (CBM) [1] is one of the four major experiment that are planned to be performed at the Facility for Anti-proton and Ion Research (FAIR)[2] at Darmstadt, Germany. The CBM experiment aims to explore the phase diagram of nuclear matter [3]. The CBM experiment at FAIR is a fixed target heavy-ion experiment which includes a Micro Vertex Detector (MVD), a Silicon Tracking System (STS), a Ring Imaging Cherenkov Detector (RICH), a Muon Chamber System (MUCH), a Transition Radiation Detector (TRD), a Time-of-Flight detection system (TOF) using resistive plate chambers, a Shashlik type Electromagnetic Calorimeter (ECAL) and a Projectile Spectator Detector (PSD) [4]. The experimental strategy is to perform systematic and multi- differential measurements of all particles produced in nuclear collisions with unprecedented precision and statistics. The CBM detector will identify hadrons and leptons in nuclear collisions with up to 1000 charged particles in CBM acceptance at collision rates up to 10 MHz. The experiment will be optimized in particular for the detection of rare probes, like hadronic decays of D mesons and leptonic decays of light vector mesons, that can yield information on the dense phase of the collisions. The challenge is to accomplish in this environment high-resolution charged particle tracking, momentum measurement and secondary vertex selection with a silicon tracking (STS) and vertex detection system (MVD), the central component of the CBM detector. The system requirements include a very low material budget, radiation tolerant sensors with high spatial resolution, and a fast readout to cope with free streaming data.

Journal of Physics: Conference Series **503** (2014) 012028 doi:10.1088/1742-6596/503/1/012028

2. The Silicon Tracking System

The Silicon Tracking System (STS) is the core of the CBM experiment. It is located in the field of a large aperture dipole magnet. The main task of the STS is to provide track reconstruction and momentum information of the charged particles. The multiplicity of the charged particles is up to 700 per event covering the polar angles $2.5° < \Theta < 25°$.

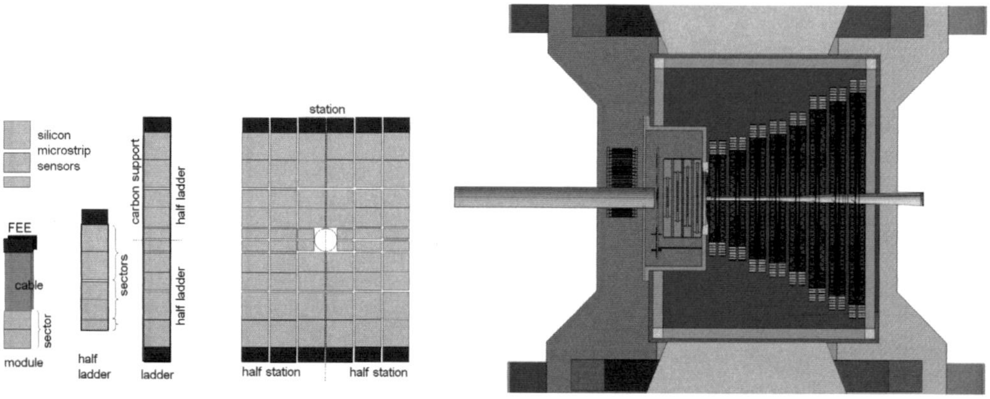

(a) (b)

Figure 1. (a) Conceptual design of STS with ladder structure and modules. (b) Engineering model of STS (Green) + MVD (Orange, see section 5)

The STS will comprise of 8 stations in ladder structure having silicon microstrip detectors [see Fig. 1(a)]. The 8 STS stations will be located downstream from the target at a distance of 30 cm to 100 cm. The required momentum resolution is of the order of 1% and single hit spatial resolution is of the order of 25 μm. This performance requires an ultra low material budget, achieved by micro-thin cables which connect the sensors to the front-end electronics located outside the active area of the detector. The conceptual and engineering design of the stations is shown in Fig.1(a) and Fig.1(b). In total 1220 sensors of three sizes will be connected to more than 14000 read-out ASICs and about 2.1 million channels are to be read. The STS consists of total 106 ladders of 8 different types. Each Ladder consists of different modules, the STS comprises about 900 modules of 25 different types. The active volume of the STS is built employing 300 μm thick double-sided silicon microstrip sensors mounted onto lightweight carbon fiber support ladders and read over ultra-light micro thin cables. STS is designed for hit finder efficiency close to 100% and track reconstruction $\geq$ 95% for momenta $\geq 1GeV/c$. The double sided micro-strip sensors are 300 μm thick and radiation hard upto a fluence of 1×10^{14} n_{eq}/cm^2 compatible to CBM physics program for SIS100/300 at FAIR.

3. Hit reconstruction in STS

Simulations have been performed that include the complete chain of physical processes caused by the charged particles traversing the detector: from charge separation in bulk silicon to digitization as an output. The first step of STS hit reconstruction is performed by the cluster finder algorithms. A cluster is a group of adjacent fired strips in a sensor with a common time stamp. A constant signal threshold is applied for every channel. The total charge of a cluster is defined as the sum of the single strip signals. The cluster position is given by the center-of-gravity method [5]. The cluster size distribution itself is presented in the Fig.2 for a threshold of 4000 electrons set in the read-out electronics. Due to inclination angles of tracks caused by the outwards bending magnetic field the cluster size increases towards the large polar angles.

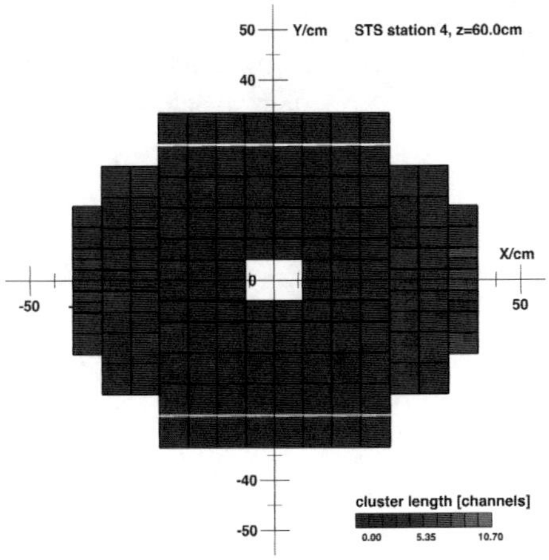

(a)

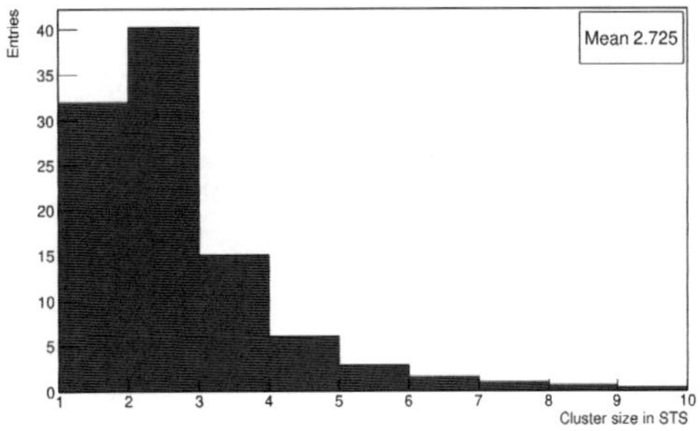

(b)

Figure 2. (a) Cluster size distribution in the microstrip sensors of STS station 4 for a threshold of 4000 electrons applied in the read-out electronics. (b) Distribution of cluster sizes for the entire STS.

4. Track reconstruction

The track finding in the STS detector system which is operated in an inhomogeneous magnetic field is based on the Cellular Automaton (CA) method [6]. Subsequent track and vertex fitting makes use of the Kalman Filter (KF). At first a set of track segments (tracklets) from hits on the neigbouring stations is created using the algorithms. A set of cuts which reflects the geometrical acceptance of tracks in STS (for e.g., forward tracks with mimimum 4 hits and momemta exceeding 100 MeV/c), is applied to create tracks with enough hits to be reconstructed. Then the Kalman Filter based track fitting procedure is used and a χ^2 fit is calculated to combine and reject tracks accordingly. The propagation of the tracks in an inhomogeneous magnetic field is described by a complex formula [7]. An example of UrQMD generated central Au+Au collisons at 25 AGeV projectile energy reconstructed with the STS is visualized in Fig. 3. The reference primary tracks (tracks from the partcle produced close to interaction point)

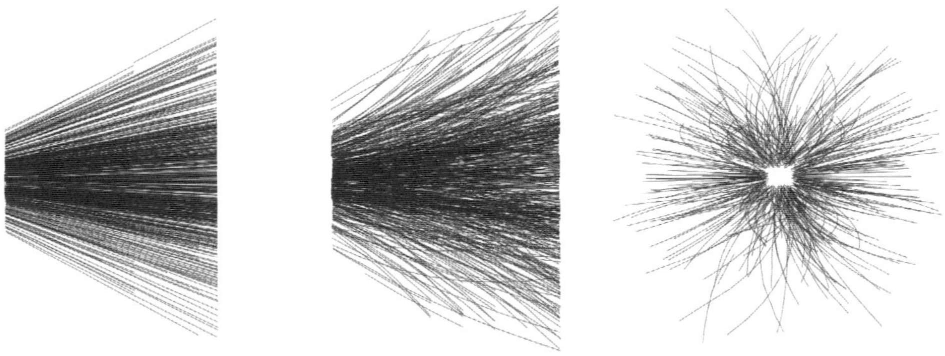

YZ plane XZ plane YX plane

Figure 3. Display of reconstructed tracks from a central Au+Au collision at 25 AGeV projectile energy, shown in three different projections.

Journal of Physics: Conference Series **503** (2014) 012028 doi:10.1088/1742-6596/503/1/012028

are reconstructed with efficiencies up to 96% depending on the particles momentum and the reconstrcution efficiency of the secondary tracks (tracks from the decayed products) reaches up to 90% shown in Fig.4(a). The momentum resolution obtained for primary tracks is shown in Fig.4(b).

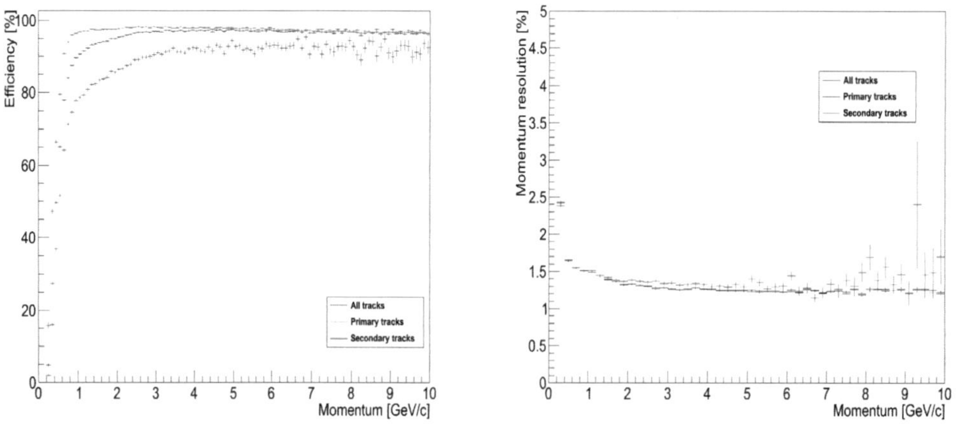

(a) (b)

Figure 4. (a) Track reconstruction efficiency and (b) momentum resolution in the STS as a function of the momentum for all tracks in central Au+Au at 25 AGeV projectile energy.

5. Particle identification

An important task of the STS is to precisely reconstruct weak decay topologies in order to measure strange and multi-strange hyperons. For this purpose, a dedicated software package (KF Particle) based on the Kalman Filter procedure was developed which reconstructs secondary vertices of track pairs or track multiplets with as high precision allowed by the high granularity of the STS. Geometrical (distance of closest approach) and topological (back-pointing of the mother track to the collision vertex) cuts allow to substantially reduce the random combinatorial background from primary tracks. The procedure can be repeated to reconstruct entire decay chains. In Fig. 5 simulated invariant-mass spectra for the decays are shown for $\Lambda \rightarrow p\pi^-$, and $\Omega^- \rightarrow \Lambda K^-$ from 1×10^6 fully reconstructed central Au+Au collisions events at 25 AGeV. In both cases, a clear separation of signal from combinatorial background is obtained; the experiment is even sensitive to the rare Ω^- close to its production threshold. The same measurement technique will be applied for D mesons, but is much more challenging because of their extremely low multiplicities in the order of 10^{-5} per collision and short decay lengths ($c\tau = 312$ μm for $D^{\pm}$ and 123 μm for D^0). The STS detector alone is not capable of resolving displaced vertices on this scale. For the measurement of open charm, it will thus be operated together with a dedicated Micro Vertex Detector (MVD), built from silicon pixel sensors and located between the target and the STS (see Fig. 1(b)). Tracks are first reconstructed in the STS and then extrapolated towards the MVD, where the corresponding high-precision pixel hits are associated. The combined MVD+STS system allows a reconstruction of displaced vertices with a precision of about 50 μm efficiently along the beam direction. This is sufficient to reconstruct the decays $D^0 \rightarrow K^-\pi^+$ and $D^+ \rightarrow K^-\pi^+\pi^+$ and their charge conjugates as shown in Fig. 6. Detailed results concerning the tracking efficiencies and the physics performance concerning the CBM physics case could be found in the STS technical design report [4].

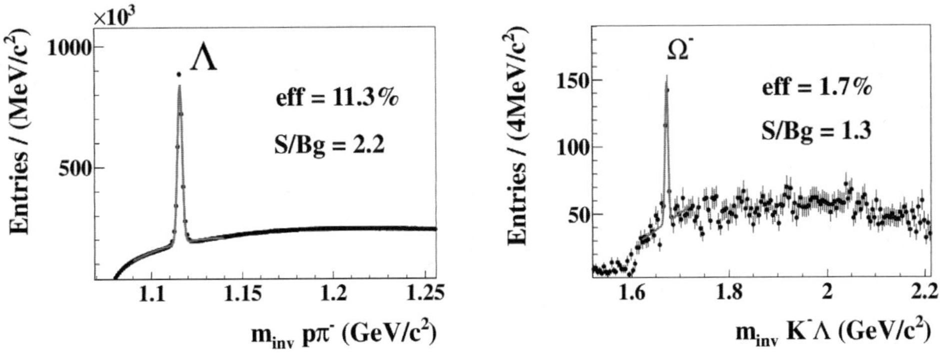

Figure 5. Invariant mass spectra for the hyperon decays: $\Lambda \to p\pi^-$(left)and $\Omega^- \to \Lambda K^-$ (right) central Au+Au collisions at 25 AGeV. p, π^-, K^- were identified by the TOF detector (time resolutionis assumed to be 80 ps). The simulated statistics results to 1×10^6 events at 25 AGeV.

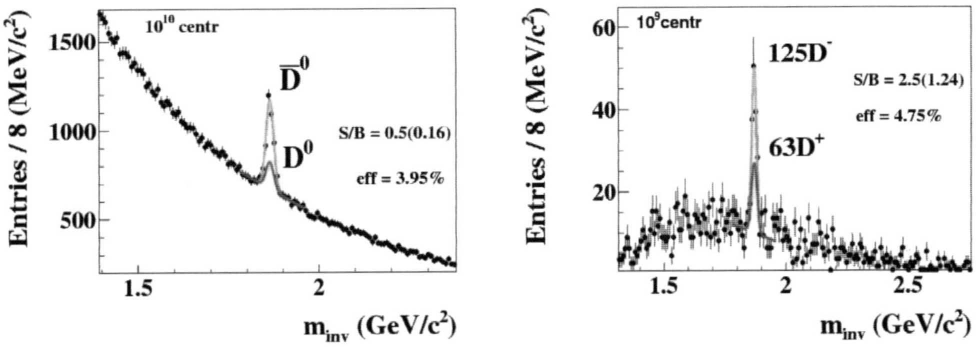

Figure 6. Invariant-mass spectra for the decays $D^0 \to K^-\pi^+$, $D^+ \to K^-\pi^+\pi^+$ and their charge conjugates for 10^{10} and 10^9 central Au+Au collisions at 25 AGeV.

6. Conclusion

The results of the simulations indicate that the STS is capable of handling high multiplicity of tracks in the CBM collision environment. The physics performance studies yield that the STS can reconstruct rare probes with good detection efficiency. The track reconstruction algorithms are in continuos development and we expect even better results in future.

References

[1] CBM Experiment: http://www.fair-center.eu/for-users/experiments/cbm.html
[2] Facility for Anti-proton and Ion Research GmbH, Darmstadt : http://www.fair-center.eu/
[3] Hoehne C, 2013 *J.Phys.* **420** 012016
[4] Heuser J, Senger P *et al.*, 2013 *Technical Design Report for the CBM Silicon Tracking System (STS)* http://repository.gsi.de/record/54798
[5] Bartsch V *et al.*, 2003 *Nucl. Instr, Meth. Phys. Res.* **A497** 389
[6] Kisel I, 2006 *Nucl. Instr, Meth. Phys. Res.* **A566** 85
[7] Gorbunov S and Kisel I, 2005 *CBM-SOFT-NOTE-2005-001* 18

Journal of Physics: Conference Series **503** (2014) 012029 doi:10.1088/1742-6596/503/1/012029

Resonances $f_0(1370)$, $f_0(1500)$ and $f_0(1710)$ within the extended Linear Sigma Model

Stanislaus Janowski and Francesco Giacosa

Institute for Theoretical Physics, Goethe-University, Max-von-Laue-Str. 1, 60438 Frankfurt am Main, Germany

Abstract. In the framework of the extended Linear Sigma Model (eLSM) we investigate the masses and decays of the three scalar-isoscalar resonances $f_0(1370)$, $f_0(1500)$ and $f_0(1710)$. The degrees of freedom of the eLSM are (pseudo)scalars and axial(vectors) as well as the scalar glueball. Although still preliminary, present results, based on the physical masses of the above mentioned resonances, show that $f_0(1710)$ is the predominantly glueball state. However, acceptable decays for this resonance can be obtained only for a (very) large gluon condensate.

1. Introduction

In this work we study the scalar-isoscalar resonances $f_0(1370)$, $f_0(1500)$ and $f_0(1710)$ [1] in the framework of the extended Linear Sigma Model (eLSM) [2, 3, 4, 5]. The eLSM is built accordingly to two fundamental properties of QCD: chiral symmetry and dilatation invariance, the former being spontaneously broken by a Mexican-hat potential, the latter being explicitly broken in such a way to mimic the trace anomaly of QCD [6, 7]. As a consequence of these two requirements, the eLSM Lagrangian contains only a finite number of terms. Moreover, (axial)vector d.o.f. are included from the very beginning in the effective model. The inclusion of the latter in a chiral invariant way makes the model more complete and has a strong influence on the phenomenology.

In Ref. [2] the eLSM has been first studied, both in the baryonic and mesonic sectors, in presence of two-flavours, $N_f = 2$. The inclusion of the dilaton/glueball field for $N_f = 2$ has been carried out in Ref. [3], where it has been shown that the glueball is predominantly contained in either $f_0(1500)$ or $f_0(1710)$ (the former case being favoured). Still, the study was not complete because scalar mesons containing the s-quark were not included. A detailed study of the eLSM for three flavours, $N_f = 3$, has been performed in Ref. [4], where a very good description of various meson masses and decays was achieved. Yet, the dilaton, although formally included to justify the form of the Lagrangian, was inert (i.e., with zero width, in agreement with large-N_c limit). The scalar-isoscalar system was studied with the nonstrange quarkonium field $\sigma_N \equiv (\bar{u}u + \bar{d}d)/\sqrt{2}$ and with the hidden-strange quarkonium field $\sigma_S \equiv \bar{s}s$ only. Ref. [4] shows through a fit to experimental data that these quark-antiquark states are heavy (above 1 GeV); similarly, the scalar isovector and isodoublet states are identified with the heavy resonances $a_0(1450)$ and $K_0^\star(1430)$, respectively. Consequently, the scalar mesons below 1 GeV shall be interpreted differently: tetraquarks or molecular/dynamically generated states are the most prominent options, e.g. Refs. [8, 9] and refs. therein.

In the scalar-isoscalar sector a three-body mixing problem must be solved: the three bare fields σ_N, σ_S, and, in addition, the dilaton $G \equiv gg$ mix and generate the physical resonances $f_0(1370)$, $f_0(1500)$ and $f_0(1710)$. The aim is to determine the mixing matrix in such a way that the masses and the decays of $f_0(1370)$, $f_0(1500)$ and $f_0(1710)$ are in agreement with the experimental results listed in Ref. [1]. Such a mixing problem was investigated in a variety of phenomenological models, see Ref. [10, 11, 12] and refs. therein. However, a full calculation involving a *full* chiral approach has not yet been achieved for $N_f = 3$. In Ref. [5] the first step toward this direction has been performed, but the attention was focused on the masses only (no decays were calculated). Based on the parameter determination obtained in Ref. [4], a correct description of the masses $f_0(1370)$, $f_0(1500)$ and $f_0(1710)$ implies that the predominant glueball content is located in the resonance $f_0(1710)$ (and not in $f_0(1500)$).

In this work we continue the study of this system by calculating, for the first time, the decay widths of the scalar-isoscalar states. It is shown that the mixing matrix of Ref. [5] implies too large decay widths for $f_0(1500)$ and $f_0(1710)$, and must be discarded. Therefore, we search for other solutions. The system is extremely dependent on subtle issues such as constructive and destructive interference effects; for this reason, a fit (which is the most straightforward thing to do) could not yet deliver acceptable results. Anyhow, an interesting partial solution could be obtained by a simple 'guess and try' procedure. In this solution, the gluon condensate is very large and the glueball is to a very good extent described by the resonance $f_0(1710)$ only. The phenomenology of $f_0(1710)$ and $f_0(1370)$ can be qualitatively described, but the kaon-kaon decay of $f_0(1500)$ is still too large. Future studies will show if this novel solution is phenomenologically acceptable or not.

This proceeding is organized as follows: In Sec. 2 we present the effective Lagrangian of the eLSM, in Sec. 3 we discuss the results and in Sec. 4 we provide a summary and an outlook for work in progress.

2. The eLSM Lagrangian

For the purpose of studying the mixing behavior in vacuum of the scalar-isoscalar mesons below 2 GeV we use eLSM Lagrangian developed in Refs. [2, 3, 4]:

$$
\begin{aligned}
\mathcal{L} = {} & \mathcal{L}_{dil} + \mathrm{Tr}[(D^{\mu}\Phi)^{\dagger}(D_{\mu}\Phi)] - m_0^2\left(\frac{G}{G_0}\right)^2 \mathrm{Tr}[\Phi^{\dagger}\Phi] - \lambda_1(\mathrm{Tr}[\Phi^{\dagger}\Phi])^2 - \lambda_2\mathrm{Tr}[(\Phi^{\dagger}\Phi)^2] \\
& + c_1(\det\Phi - \det\Phi^{\dagger})^2 + \mathrm{Tr}[H(\Phi^{\dagger}+\Phi)] + \mathrm{Tr}\left[\left(\frac{m_1^2}{2}\left(\frac{G}{G_0}\right)^2 + \Delta\right)(L^{\mu 2} + R^{\mu 2})\right] \\
& - \frac{1}{4}\mathrm{Tr}\left[L^{\mu\nu 2} + R^{\mu\nu 2}\right] + \frac{h_1}{2}\mathrm{Tr}[\Phi^{\dagger}\Phi]\mathrm{Tr}[L_{\mu}L^{\mu} + R_{\mu}R^{\mu}] + h_2\mathrm{Tr}[\Phi^{\dagger}L_{\mu}L^{\mu}\Phi + \Phi R_{\mu}R^{\mu}\Phi^{\dagger}] \\
& + 2h_3\mathrm{Tr}[\Phi R_{\mu}\Phi^{\dagger}L^{\mu}] + \ldots\ , \qquad (1)
\end{aligned}
$$

where $D^{\mu}\Phi = \partial^{\mu}\Phi - ig_1(L^{\mu}\Phi - \Phi R^{\mu})$. The nonstrange $\sigma_N \equiv (\bar{u}u + \bar{d}d)/\sqrt{2}$ and the strange $\sigma_S \equiv \bar{s}s$ bare quark-antiquark mesons are contained in the (pseudo)scalar multiplet

$$
\Phi = \frac{1}{\sqrt{2}}\begin{pmatrix} \frac{(\sigma_N + a_0^0) + i(\eta_N + \pi^0)}{\sqrt{2}} & a_0^+ + i\pi^+ & K_0^{\star +} + iK^+ \\ a_0^- + i\pi^- & \frac{(\sigma_N - a_0^0) + i(\eta_N - \pi^0)}{\sqrt{2}} & K_0^{\star 0} + iK^0 \\ K_0^{\star -} + iK^- & \bar{K}_0^{\star 0} + i\bar{K}^0 & \sigma_S + i\eta_S \end{pmatrix}. \qquad (2)
$$

The matrices L_{μ} and R_{μ} describe (axial)vector fields, see Ref. [4] for details. The scalar glueball $G \equiv gg$ is described by the dilaton Lagrangian $\mathcal{L}_{dil}$, in which a logarithmic term with the energy scale Λ mimics the trace anomaly of the pure Yang-Mills sector of QCD [6, 7]:

$$\mathcal{L}_{dil} = \frac{1}{2}(\partial_\mu G)^2 - \frac{1}{4}\frac{m_G^2}{\Lambda^2}\left(G^4 \ln\left|\frac{G}{\Lambda}\right| - \frac{G^4}{4}\right) . \tag{3}$$

The three scalar fields σ_N, σ_S, and G condense, leading to the following shifts: $\sigma_N \to \sigma_N + \phi_N$, $\sigma_S \to \sigma_S + \phi_S$, and $G \to G + G_0$. As a consequence, bilinear mixing terms $\sim \sigma_N\sigma_S$, $G\sigma_N$ and $G\sigma_S$ arise. The corresponding potential reads:

$$V(\sigma_N, G, \sigma_S) = \frac{1}{2}(\sigma_N, G, \sigma_S)\begin{pmatrix} m_{\sigma_N}^2 & z_{G\sigma_N} & z_{\sigma_S\sigma_N} \\ z_{G\sigma_N} & M_G^2 & z_{G\sigma_S} \\ z_{\sigma_S\sigma_N} & z_{G\sigma_S} & m_{\sigma_S}^2 \end{pmatrix}\begin{pmatrix} \sigma_N \\ G \\ \sigma_S \end{pmatrix} \tag{4}$$

where $z_{G\sigma_N} = -2m_0^2\phi_N/G_0$, $z_{G\sigma_S} = -2m_0^2\phi_S/G_0$ and $z_{\sigma_N\sigma_S} = 2\lambda_1\phi_N\phi_S$. The physical states are obtained upon diagonalization:

$$\begin{pmatrix} f_0(1370) \\ f_0(1500) \\ f_0(1710) \end{pmatrix} = B \begin{pmatrix} \sigma_N \equiv \left(\bar{u}u + \bar{d}d\right)/\sqrt{2} \\ G \equiv gg \\ \sigma_S \equiv \bar{s}s \end{pmatrix} . \tag{5}$$

The aim is to determine the mixing matrix B.

3. Results

We use the parameters determined in Ref. [4]. These parameters allow for a correct description of a variety of vacuum properties of mesons up to 1.5 GeV. However, four parameters could not be uniquely determined in Ref. [4]: these are m_G, Λ, λ_1 and h_1. In Ref. [5] three of them (m_G, Λ and λ_1) were determined in order to obtain the measured experimental masses of the resonances $f_0(1370)$, $f_0(1500)$ and $f_0(1710)$: $M_{f_0(1370)} =$ (1200-1500) MeV, $M_{f_0(1500)} = (1505 \pm 6)$ MeV and $M_{f_0(1710)} = (1720 \pm 6)$ MeV [1]. A solution in which $f_0(1500)$ is predominantly gluonic could not be found. On the contrary, the masses could be well described for $m_G = 1.580$ GeV, $\Lambda = 0.93$ GeV, and $\lambda_1 = 2.03$. The resulting mixing matrix reads:

$$B = \begin{pmatrix} 0.92 & -0.39 & 0.05 \\ -0.22 & -0.40 & 0.89 \\ -0.33 & -0.83 & -0.45 \end{pmatrix} . \tag{6}$$

We have now tested this scenario by evaluating the decay widths (the corresponding mathematical expressions will be presented in a forthcoming publication [13]). For $h_1 = 0$ (large-N_c limit), one finds the decay widths into kaons and pions: $\Gamma_{f_0(1710)\to\pi\pi} = 0.83$ GeV, $\Gamma_{f_0(1710)\to KK} = 0.42$ GeV, $\Gamma_{f_0(1500)\to\pi\pi} = 0.22$ GeV, $\Gamma_{f_0(1500)\to KK} = 1.14$ GeV, $\Gamma_{f_0(1370)\to\pi\pi} = 1.78$ GeV, $\Gamma_{f_0(1370)\to KK} = 0.89$ GeV. These results are clearly *too large* and cannot be cured by varying the only remaining free parameter h_1 (which should anyhow be small). Thus, the decay widths do not support this scenario as physical. Note, such a large decay width of the predominantly glueball state is in agreement with the study of Ref. [14].

The search for an acceptable solution is extremely difficult due to interference effects in the decay amplitudes. A direct fit to the known decay widths was so far not successful. This is why we have searched a solutions by trying to guess the right area of the parameter space. First, we use as an input the bare glueball mass $m_G = 1.7$ GeV in agreement with lattice QCD [15]. Then, due to the fact that $f_0(1710)$ was too broad in the previous solution, we have increased the value of the dimensionful parameter Λ: for $\Lambda \simeq 2$ GeV the resonance $f_0(1710)$ is sufficiently narrow. By further tuning $\lambda_1 \simeq -10$ and $h_1 \simeq -5$, we obtain the mixing matrix

$$B = \begin{pmatrix} 0.90 & -0.05 & 0.41 \\ -0.42 & -0.03 & 0.90 \\ -0.04 & -0.99 & -0.05 \end{pmatrix} . \tag{7}$$

The resonance $f_0(1710)$ is (almost) a pure glueball. The masses that are determined by these parameters are acceptable: $M_{f_0(1370)} = 1.06$ GeV, $M_{f_0(1500)} = 1.48$ MeV and $M_{f_0(1710)} = 1.70$ GeV. The resulting decay widths are: $\Gamma_{f_0(1710)\to\pi\pi} = 0.082$ GeV, $\Gamma_{f_0(1710)\to KK} = 0.064$ GeV, $\Gamma_{f_0(1500)\to\pi\pi} = 0.14$ GeV, $\Gamma_{f_0(1500)\to KK} = 0.13$ GeV, $\Gamma_{f_0(1370)\to\pi\pi} = 0.12$ GeV, $\Gamma_{f_0(1370)\to KK} = 0.07$ GeV.

Thus, while the decays of $f_0(1370)$ and $f_0(1710)$ are at least in qualitative agreement with the experiment, this is not the case for $f_0(1500)$: the decays are still too large. Note also that the very large value of Λ implies a large gluon condensate: lattice results [16] suggest that $\Lambda \lesssim 0.6$ GeV, see the discussion in Ref. [3]. Thus, at this level this solution is not yet conclusive, but can point to an interesting direction where to look for it: large bare glueball mass in agreement with lattice (1.7 GeV) and a large value of the gluon condensate.

4. Conclusions and outlook

In this work we have studied the masses and the decays of the resonances $f_0(1370)$, $f_0(1500)$ and $f_0(1710)$ within the eLSM. Presently, the favoured glueball seems to be $f_0(1710)$, in agreement with Refs. [12, 17], but further studies are needed. Namely, decay widths which are -at least qualitatively- in agreement with data could only be found for a large (eventually too large!) value of the gluon condensate.

Another possibility is an improvement of the underlying effective model of Ref. [4], by studying the influence of a quadratic mass term in the (pseudo)scalar sector. This is a minimal change of Ref. [4], which however can have interesting phenomenological implications due to the fact that the strange current quark mass is not negligible. For value of the gluon condensate in agreement with lattice, a not too broad glueball can only be found if destructive interferences between the different amplitudes occur. This is why an improved numerical analysis, which allows to study in detail the whole parameter space, would be also helpful.

More in general, one can extend the study of glueballs with other quantum numbers. In Ref. [18] the pseudoscalar glueball was investigated within the eLSM. A similar program can be carried out for a tensor glueball with a mass of about 2.2 GeV, see e.g. Ref. [19] and refs. therein, as well as heavier glueballs, such as the vector and pseudovector glueball states, with lattice predicted masses of about 3.8 and 3 GeV, respectively.

Acknowledgments

The authors thank I. Mishustin, D. Parganlija, and D. H. Rischke for useful discussions. S. J. acknowledges support from H-QM and HGS-HIRe.

References

[1] J. Beringer *et al.* (Particle Data Group), Phys. Rev. D **86**, 010001 (2012).
[2] S. Gallas, F. Giacosa and D. H. Rischke, Phys. Rev. D **82**, 014004 (2010) [arXiv:0907.5084 [hep-ph]]; D. Parganlija, F. Giacosa and D. H. Rischke, Phys. Rev. D **82**, 054024 (2010) [arXiv:1003.4934 [hep-ph]].
[3] S. Janowski, D. Parganlija, F. Giacosa and D. H. Rischke, Phys. Rev. D **84**, 054007 (2011) [arXiv:1103.3238 [hep-ph]]; S. Janowski, Acta Phys. Polon. Supp. **5**, 1071 (2012) [arXiv:1208.5397 [hep-ph]].
[4] D. Parganlija, P. Kovacs, G. Wolf, F. Giacosa and D. H. Rischke, Phys. Rev. D **87**, 014011 (2013) arXiv:1208.0585 [hep-ph]].
[5] S. Janowski, Acta Phys. Polon. Supp. **6**, 899 (2013) [arXiv:1306.3155 [hep-ph].
[6] C. Rosenzweig, A. Salomone and J. Schechter, Phys. Rev. D **24**, 2545 (1981); A. Salomone, J. Schechter and T. Tudron, Phys. Rev. D **23**, 1143 (1981); C. Rosenzweig, A. Salomone and J. Schechter, Nucl. Phys. B

206, 12 (1982) [Erratum-ibid. B **207**, 546 (1982)]; H. Gomm and J. Schechter, Phys. Lett. B **158**, 449 (1985); R. Gomm, P. Jain, R. Johnson and J. Schechter, Phys. Rev. D **33**, 801 (1986).

[7] A. A. Migdal and M. A. Shifman, Phys. Lett. B **114**, 445 (1982).

[8] R. L. Jaffe, Phys. Rev. D **15**, 267 (1977); R. L. Jaffe, Phys. Rev. D **15**, 281 (1977); L. Maiani, F. Piccinini, A. D. Polosa and V. Riquer, Phys. Rev. Lett. **93**, 212002 (2004) [arXiv:hep-ph/0407017]; F. Giacosa, Phys. Rev. D **74**, 014028 (2006) [arXiv:hep-ph/0605191]; A. H. Fariborz, R. Jora and J. Schechter, Phys. Rev. D **72**, 034001 (2005) [arXiv:hep-ph/0506170]; M. Napsuciale and S. Rodriguez, Phys. Rev. D **70**, 094043 (2004); F. Giacosa, G. Pagliara, mesons, Nucl. Phys. A **833**, 138-155 (2010) [arXiv:0905.3706 [hep-ph]]; F. Giacosa, Phys. Rev. D **75**, 054007 (2007) [arXiv:hep-ph/0611388].

[9] E. van Beveren, T. A. Rijken, K. Metzger, C. Dullemond, G. Rupp and J. E. Ribeiro, Z. Phys. C **30**, 615 (1986) [arXiv:0710.4067 [hep-ph]]; N. A. Tornqvist, Z. Phys. C **68**, 647 (1995) [arXiv:hep-ph/9504372]; M. Boglione and M. R. Pennington, Phys. Rev. D **65**, 114010 (2002) [arXiv:hep-ph/0203149]; E. van Beveren, D. V. Bugg, F. Kleefeld and G. Rupp, Phys. Lett. B **641**, 265 (2006) [arXiv:hep-ph/0606022]; J. R. Pelaez, Phys. Rev. Lett. **92**, 102001 (2004) [arXiv:hep-ph/0309292]; J. A. Oller and E. Oset, Nucl. Phys. A **620**, 438 (1997) [Erratum-ibid. A **652**, 407 (1999)] [arXiv:hep-ph/9702314].

[10] C. Amsler and N. A. Tornqvist, Phys. Rept. **389**, 61 (2004); E. Klempt and A. Zaitsev, Phys. Rept. **454**, 1 (2007) [arXiv:0708.4016 [hep-ph]]; F. Giacosa, Phys. Rev. D **80** , 074028 (2009) [arXiv:0903.4481 [hep-ph]].

[11] C. Amsler and F. E. Close, Phys. Rev. D **53**, 295 (1996) [arXiv:hep-ph/9507326].

[12] W. J. Lee and D. Weingarten, Phys. Rev. D **61**, 014015 (2000) [arXiv:hep-lat/9910008]; F. Giacosa, T. Gutsche, V. E. Lyubovitskij and A. Faessler, Phys. Rev. D **72**, 094006 (2005) [arXiv:hep-ph/0509247]; H. Y. Cheng, C. K. Chua and K. F. Liu, Phys. Rev. D **74**, 094005 (2006) [arXiv:hep-ph/0607206].

[13] S. Janowski, F. Giacosa and D. H. Rischke, in preparation.

[14] J. R. Ellis and J. Lanik, Phys. Lett. B **150** (1985) 289.

[15] C. Morningstar and M. J. Peardon, AIP Conf. Proc. **688**, 220 (2004) [arXiv:nucl-th/0309068]; C. J. Morningstar and M. J. Peardon, Phys. Rev. D **60** (1999) 034509 [hep-lat/9901004]; M. Loan, X. Q. Luo and Z. H. Luo, Int. J. Mod. Phys. A **21**, 2905 (2006) [arXiv:hep-lat/0503038]; Y. Chen *et al.*, Phys. Rev. D **73**, 014516 (2006) [arXiv:hep-lat/0510074]; E. Gregory, A. Irving, B. Lucini, C. McNeile, A. Rago, C. Richards and E. Rinaldi, arXiv:1208.1858 [hep-lat].

[16] J. Kripfganz, Phys. Lett. B **101**, 169 (1981); A. Di Giacomo and G. C. Rossi, Phys. Lett. B **100**, 481 (1981); A. Di Giacomo and G. Paffuti, Phys. Lett. B **108**, 327 (1982); E. M. Ilgenfritz and M. Muller-Preussker, Phys. Lett. B **119**, 395 (1982); S. s. Xue, Phys. Lett. B **191**, 147 (1987); M. Campostrini, A. Di Giacomo and Y. Gunduc, Phys. Lett. B **225**, 393 (1989); A. Di Giacomo, H. Panagopoulos and E. Vicari, Nucl. Phys. B **338**, 294 (1990); X. D. Ji, arXiv:hep-ph/9506413; G. Boyd and D. E. Miller, arXiv:hep-ph/9608482.

[17] L. -C. Gui, Y. Chen, G. Li, C. Liu, Y. -B. Liu, J. -P. Ma, Y. -B. Yang and J. -B. Zhang, Phys. Rev. Lett. **110** (2013) 021601 [arXiv:1206.0125 [hep-lat]].

[18] W. I. Eshraim, S. Janowski, F. Giacosa and D. H. Rischke, Phys. Rev. D **87**, 054036 (2013) [arXiv:1208.6474 [hep-ph]]; W. I. Eshraim, S. Janowski, A. Peters, K. Neuschwander and F. Giacosa, Acta Phys. Polon. Supp. **5**, 1101 (2012) [arXiv:1209.3976 [hep-ph]]; W. I. Eshraim and S. Janowski, J. Phys. Conf. Ser. **426**, 012018 (2013) [arXiv:1211.7323 [hep-ph]]; W. I. Eshraim and S. Janowski, PoS ConfinementX , 118 (2012) [arXiv:1301.3345 [hep-ph]].

[19] F. Giacosa, T. Gutsche, V. E. Lyubovitskij, A. Faessler, Phys. Rev. D **72**, 114021 (2005) [hep-ph/0511171]; L. Burakovsky and J. T. Goldman, Phys. Rev. D **57**, 2879 (1998) [hep-ph/9703271].

Journal of Physics: Conference Series **503** (2014) 012030 doi:10.1088/1742-6596/503/1/012030

Experimental overview of the $\bar{\text{P}}$ANDA experiment

Elisa Fioravanti for the $\bar{\text{P}}$ANDA Collaboration
INFN Ferrara, Via Saragat 1, 44100 Ferrara, Italy
E-mail: fioravanti@fe.infn.it

Abstract. The physics program of the $\bar{\text{P}}$ANDA (anti-Proton ANhiliation ar DArmstadt) experiment will address various questions related to the strong interactions by employing a multi-purpose detector system at the High Energy Storage Ring (HESR) for anti-protons of the upcoming Facility for Anti-proton and Ion Research (FAIR). The excellent antiproton beam resolution of $\Delta p/p \sim 10^{-5}$ and the high luminosity $\mathcal{L} = 2\times10^{32}\text{cm}^{-2}\text{s}^{-1}$ will allow the precise measurement of the charmonium and open charm spectroscopy, the search for exotic hadrons like multiquarks, glueballs and hybrids, the study of in-medium modifications of hadrons and the nucleon structure.

1. The FAIR facility

The recently approved FAIR facility, which will be built as a major upgrade of the existing GSI laboratory in Germany, will provide antiproton beams of the highest quality in terms of intensity and resolution. The FAIR facility consists of two synchrotron rings, called SIS100 and SIS300, housed in the same tunnel, which will provide proton and ion beams. The SIS100, a 100 T·m proton ring, will feed the radioactive ion and antiproton beam lines for experiments to be carried out in the HESR, the Collector and Cooler rings (CR) and the New Experimental Storage Ring (NESR). The SIS300 will deliver high energy ion beams for the study of ultra relativistic heavy ion collisions.

1.1. The High-Energy Storage Ring

The antiproton beam will be produced by a primary proton beam from the SIS100. The $\bar{p}$ production rate will be 2×10^7/s. After 5×10^5 $\bar{p}$ have been produced they will be transferred to the HESR, where the $\bar{\text{P}}$ANDA experiment, in the $\bar{p}$ momentum range from 1 GeV/c to 15 GeV/c can be performed. Two modes of operation are foreseen: the high intensity mode, where with a beam momentum spread $\Delta p/p$=10^{-4} a luminosity of 2×10^{32} $\text{cm}^{-1}\text{s}^{-1}$ will be available, and the high resolution mode, where the luminosity requirement will be released to 10^{31} $\text{cm}^{-2}\text{s}^{-1}$ to have a maximum momentum precision of 10^{-5}.

2. The $\bar{\text{P}}$ANDA Physics Program

$\bar{\text{P}}$ANDA will use the antiproton beam[1] at the HESR with momentum range between 1.5 GeV/c and 15 GeV/c, corresponding to total center-of-mass energies in the antiproton-proton system between 2.25 GeV and 5.5 GeV. The $\bar{\text{P}}$ANDA experiment aims at exploring hadronic matter by means of the gluon rich environment of the $\bar{p}p$ annhilation which allows to access a wide range of

[1] The antiprotons are created by colliding high intensity protons with a copper target

final states. The 4π acceptance of the detector for both charged and neutral particles, together with the envisaged high quality of the antiproton beam, will create an ideal environment to collect high statistics data to address many open problems related with the strong interaction. $\bar{\text{P}}$ANDA will perform a complete program of hadron spectroscopy to test many unclear aspects of Quantum Chromo Dynamics (QCD) (Fig. 1). The aim is to investigate both the dynamics of the interaction, and the characteristics of new forms of matter such as exotic states in the charm energy range. Furthermore, particles properties, when produced inside the nuclear medium, and the structure of hadrons, will be investigated.
In this paper we will discuss the following aspects of the $\bar{\text{P}}$ANDA physics program [1]: the charmonium spectroscopy, the search for gluonic excitations (hybrids and glueballs), the study of hadrons in nuclear matter, the open charm spectroscopy and the hypernuclear physics.

2.1. Charmonium spectroscopy
Experimentally charmonium has been studied in e^+e^- and $\bar{p}p$ experiments. In e^+e^- annihilations direct charmonium formation is possible only for states with the quantum numbers of the photon $J^{PC} = 1^{--}$. Precise measurements of the masses and widths of these states can be obtained from the energy of the electron and positron beams, which are known with good accuracy. All other states can be reached by means of other production mechanisms. On the other hand, all $c\bar{c}$ states can be directly formed in $\bar{p}p$ annihilations, through the coherent annihilation of the three quarks in the proton with the three antiquarks in the antiproton. With this method the masses and widths of all charmonium states can be measured with excellent accuracy, determined by the very precise knowledge of the initial $\bar{p}p$ state and not limited by the resolution of the detector.
The spectrum of charmonium states consists of eight narrow states below the open charm threshold (3.73 GeV) and several tens of states above the threshold. All eight states below $D\bar{D}$ threshold are well established, but whereas the triplet states are measured with very good accuracy, the same cannot be said for the singlet states. For instance, the η_c ws discovered almost thirty years ago but the mass and width measurements are far from satisfactory [2]. An additional topic is the study of the $\eta_c(2S)$, its properties need to be measured with good accuracy. Another open question is the spin-dependence of the $q\bar{q}$ confinement potential. With this aim, a precise measurement of the mass, width and decay channels of the singlet-P resonance, h_c, is of extreme importance.
The region above threshold is rich in interesting new physics. A lot of new states have recently been discovered at the B-factories, mainly in the hadronic decays of the meson: these new states (X, Y, Z) are associated with charmonium because they decay predominantly into charmonium states but their interpretation is far from obvious. For instance one of the most established among the XYZ states is the narrow X(3872) discovered by Belle [3]. For the width of this state only an upper limit has been established, and $\bar{\text{P}}$ANDA will be ideal to perform this measurement. An additional state is the Y(4260) discovered by BaBar [4], for which the interpretation is not so obvious.
At full luminosity $\bar{\text{P}}$ANDA will be able to collect several thousand $c\bar{c}$ states per day. By means of fine scans it will be possible to measure masses with accuracies of the order of 100 KeV and widths to 10% or better. The entire energy region below and above open charm threshold will be explored.

2.2. Gluonic excitations
An important item in the $\bar{\text{P}}$ANDA physics program is the search for gluonic excitations, i.e. hadrons in which the gluons can act as principal components. The gluonic hadrons fall into two categories: glueballs, states of pure glue, and hybrids, which consist of $q\bar{q}$ pair and excited glue. The additional degrees of freedom carried by gluons allow these hybrids and glueballs to have

J^{PC} exotic quantum numbers: in this case mixing effects with nearby $q\bar{q}$ states are excluded and this makes their experimental identification easier.
Antiproton-proton annihilations provide a very favorable environment in which look for gluonic hadrons. Two particles, first seen in πN scattering [5] with exotic quantum numbers $J^{PC} = 1^{-+}$, $\pi_1(1400)$ [6] and $\pi_1(1600)$ [7], are clearly seen in $\bar{p}p$ annihilation at rest. On the other hand a narrow state at 1500 MeV/c^2 discovered in $\bar{p}p$ annihilations by the Crystal Barrel experiment [8], is considered the best candidate for the glueball ground state ($J^{PC} = 0^{++}$), even though the mixing with nearby $q\bar{q}$ states makes this interpretation difficult. So far the experimental search for glueballs and hybrids has been mainly carried out in the mass region below 2.2 GeV/c^2. $\bar{\text{P}}$ANDA will extend the search to higher masses, and in particular to the charmonium mass region, where light quark states form a structure-less continuum and heavy quark states are far fewer in number. Therefore exotic hadrons in this mass region could be resolved and identified unambiguously.

2.3. Hadrons in nuclear matter

The study of medium modifications of hadrons embedded in hadronic matter is aimed at understanding the origin of hadron masses in the context of spontaneous chiral symmetry breaking in QCD and its partial restoration in a hadronic environment. So far experiments have been focused on the light quark sector: evidence of mass changes for pions and kaons have been deduced by the study of deep bound pionic atoms [9] and of K meson production in proton-nucleus and heavy-ion collisions [10].
The high-intensity $\bar{p}$ beam of up to 15 GeV/c will allow an extension of this program to the charm sector both for hadrons with hidden and open charm. The in-medium masses of these states are expected to be affected primarily by the gluon condensate. Recent theoretical calculations predict small mass shifts (5-10 MeV/c^2) for the low-lying charmonium states [11] and more consistent effects for the χ_{cJ} (40 MeV/c^2), ψ' (100 MeV/c^2) and $\psi(3770)$ (140 MeV/c^2) [12]. D mesons, on the other hand, offer the unique opportunity to study the in-medium dynamics of a system with a single light quark. Recent theoretical calculations agree in the prediction of a mass splitting for D mesons in nuclear matter but, unfortunately, they disagree in sign and size of the effect.
Another study which can be carried out in $\bar{\text{P}}$ANDA is the measurement of J/ψ and D meson production cross sections in $\bar{p}$ annihilation on a series of nuclear targets. The comparison of the resonant J/ψ yield obtained from $\bar{p}$ annihilation on protons and different nuclear targets allows to deduce the J/ψ-nucleus dissociation cross section, a fundamental parameter to understand J/ψ suppression in relativistic heavy ion collisions interpreted as a signal for quark-gluon plasma formation.

2.4. Open charm physics

The HESR running at full luminosity and at $\bar{p}$ momenta larger than 6.4 GeV/c would produce a large number of D meson pairs. The high yield (100 charm pairs per second around the $\psi(4040)$) and the well defined production kinematics of D meson pairs, would allow to carry out a significant charmed meson spectroscopy program which would include, for example, the rich D and D_s meson spectra.
The B-factory experiments have discovered several new resonances in the D and D_s sectors, where two of them are extremely narrow: the $D^*_{sJ}(2317)$ [13] and the $D_{sJ}(2457)$ [14]. These new states appear at unexpected locations, since their masses are more than 140 MeV/c^2 lower than expected from potential models. This has given rise to speculations about their nature. It is important to verify these findings by means of new measurements. Threshold pair production can be employed for precision measurements of the mass and the width of the narrow excited D states.

Journal of Physics: Conference Series **503** (2014) 012030 doi:10.1088/1742-6596/503/1/012030

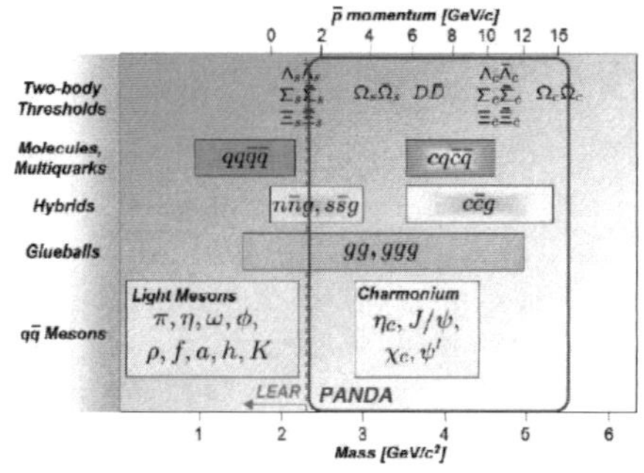

Figure 1. Mass range of hadrons accessible at $\bar{\text{P}}$ANDA.

2.5. Hypernuclei

The hypernuclei are systems where one (or more) nucleon is replaced by one (or more) hyperon(s). They are promising laboratories to study the hyperon-nucleon and hyperon-hyperon interactions. In 50 years of study very few double hypernuclei are presently known. With a dedicate setup of $\bar{\text{P}}$ANDA and its antiproton beam, a copious production of double Λ-hypernuclei is expected to be observed, providing a precision investigation of the $\Lambda - \Lambda$ interaction. For this purpose, antiprotons at a moderate momentum of 3 GeV/c will interact with a primary target to produce large numbers of $\Xi\bar{\Xi}$ pairs. The $\bar{\Xi}$ decay provides a unique signature for the production. The corresponding Ξ particle will be stopped and captured in a secondary target. In case the Ξ is absorbed inside the nucleus, it will yield a Λ pair with very small relative energy. A double hypernucleus can be formed as well and its decay could be observed with spectroscopic precision by using Germanium detectors.

3. The $\bar{\text{P}}$ANDA Detector

In order to accomplish the physics program discussed above, the $\bar{\text{P}}$ANDA detector must fulfill a number of requirements: it must provide full solid angle coverage, it must be able to handle high rates with good particle identification and momentum resolution for γ, e, μ, π, K and p. Additional requirements are vertex reconstruction capability, efficient lepton identification and excellent calorimetry, both in terms of resolution and of sensitivity to low energy showers. The schematic view of the detector is shown in Fig. 2. The antiprotons circulating in the HESR hit an internal hydrogen target (either pellet or cluster jet). The apparatus consists of a central detector, called Target Spectrometer (TS) and a Forward Spectrometer (FS). The TS, for the measurement of particles emitted at laboratory angles larger than 5°, will be located inside a solenoidal magnet which provides a field of 2 T. Its main components will be a micro vertex silicon detector, a central tracker (straw tube detector), an inner time-of-flight telescope, a cylindrical DIRC (Detector of Internally Reflected Light) and electromagnetic calorimeter, a set of muon counters and of multiwire drift chambers. The FS will detect particles emitted at polar angles below 10° in the horizontal and 5° in the vertical direction. It will consist of a 2 T·m dipole magnet, with tracking detectors, Cherenkov and time-of-flight detectors, electromagnetic and hadron calorimeters.

4. Conclusions

The availability of high-intensity, cooled antiproton beams at FAIR will make it possible to perform a very rich experimental program. The $\bar{\text{P}}$ANDA experiment will perform high-precision hadron spectroscopy from $\sqrt{s}$ =2.25 GeV to $\sqrt{s}$ =5.5 GeV and produce a wealth of new results:

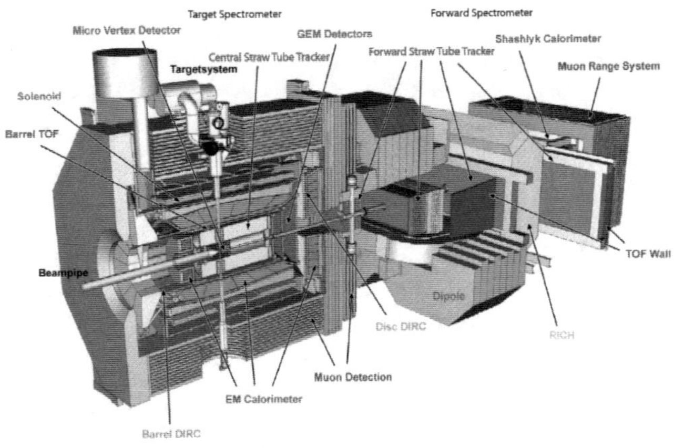

Figure 2. Schematic view of the $\bar{\text{P}}$ANDA detector.

- precision measurement of the parameters of all charmonium states, both below and above open charm threshold, with the possible discovery of the missing states which will lead to a full understanding of the charmonium spectrum;
- the observation/discovery of glueballs and hybrids, particularly in the mass range between 3 and 5 GeV/c^2, yielding new insights into the structure of the QCD vacuum;
- the measurement of mass shifts of charmonium and open charm mesons in nuclear matter, related to the partial restoration of QCD chiral symmetry in a dense nuclear medium;
- open charm spectroscopy (D and D_s spectra);
- the hypernuclear physics.

The $\bar{\text{P}}$ANDA physics program includes other topics, which are not explained in this paper, for instance, the study of the structure of the nucleon using electromagnetic processes and electroweak physics (CP-violation).
All these new measurements will make it possible to achieve a very significant progress in our understanding of QCD and the strong interaction.

References

[1] PANDA Physics Performance Report, arXiv:0903.3905
[2] J. Beringer *et al.*, (Particle Data Group), Phys. Rev. D**86**, 010001 (2012)
[3] S. C. Choi *et al.*, Phys. Rev. Lett. **91**, 262001 (2003)
[4] B. Aubert *et al.*, Phys. Rev. Lett **95**, 142001 (2005)
[5] D. Thompson *et al.*, Phys. Rev. Lett **79**, 1630 (1997); G. Adams *et al.*, Phys. Rev. Lett **81**, 5760 (1998)
[6] A. Abele *et al.*, Phys. Lett **B 423**, 175 (1998)
[7] J. Reinnarth *et al.*, Nucl. Phys. **A 692**, 268c (2001)
[8] C. Amsler *et al.*, Phys. Lett **B 355**, 425 (1995)
[9] H. Geissel *et al.*, Phys. Rev. Lett **88**, 122301 (2002), H. Geissel *et al.*, Phys. Lett **B 549**, 64 (2002); H. Geissel *et al.*, Phys. Rev. Lett **92**, 072302 (2002)
[10] M. Nekipelov *et al.*, Phys. Lett **B 540**, 207 (2002); Z. Rudy *et al.*, Eur. Phys. J. **A 15**, 303 (2002); F. Laue *et al.*, Phys. Rev. Lett **82**, 1640 (1999); P. Crochet *et al.*, Phys. Lett **B 486**, 6 (2000); K. Wisniewski *et al.*, Eur. Phys. J. **A 9**, 515 (2000)
[11] W. Klingl *et al.*, Nucl. Phys. **A 624**, 527 (1997)
[12] S. Lee and C. Ko, Phys. Rev. **C 67**, 038202 (2003); S. Lee, Proceedings of Int. Workshop "Hadron 2003", edited by H.K.E. Klempt and H.Orth., Aschaffenburg, 2004.
[13] B. Aubert *et al.*, Phys. Rev. Lett **90**, 242001 (2003)
[14] P. Krokovny *et al.*, Phys. Rev. Lett **91**, 262002 (2003)

Investigation of light and heavy tetraquark candidates using lattice QCD

Marc Wagner[1], Abdou Abdel-Rehim[2], Constantia Alexandrou[2,3], Mattia Dalla Brida[4], Mario Gravina[3], Giannis Koutsou[2], Luigi Scorzato[5], Carsten Urbach[6]

[1] Goethe-Universität Frankfurt am Main, Institut für Theoretische Physik, Max-von-Laue-Straße 1, D-60438 Frankfurt am Main, Germany
[2] Computation-based Science and Technology Research Center, Cyprus Institute, 20 Kavafi Street, Nicosia 2121, Cyprus
[3] Department of Physics, University of Cyprus, P.O. Box 20537, 1678 Nicosia, Cyprus
[4] School of Mathematics, Trinity College Dublin, Dublin 2, Ireland
[5] INFN – gruppo collegato di Trento, via sommarive 14, 38123 – Trento, Italy
[6] Helmholtz-Institut für Strahlen- und Kernphysik (Theorie) and Bethe Center for Theoretical Physics, Universität Bonn, D-53115 Bonn, Germany

E-mail: `mwagner@th.physik.uni-frankfurt.de`

Abstract. We review the status of an ongoing long-term lattice investigation of the spectrum and structure of tetraquark candidates. We focus on the light scalar meson $a_0(980)$. First steps regarding the study of a possibly existing $cc\bar{c}\bar{c}$ tetraquark are also outlined.

1. Introduction

The nonet of light scalar mesons formed by $\sigma \equiv f_0(500)$, $\kappa \equiv K_0^*(800)$, $a_0(980)$ and $f_0(980)$ is poorly understood [1]. Compared to expectations based on a standard quark antiquark picture all nine states are rather light and their ordering is inverted. This can, however, naturally be explained assuming a tetraquark structure, which is also supported by certain decay channels, e.g. $a_0(980) \to K + \bar{K}$.

Here we report about the status of an ongoing long-term project with the aim to study such tetraquark candidates using lattice QCD. We mainly focus on the $a_0(980)$. In section 2 we summarize recently published results obtained with Wilson twisted mass quarks, where diagrams with closed fermion loops (also called "singly disconnected diagrams") have been neglected [3, 4, 5, 6, 7]. In section 3 we discuss latest technical advances obtained with clover improved Wilson quarks, in particular the inclusion of diagrams with closed fermion loops. Finally, in section 4 we discuss first steps of a study of a possibly existing $cc\bar{c}\bar{c}$ tetraquark, which has recently been predicted using a coupled system of covariant Bethe-Salpeter equations [8].

2. Wilson twisted mass quarks, diagrams with closed fermion loops neglected

2.1. Lattice setup

We use gauge link configurations with 2+1+1 dynamical quark flavors generated by the ETM Collaboration [9, 10]. We consider several ensembles with lattice spacing $a \approx 0.086\,\mathrm{fm}$. The ensembles differ in the volume $(L/a)^3 \times (T/a) = 20^3 \times 48, \ldots, 32^3 \times 64$ and the unphysically heavy u/d quark mass corresponding to $m_\pi \approx 280\,\text{MeV} \ldots 460\,\text{MeV}$.

For the computations presented in this section we have ignored diagrams with closed fermion loops, which are technically rather challenging (cf. section 3.2). Physical consequences are discussed in detail in [4].

2.2. Four-quark creation operators

$a_0(980)$ has quantum numbers $I(J^P) = 1(0^+)$. As usual in lattice QCD we extract the low lying spectrum in that sector by studying the asymptotic exponential behavior of correlation functions $C_{jk}(t) = \langle (\mathcal{O}_j(t))^\dagger \mathcal{O}_k(0) \rangle$ [11]. $\mathcal{O}_j$ and $\mathcal{O}_k$ denote suitable creation operators, i.e. operators generating the $a_0(980)$ quantum numbers, when applied to the vacuum state.

Assuming that the experimentally measured $a_0(980)$ with mass $980 \pm 20\,\text{MeV}$ is a rather strongly bound four quark state, suitable creation operators to excite such a state are

$$\mathcal{O}_{a_0(980)}^{K\bar{K}\text{ molecule}} = \sum_{\mathbf{x}} \Big(\bar{s}(\mathbf{x})\gamma_5 u(\mathbf{x})\Big)\Big(\bar{d}(\mathbf{x})\gamma_5 s(\mathbf{x})\Big) \tag{1}$$

$$\mathcal{O}_{a_0(980)}^{\text{diquark}} = \sum_{\mathbf{x}} \Big(\epsilon^{abc}\bar{s}^b(\mathbf{x})C\gamma_5\bar{d}^{c,T}(\mathbf{x})\Big)\Big(\epsilon^{ade}u^{d,T}(\mathbf{x})C\gamma_5 s^e(\mathbf{x})\Big). \tag{2}$$

The first operator has the spin/color structure of a $K\bar{K}$ molecule. The second resembles a bound diquark antidiquark pair, where spin coupling via $C\gamma_5$ corresponds to the lightest diquarks/antidiquarks (cf. e.g. [1, 12, 13]). Further low lying states with $a_0(980)$ quantum numbers, which need to be considered are the two-particle states $K + \bar{K}$ and $\eta_s + \pi$. Suitable creation operators to resolve these states are

$$\mathcal{O}_{a_0(980)}^{K+\bar{K}\text{ two-particle}} = \Big(\sum_{\mathbf{x}} \bar{s}(\mathbf{x})\gamma_5 u(\mathbf{x})\Big)\Big(\sum_{\mathbf{y}} \bar{d}(\mathbf{y})\gamma_5 s(\mathbf{y})\Big) \tag{3}$$

$$\mathcal{O}_{a_0(980)}^{\eta_s+\pi\text{ two-particle}} = \Big(\sum_{\mathbf{x}} \bar{s}(\mathbf{x})\gamma_5 s(\mathbf{x})\Big)\Big(\sum_{\mathbf{y}} \bar{d}(\mathbf{y})\gamma_5 u(\mathbf{y})\Big). \tag{4}$$

2.3. Numerical results

Figure 1(left) shows effective mass plots for a 2×2 correlation matrix with a $K\bar{K}$ molecule operator (1) and a diquark-antidiquark operator (2) (lattice volume $(L/a)^3 \times (T/a) = 20^3 \times 48$, pion mass $m_\pi \approx 341\,\text{MeV}$). The corresponding two plateaus are around 1100 MeV and, therefore, consistent both with the expectation for possibly existing $a_0(980)$ tetraquark states and with two-particle $K + \bar{K}$ and $\eta_s + \pi$ states, where both particles are at rest ($2 \times m(K) \approx 1198\,\text{MeV}$; $m(\eta_s) + m(\pi) \approx 1115\,\text{MeV}$ in our lattice setup).

Increasing this correlation matrix to 4×4 by adding $K + \bar{K}$ two-particle and $\eta_s + \pi$ two-particle operators (eqs. (3) and (4)) yields the effective mass results shown in Figure 1(right). Two additional states are observed, whose plateaus are around $1500\,\text{MeV} \ldots 2000\,\text{MeV}$.

From these plots one can conclude that the two low-lying states are the expected two-particle $K + \bar{K}$ and $\eta_s + \pi$ states, while an additional stable $a_0(980)$ tetraquark state does not exist. The second and third excitation can also be interpreted as two-particle states, which have non-vanishing relative momentum. A detailed discussion of these plots and corresponding arguments can be found in [4].

Qualitatively identical results are found, when using other ensembles with light quark masses and spacetime volumes as mentioned in section 2.1.

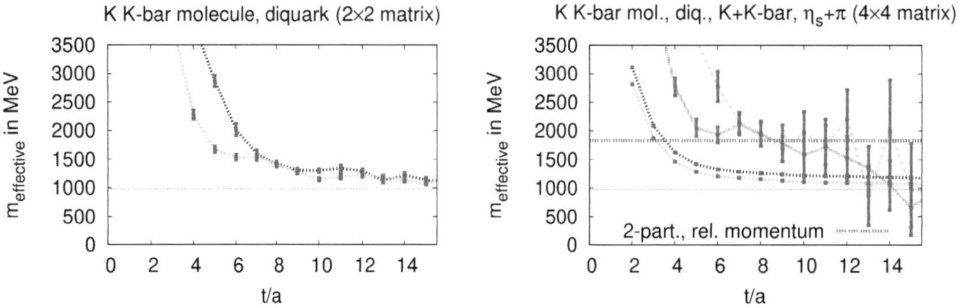

Figure 1. Effective masses as functions of the temporal separation. **(left)** 2×2 correlation matrix (operators: $K\bar{K}$ molecule, diquark). **(right)** 4×4 correlation matrix (operators: $K\bar{K}$ molecule, diquark, $K + \bar{K}$ two-particle, $\eta_s + \pi$ two-particle).

3. Clover improved Wilson quarks, diagrams with closed fermion loops included

3.1. Lattice setup

Recently we started to perform similar computations using gauge link configurations generated by the PACS-CS Collaboration [14] with 2+1 flavors of clover improved Wilson sea quarks, a lattice spacing $a \approx 0.09\,\text{fm}$, a volume $(L/a)^3 \times T/a = 32^3 \times 64$ and a u/d quark mass corresponding to $m_\pi \approx 300\,\text{MeV}$. A significant advantage compared to Wilson twisted mass quarks is that parity and isospin are exact symmetries. For example there is no pion and kaon mass splitting and $P = +$ and $P = -$ states are separated by quantum numbers (these problems in the context of Wilson twisted mass quarks and the $a_0(980)$ are discussed in detail in [4]).

3.2. Diagrams with closed fermion loops

A major improvement compared to our previous results presented in section 2.3 is that this time diagrams with closed fermion loops are included, i.e. where strange quark propagators start and end at the same timeslice. Ignoring such diagrams introduces a systematic error, which is hard to quantify. With these new computations this systematic error has been eliminated. Another important consequence of the inclusion of closed fermion loop is that quark-antiquark and four-quark trial states can have non-vanishing overlap. This allows to study a larger correlation matrix containing not only the four-quark operators (1) to (4), but also a quark-antiquark operator

$$\mathcal{O}^{q\bar{q}}_{a_0(980)} \quad = \quad \sum_{\mathbf{x}} \Big(\bar{d}(\mathbf{x})u(\mathbf{x})\Big). \tag{5}$$

Such a correlation matrix will enable us to make stronger statements about the structure of states from the $a_0(980)$ sector. The techniques we use to compute these diagrams with closed fermion loops are described in [7].

3.3. First numerical results

At the moment only the computation of a 2×2 correlation matrix containing the $q\bar{q}$ operator (eq. (5)) and the four-quark $K\bar{K}$ molecule operator (eq. (1)) has been finished. The effective masses corresponding to the diagonal element of the $q\bar{q}$ operator is plotted in Figure 2, left. The plot on the right shows the two effective masses obtained by solving a generalized eigenvalue problem using the full 2×2 correlation matrix. All effective masses are around 1000 MeV and, therefore, consistent with $2 \times m(K)$, with $m(\eta) + m(\pi)$ and also with $m(a_0(980))$.

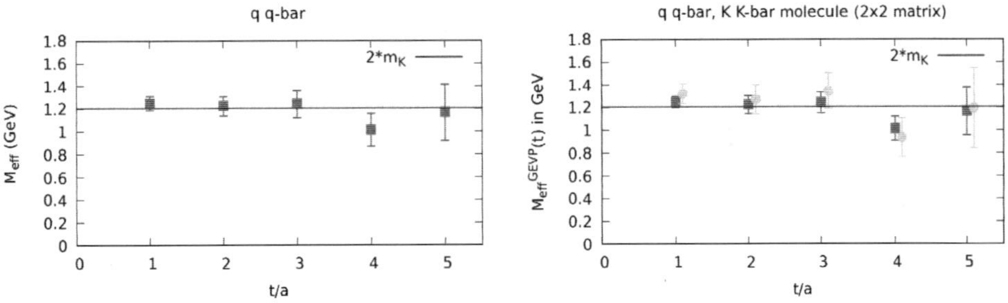

Figure 2. Effective masses as functions of the temporal separation. **(left)** $q\bar{q}$ operator. **(right)** 2×2 correlation matrix (operators: $q\bar{q}$, $K\bar{K}$ molecule).

For statements regarding the structure of the observed states we need to include also the four-quark operators (2) to (4), which is currently in progress. This will allow us to address similar questions as in section 2.3, i.e. to answer, whether there is in addition to the two-particle $K + \bar{K}$ and $\eta + \pi$ states also a third (bound) state near the mass of $a_0(980)$. We also intend to study the volume dependence of the two-particle spectrum using "Lüscher's method" (cf. e.g. [15, 16, 17]) or improved techniques taking e.g. coupled channel scattering into account [20, 21, 22]. Such computations are very challenging using lattice QCD, but first results have recently been published, e.g. for the κ and positive parity D mesons [18, 19].

4. Lattice investigation of a possibly existing $\bar{c}c\bar{c}c$ tetraquark

Recently a $\bar{c}c\bar{c}c$ tetraquark has been predicted using a coupled system of covariant Bethe-Salpeter equations [8]. It is expected to be a mesonic molecule-like state formed by two η_c mesons with a mass of $m(\bar{c}c\bar{c}c) = (5.3 \pm 0.5)\,\text{GeV}$. Comparing this mass to $2 \times m(\eta_c) = 6.0\,\text{GeV}$ indicates that the binding energy could be quite large, $\Delta E = m(\bar{c}c\bar{c}c) - 2 \times m(\eta_c) \approx -(0.7 \pm 0.5)\,\text{GeV}$. The uncertainty quoted in [8] is, however, of the same order of magnitude. Investing such a $\bar{c}c\bar{c}c$ tetraquark using lattice methods could provide an independent confirmation of its existence.

We use the same setup as discussed for the $a_0(980)$ meson in section 3. In a first attempt we consider a single creation operator of molecule type, which models a bound $\eta_c\eta_c$ state,

$$\mathcal{O}_{\bar{c}c\bar{c}c}^{\eta_c\eta_c\text{ molecule}} = \sum_{\mathbf{x}} \Big(\bar{c}(\mathbf{x})\gamma_5 c(\mathbf{x})\Big)\Big(\bar{c}(\mathbf{x})\gamma_5 c(\mathbf{x})\Big). \tag{6}$$

Diagrams with closed fermion loops are ignored.

In Figure 3 we compare the corresponding effective mass (blue points) with two times the η_c effective mass (red points). We do not observe any indication of a four quark $\bar{c}c\bar{c}c$ state, which is lighter than $2 \times m(\eta_c)$. On the other hand the effective mass corresponding to the creation operator (6) still seems to slightly decrease at large temporal separations and does not exhibit a clear plateau. This signals a trial state $\mathcal{O}_{\bar{c}c\bar{c}c}^{\eta_c\eta_c\text{ molecule}}|\Omega\rangle$, which has a poor ground state overlap. The ground state could either be two unbound η_c (then the plateau is expected at $2 \times m(\eta_c)$) or still a $\bar{c}c\bar{c}c$ bound state, however, with a structure quite different from (6).

Currently we explore the latter alternative. While the creation operator (6) places the two η_c mesons essentially on top of each other, they could be rather far separated, e.g. by a distance up to $r \approx 1\,\text{fm}$ (for a recent study of this distance for heavy-heavy-light-light tetraquarks cf. [23]). To this end we employ corresponding more general molecule type $\bar{c}c\bar{c}c$ creation operators. Computations are in progress.

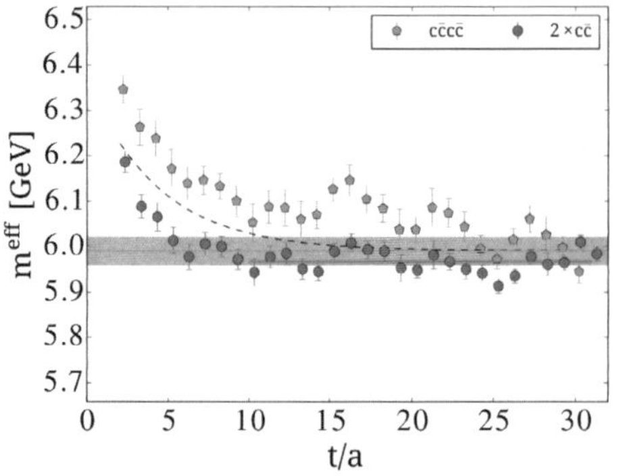

Figure 3. Effective mass as functions of the temporal separation for the $\eta_c\eta_c$ molecule operator (blue points) and the η_c meson multiplied by a factor 2 (red points).

Acknowledgments

We acknowledge helpful discussion with Christian Fischer. M.W. acknowledges support by the Emmy Noether Programme of the DFG (German Research Foundation), grant WA 3000/1-1. M.G. acknowledges support by the Marie-Curie European training network ITN STRONGnet grant PITN-GA-2009-238353. M.D.B. is funded by the Irish Research Council and is grateful for the hospitality at the University of Cyprus and Cyprus Institute, where part of this work was carried out. L.S. acknowledges support by STRONGnet. This work was supported in part by the Helmholtz International Center for FAIR within the framework of the LOEWE program launched by the State of Hesse and by the DFG and the NSFC through funds provided to the Sino-German CRC 110 "Symmetries and the Emergence of Structure in QCD".

References

[1] R. L. Jaffe, Phys. Rept. **409**, 1 (2005) [Nucl. Phys. Proc. Suppl. **142**, 343 (2005)] [arXiv:hep-ph/0409065].
[2] J. Beringer et al. [Particle Data Group], Phys. Rev. D86, 010001 (2012).
[3] J. O. Daldrop *et al.* [ETM Collaboration], PoS LATTICE **2012**, 161 (2012) [arXiv:1211.5002 [hep-lat]].
[4] C. Alexandrou *et al.* [ETM Collaboration], JHEP **1304**, 137 (2013) [arXiv:1212.1418].
[5] M. Wagner *et al.* [ETM Collaboration], PoS ConfinementX , 108 (2012) [arXiv:1212.1648 [hep-lat]].
[6] M. Wagner *et al.* [ETM Collaboration], arXiv:1302.3389 [hep-lat].
[7] M. Wagner *et al.*, arXiv:1309.0850 [hep-lat].
[8] W. Heupel, G. Eichmann and C. S. Fischer, Phys. Lett. B **718**, 545 (2012) [arXiv:1206.5129 [hep-ph]].
[9] R. Baron *et al.* [ETM Collaboration], JHEP **1006**, 111 (2010) [arXiv:1004.5284 [hep-lat]].
[10] R. Baron *et al.* [ETM Collaboration], Comput. Phys. Commun. **182**, 299 (2011) [arXiv:1005.2042 [hep-lat]].
[11] J. Weber *et al.*, arXiv:1310.1760 [hep-lat].
[12] C. Alexandrou, P. .de Forcrand and B. Lucini, Phys. Rev. Lett. **97**, 222002 (2006) [hep-lat/0609004].
[13] M. Wagner and C. Wiese [ETM Collaboration], JHEP **1107**, 016 (2011) [arXiv:1104.4921 [hep-lat]].
[14] S. Aoki *et al.* [PACS-CS Collaboration], Phys. Rev. D **79**, 034503 (2009) [arXiv:0807.1661 [hep-lat]].
[15] M. Lüscher, Commun. Math. Phys. **105**, 153 (1986).
[16] M. Lüscher, Nucl. Phys. B **354**, 531 (1991).
[17] M. Lüscher, Nucl. Phys. B **364**, 237 (1991).
[18] C. B. Lang *et al.*, Phys. Rev. D **86**, 054508 (2012) [arXiv:1207.3204 [hep-lat]].
[19] D. Mohler, S. Prelovsek and R. M. Woloshyn, Phys. Rev. D **87**, no. 3, 034501 (2013) [arXiv:1208.4059 [hep-lat]].
[20] V. Bernard *et al.*, JHEP **1101**, 019 (2011) [arXiv:1010.6018 [hep-lat]].
[21] M. Doring *et al.*, Eur. Phys. J. A **48**, 114 (2012) [arXiv:1205.4838 [hep-lat]].
[22] M. Doring, plenary talk at "31st International Symposium on Lattice Field Theory" (2013).
[23] P. Bicudo and M. Wagner, Phys. Rev. D **87**, 114511 (2013) [arXiv:1209.6274 [hep-ph]].

First results of proton antiproton annihilation into a pion pair at large scattering angles within the handbag approach

A T Goritschnig[1], S Kofler[2] and W Schweiger[3]

Institute for Physics, Karl-Franzens-Universität Graz, Universitätsplatz 5, 8010 Graz, Austria

E-mail: [1]alexander.goritschnig@uni-graz.at

E-mail: [2]stefan.kofler@uni-graz.at

E-mail: [3]wolfgang.schweiger@uni-graz.at

Abstract. We propose to describe the process $p\,\bar{p} \to \pi^+\,\pi^-$ in a perturbative, QCD motivated framework in which a hard $ud\,\bar{u}\bar{d} \to d\,\bar{d}$ annihilation factorizes from soft transition distribution amplitudes. We advocate that the scale allowing for this factorization is the large transverse momentum transfer. In our simplified model, in which the proton is considered as a (scalar)diqark-quark system, a transition distribution amplitude describes the non-perturbative transition of the proton to the meson by emission of a scalar, isoscalar ud-diquark and absorption of an antiquark (analogously for $\bar{p} \to \pi^-$). We model the transition distribution amplitudes as an overlap of light-cone wave functions and present first results for the differential cross section. This process will be measured by the $\bar{\text{P}}$ANDA experiment at GSI-FAIR.

1. Introduction

The $\bar{\text{P}}$ANDA detector at FAIR (Facility for Antiproton and Ion Research) in Darmstadt is ideally suited for the study of exclusive channels in proton-antiproton collisions. Studying these reactions will open a new area of looking into the interior of hadrons and thus bring new perspectives to the question on how hadrons can be understood in terms of the fundamental degrees of freedom of QCD, i.e. in terms of quarks and gluons. One possibility is to investigate the process $p\,\bar{p} \to \pi^+\,\pi^-$. Pion production is particularly interesting and illuminating because one has well established constraints at hand to find realistic light-cone wave functions (LCWFs) which serve as model input. We adapt the description of heavy D-meson production within the handbag approach, cf. Ref. [1], to the case of light meson production. We advocate that for large transverse momentum transfer and for restricted parton virtualities and intrinsic transverse momenta the process amplitude factorizes into a (hard) partonic subprocess and (soft) hadronic transition matrix elements. The hadronic transition matrix elements can be parameterized by transition distribution amplitudes (TDAs) which have been introduced in a series of papers, cf. Refs. [2]-[5]. To model the TDAs we extend the formalism of LCWF-overlap, developed in Ref. [6].

Journal of Physics: Conference Series **503** (2014) 012032 doi:10.1088/1742-6596/503/1/012032

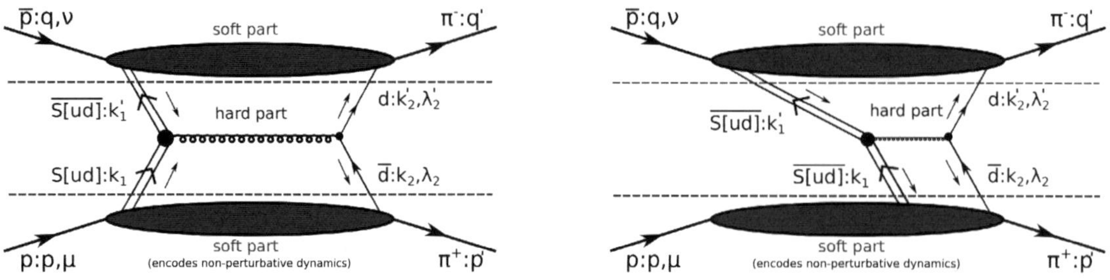

Figure 1. The double-handbag mechanism for $p\,\bar{p} \to \pi^+\,\pi^-$. Left: DGLAP-contribution. Right: ERBL-contribution (one of two). The blobs contain the non-perturbative dynamics of the process and are parameterized by transition distribution amplitudes.

2. The Handbag Mechanism

The assignments of the momenta of the incoming baryons and outgoing mesons can be read off from Fig. 1. The reference frame (symmetric center-of-momentum system (CMS)) is aligned along the z-axis of the 3-vector of the average momentum $\bar{\mathbf{p}}$ defined by $\bar{p} \equiv \frac{1}{2}(p + p')$. The momentum transfer is defined by $\Delta \equiv p' - p = q - q'$. Symmetric CMS means that the transverse component of the momentum transfer is symmetrically shared between the incoming and outgoing particles. In light-cone (LC) coordinates[1] the incoming proton and the outgoing π^+ momenta are written as

$$p = \left[(1+\xi)\,\bar{p}^+, \frac{M^2 + \mathbf{\Delta}_\perp^2/4}{2\,(1+\xi)\,\bar{p}^+}, -\frac{\mathbf{\Delta}_\perp}{2}\right] \quad \text{and} \quad p' = \left[(1-\xi)\,\bar{p}^+, \frac{m^2 + \mathbf{\Delta}_\perp^2/4}{2\,(1-\xi)\,\bar{p}^+}, \frac{\mathbf{\Delta}_\perp}{2}\right], \tag{1}$$

where M denotes the proton mass and m the π^+ mass. The so called "skewness parameter" ξ is defined by $\xi \equiv -\frac{\Delta^+}{2\bar{p}^+}$ and gives the relative momentum transfer between the proton and the π^+ in the longitudinal plus direction[2]. The antiproton and the π^- momenta are parameterized in an analogous way.

Similar to wide-angle Compton scattering, cf. Ref. [7], we argue that for large CMS scattering angles, i.e. large Mandelstam s, t and u ($t, u \geq 1$ GeV^2), the process can be described by a double-handbag mechanism. This double-handbag mechanism is shown in Fig. 1. Given the large scattering angles, that provides us with a large energy scale, it is assumed that the $p\,\bar{p} \to \pi^+\,\pi^-$ amplitude factorizes into a hard-scattering kernel on the partonic level and soft hadronic $p \to \pi^+$ and $\bar{p} \to \pi^-$ transition matrix elements. We stress that there is no rigorous proof for factorization for our process. It is, at most, a first step in this direction. In the handbag mechanism only the minimal number of constituents which are required to transform the (initial) $p\bar{p}$- to the (final) $\pi^+\pi^-$-pair actively take part in the partonic subprocess. Since we consider the proton to be a bound state of a scalar diquark, denoted by $S[ud]$, and a u-quark the active constituents are the $S[ud]$-diquarks and d-quarks on the proton side and the corresponding antiparticles on the antiproton side.

Using the physically plausible assumptions of restricted parton virtualities and restricted

[1] Any four-vector $a = (a^0, a^1, a^2, a^3)$ is written in LC-coordinates as $a = [a^+, a^-, \mathbf{a}_\perp]$ with $a^\pm = \frac{a^0 \pm a^3}{\sqrt{2}}$ and $\mathbf{a}_\perp = (a^1, a^2)$.
[2] Note that in our reference frame and for the pion production $\xi < 0$.

intrinsic transverse momenta, cf. Refs. [1, 7, 8], the hadronic amplitude can be written as

$$\begin{aligned}\mathscr{M}_{\mu\nu} &= \int \mathrm{d}\bar{x} \int \mathrm{d}\bar{x}' \; \widetilde{H}\left(\bar{x},\bar{x}'\right) \\ &\times \bar{p}^+ \int \frac{\mathrm{d}z^-}{2\pi} e^{i\bar{x}\bar{p}^+z^-} \langle \pi^+ : p' | \Psi^d\left(-\frac{z^-}{2}\right) \phi^{S[ud]}\left(\frac{z^-}{2}\right) |p : p, \mu\rangle \\ &\times \bar{q}^- \int \frac{\mathrm{d}z'^+}{2\pi} e^{i\bar{x}'\bar{q}^-z'^+} \langle \pi^- : q' | \, \phi^{S[ud]\dagger}\left(\frac{z'^+}{2}\right) \bar{\Psi}^d\left(-\frac{z'^+}{2}\right) |\bar{p} : q, \nu\rangle \,.\end{aligned} \tag{2}$$

It is a convolution of a hard-scattering kernel $\widetilde{H}$ and two hadronic matrix elements with respect to the average momentum fraction defined by

$$\bar{x}^{(\prime)} \equiv \frac{\bar{k}^{+(\prime-)}}{\bar{p}^+(\bar{q}^-)} \quad \text{with} \quad \bar{k}^{+(\prime-)} = \frac{k_1^{+(\prime-)} + k_2^{+(\prime-)}}{2}. \tag{3}$$

The assignments of the active parton momenta $k_{1,2}^{(\prime)}$ can be read off Fig. 1. Since we are working in the LC-quantization framework we want to work with the independent degrees of freedom of the d-quark field operators. This can be done using the same projection techniques as in Refs. [1, 7, 8]. The amplitude thus becomes

$$\begin{aligned}\mathscr{M}_{\mu\nu} &= \frac{1}{4\bar{p}^+\bar{q}^-} \sum_{\lambda_1,\lambda_2} \int \mathrm{d}\bar{x} \int \mathrm{d}\bar{x}' \; H_{\lambda_2,\lambda_2'}\left(\bar{x},\bar{x}'\right) \frac{1}{\bar{x}-\xi} \frac{1}{\bar{x}'-\xi} \\ &\times \bar{v}\left(k_2,\lambda_2\right) \gamma^+ \bar{p}^+ \int \frac{\mathrm{d}z^-}{2\pi} e^{i\bar{x}\bar{p}^+z^-} \langle \pi^+ : p' | \Psi^d\left(-\frac{z^-}{2}\right) \phi^{S[ud]}\left(\frac{z^-}{2}\right) |p : p, \mu\rangle \\ &\times \bar{q}^- \int \frac{\mathrm{d}z'^+}{2\pi} e^{i\bar{x}'\bar{q}^-z'^+} \langle \pi^- : q' | \, \phi^{S[ud]\dagger}\left(\frac{z'^+}{2}\right) \bar{\Psi}^d\left(-\frac{z'^+}{2}\right) |\bar{p} : q, \nu\rangle \, \gamma^- u\left(k_2',\lambda_2'\right),\end{aligned} \tag{4}$$

where we have defined the hard-scattering amplitude as $H_{\lambda_2,\lambda_2'}\left(\bar{x},\bar{x}'\right) \equiv \bar{u}\left(k_2',\lambda_2'\right) \widetilde{H}\left(\bar{x}\bar{p}^+,\bar{x}'\bar{p}^+\right)$ $\times v\left(k_2,\lambda_2\right)$ and $\lambda_2^{(\prime)}$ denote the active quark helicity. Now, depending on the $\bar{x}^{(\prime)}$-region, three different partonic subprocesses contribute (see Fig. 1): $S[ud]\,\overline{S[ud]} \to d\,\bar{d}$ (left), $\overline{S[ud]} \to \overline{S[ud]}\,d\,\bar{d}$ (right) and $S[ud] \to S[ud]\,d\,\bar{d}$ (not shown). So far we have only taken a closer look on the contribution coming from the DGLAP-region alone, i.e. Fig. 1 left. In Sec. 5 we discuss the possible influence of the ERBL-region.

3. Modeling the hadronic transitions

The soft hadronic matrix elements are modeled in the spirit of an overlap of LCWFs (focusing on the $p \to \pi^+$-transition, since the $\bar{p} \to \pi^-$-transition can be treated along the same lines). For doing that we take the Fourier representation of the field operators and the Fock-state decomposition of the hadron states in LC-quantum field theory. We only take the valence Fock states of the hadrons into account, i.e. a $|p : S[ud], d\rangle$ and a $|\pi^+ : u, \bar{d}\rangle$ state for the proton and π^+, respectively. Furthermore we only consider configurations with zero orbital angular momenta between the partons such that the parton helicities sum up to their parent-hadron helicity. According to Ref. [9] we choose for the proton

$$\Psi_p = N_p \, \tilde{x} \, \exp\left[-\frac{a_p^2}{\tilde{x}(1-\tilde{x})} \tilde{\mathbf{k}}_\perp^2\right] \tag{5}$$

and according to Ref. [10] for the pion

$$\Psi_\pi = N_\pi \exp\left[-\frac{a_\pi^2}{\hat{x}(1-\hat{x})} \hat{\mathbf{k}}_\perp^2\right] \tag{6}$$

as LCWFs. $\tilde{x}$ and $\hat{x}$ denote the momentum fractions of the active $S[ud]$-diquark inside the proton and the active $\bar{d}$-quark inside the π^+, respectively. Each of the wave functions has two free parameters, a global normalization constant $N_{p,\pi}$ and an oscillator parameter $a_{p,\pi}$. For the proton they are fixed by requiring a valence Fock state probability of $P_p = 0.5$ and a r.m.s of the intrinsic transverse momentum of the active quark of $\sqrt{\langle \mathbf{k}_\perp \rangle^2} = 0.280$ GeV. This yields $N_p = 61.8\ \text{GeV}^{-2}$ and $a_p = 1.1\ \text{GeV}^{-1}$. The pion parameters are fixed by demanding the r.m.s. of the active quark to be $\sqrt{\langle \mathbf{k}_\perp \rangle^2} = 0.350$ GeV and the pion decay constant $f_\pi = 0.093$ GeV. This gives $N_\pi = 26.1\ \text{GeV}^{-2}$ and $a_\pi = 1.0\ \text{GeV}^{-1}$. The result for the LCWF overlap is shown in Fig. 2 for different values of the CMS scattering angle.

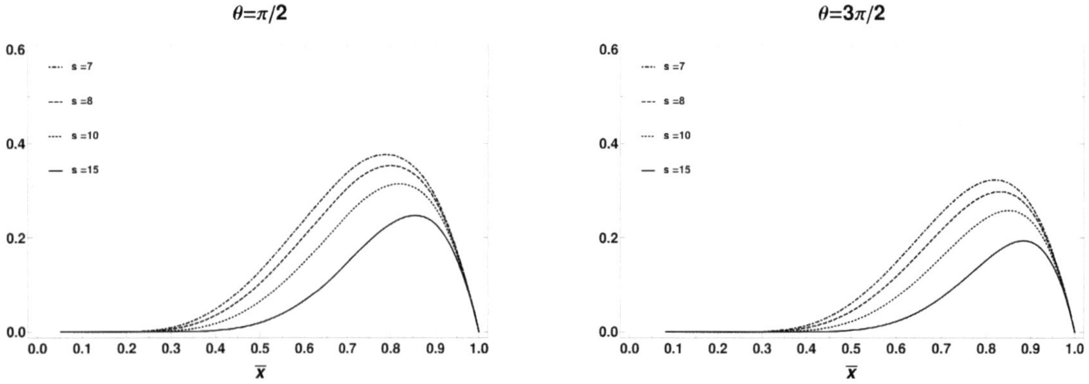

Figure 2. The wave function overlap versus the average momentum $\bar{x}$ for CMS scattering angles $\theta = \pi/2$ (left panel) and $\theta = 3\pi/2$ (right panel) and different values of Mandelstam s.

4. Hard-Scattering Amplitudes

Under the assumptions on the parton momenta discussed in Sec. 2 one can show that in the hard-scattering amplitude the momentum fractions are approximately 1, i.e. $x_{1,2} = x'_{1,2} \approx 1$. Since we treat the scalar diquarks and the d-quarks as being massless, the hard-scattering amplitude vanishes when $\lambda_2 = \lambda'_2$. The non-vanishing contributions are

$$H_{+,-} = H_{-,+} = -4\pi \frac{4}{9} \alpha_s(s) F_s(s) \sin\theta, \tag{7}$$

where $F_s(s)$, taken from Ref. [11], takes care of the composite nature of the diquarks and the fact that they should dissolve into quarks if a large amount of momentum is transfered to the diquarks.

5. Results and Discussions

The differential cross section for $p\,\bar{p} \to \pi^+\,\pi^-$ reads

$$\frac{d\sigma}{d\Omega} = \frac{1}{64\pi^2}\frac{1}{s}\sqrt{\frac{1-4m^2/s}{1-4M^2/s}}\,\sigma_0 \quad \text{with} \quad \sigma_0 \equiv \frac{1}{4}\sum_{\mu,\nu} |\mathcal{M}_{\mu\nu}|^2. \tag{8}$$

The result for the differential cross section in the CMS-angle interval $\frac{\pi}{2} \le \theta \le \frac{2\pi}{3}$ is shown in Fig. 3. Comparing our result at $s = 12.9\ \text{GeV}^2$ for a CMS scattering angle of $\pi/2$ with the experimental data of Ref. [12] reveals that our calculations in which we have only taken the

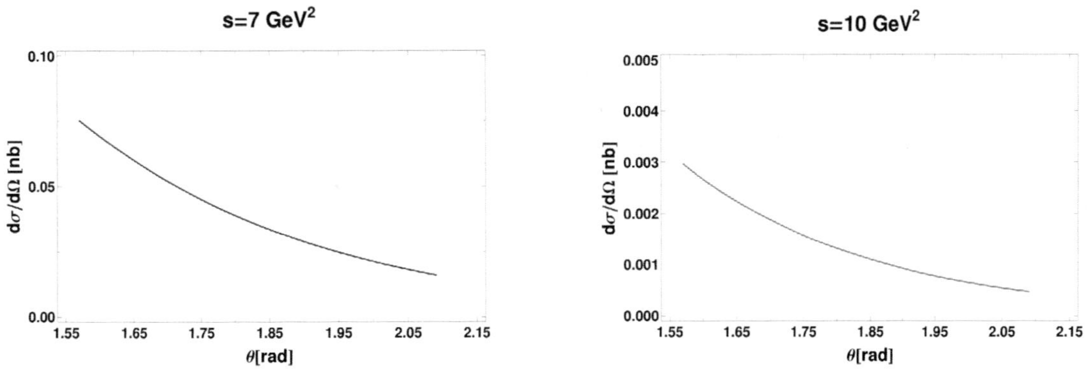

Figure 3. The differential cross section for $s = 7$ GeV^2 and 10 GeV^2 as a function of the CMS scattering angle θ.

DGLAP-region into account, predict a much smaller differential cross section. It is about three orders of magnitude smaller.
For CMS energies around 10 GeV^2 (and larger) one may think of the ERBL-region to play a minor role, since there the skewness parameter is small[3], i.e. it is about 0.1 for a CMS angle of $\theta = \frac{2\pi}{3}$ and most of the $\bar{x}^{(\prime)}$ integration region $0 \le \bar{x}^{(\prime)} \le 1$ is covered by the DGLAP-domain ($\xi \le \bar{x}^{(\prime)} \le 1$). If this would be the case, the handbag mechanism, in its present form, would fail to give a reasonable description for $p\,\bar{p} \to \pi^+\,\pi^-$. However, such a conclusion could be premature. In Ref. [3] it is argued that the DGLAP-region dominates the cross section for large values of $\xi \approx 1$, while for small (and intermediate) values of ξ the dominant contribution is due to the ERBL-region, which is mainly determined by a baryon pole contribution. Therefore it might be that the big discrepancy between our results and the experimental data can be attributed to the missing contributions from the ERBL-region. In order to draw definite conclusions we thus have to include also the ERBL-region. This is the subject of current investigations.

Acknowledgments

We want to thank the organizers of the "Fairness 2013" for providing an interesting workshop. ATG and SK are supported by the Fonds zur Förderung der wissenschaftlichen Forschung in Österreich via Grand No. J3163-N16 and FWF DK W1203-N16, respectively. We further acknowledge helpful discussions with Bernard Pire from the École Polytechnique.

References

[1] Goritschnig A, Pire B and Schweiger W 2013 *Phys. Rev. D* **87** 014017
[2] Pire B, Semenov-Tian-Shansky and Szymanowski L 2010 *Phys. Rev. D* **82** 094030
[3] Pire B, Semenov-Tian-Shansky and Szymanowski L 2011 *Phys. Rev. D* **84** 074014
[4] Lansberg J P, Pire B, Semenov-Tian-Shansky and Szymanowski L 2012 *Phys. Rev. D* **86** 114033
[5] Pire B, Semenov-Tian-Shansky and Szymanowski L 2013 *Phys. Lett. B* **724** 99
[6] Diehl M, Feldmann T, Jakob R and Kroll P 2001 *Nucl. Phys. B* **596** 33
[7] Diehl M, Feldmann T, Jakob R and Kroll P 1998 *Eur. Phys. J.* **C8** 409
[8] Goritschnig A, Kroll P and Schweiger W 2009 *Eur. Phys. J.* **A42** 43
[9] Kroll P, Quadder B and Schweiger W 1989 *Nucl. Phys. B* **316** 373
[10] Feldmann T and Kroll P 2000 *Eur. Phys. J.* **C12** 99
[11] Kroll P, Pilsner T, Schürmann M and Schweiger W (1993) *Phys. Lett. B* **316** 546
[12] Buran T et al. 1976 *Nucl. Phys. B* **116** 51

[3] An increase of the CMS scattering angle θ increases the skewness in magnitude in our kinematics.

Journal of Physics: Conference Series **503** (2014) 012033 doi:10.1088/1742-6596/503/1/012033

Non-mesonic weak decay of hypernuclei with effective field theory

A Pérez-Obiol[1], D R Entem[2], B Juliá-Díaz[1,3] and A Parreño[1]

[1] Departament d'Estructura i Constituents de la Matèria, Institut de Ciències del Cosmos (ICC), Universitat de Barcelona, Martí i Franquès 1, E–08028 Barcelona, Spain

[2] Grupo de Física Nuclear and IUFFyM, Universidad de Salamanca, E-37008 Salamanca, Spain

[3] ICFO - Institut de Ciències Fotòniques, Av. Carl Friedrich Gauss 3, E–08860 Castelldefels (Barcelona), Spain

E-mail: axel@ecm.ub.edu, entem@usal.es, bruno@ecm.ub.edu, assum@ecm.ub.edu

Abstract. The $\Lambda N \rightarrow NN$ interaction, the main responsible for the decay of heavy hypernuclei, is described within the effective field theory (EFT) framework. The EFT is developed up to $\mathcal{O}(\vec{q}^{\,2})$, where $\vec{q}$ is the transferred momentum between the interacting baryons. Numerical values for the leading order low-energy constants are obtained by two means. First, from a fit to the hypernuclear decay observables, and second, by comparing the EFT to the one-meson-exchange model describing the same interaction. All the two-pion exchange diagrams entering at the next-to-leading in the EFT have been calculated.

1. Introduction

During the last 60 years hypernuclear experiments have helped us to understand how hyperons —baryons with strangeness— and nucleons interact, and how they form the bound states we call hypernuclei. Nowadays, a few tens of hypernuclei have been characterized (see Fig. 1), and in the coming years many more should be produced at experiments such as HypHI [1] and PANDA [2] at FAIR. The lightest hyperon, the Λ, has valence quarks u, d and s, a mass of ~ 1116 MeV, zero charge and zero isospin. It decays through the weak interaction in $\sim 2.6 \cdot 10^{-10}$ seconds, and hence it is very difficult to perform ΛN scattering experiments. Therefore, hypernuclei are the main laboratories to study ΛN interactions. On the one hand, the binding energies of these hypernuclei give us information about the nuclear structure and about the strong hyperon-nucleon forces. On the other hand, the decay of hypernuclei allows us to constrain the weak ΛN interaction.

The weak $\Lambda N \rightarrow NN$ transition is the main mechanism driving the decay of heavy hypernuclei. While in free space the Λ decays mostly to a nucleon and a pion, when it is embedded in $A > 5$ hypernuclei, this mechanism becomes highly suppressed. This effect is due to the Pauli-blocking of the nucleonic medium on the final nucleon. In contrast, the final nucleons in single-, $\Lambda N \rightarrow NN$, and multi-nucleon ($\Lambda NN \rightarrow NNN$, etc.) induced decay mechanisms have enough energy to either escape the nucleus or occupy states above the Fermi energy level.

The $\Delta S = 1$ ΛN interaction has been traditionally studied with one-meson-exchange models (OME) [4]. Within this picture, the different ranges of the potential are described by the

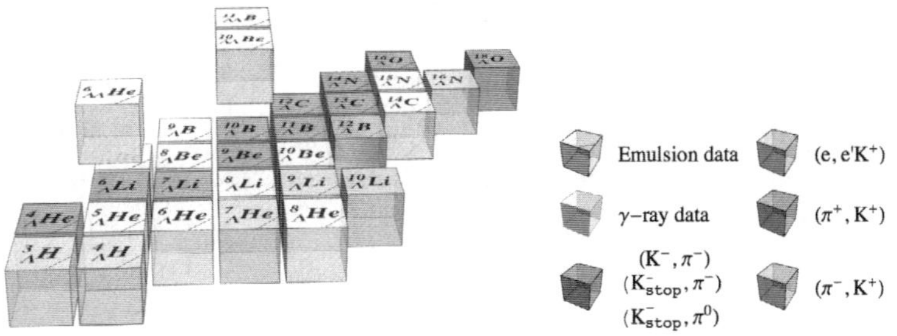

Figure 1. (Color online) Low mass region of layers $s = -1$ and $s = -2$ of the nuclear landscape. The different hypernuclei are colored according to the reactions that have produced them. In the coming years, new experiments will contribute with more hypernuclei in both layers. Original data for the Λ-hypernuclei is taken from [3].

exchange of the lightest pseudoscalar and vector mesons (π, K, η, ρ, ω and K^*). This model has been very useful to describe the hypernuclear decay data, but it is very model dependent and its connection to the underlying physics is difficult to establish. First, the non-pionic couplings and form factors are not experimentally known, and must be derived using $SU(3)$ flavor and $SU(6)$ flavor-spin symmetries, which are largely broken. Second, the heaviest mesons, which account for the short range part of the interaction, are too large to probe the physics at the ranges given by the corresponding Yukawa potentials.

In our work we describe the two-body weak $\Delta S = 1$ ΛN interaction within an effective field theory (EFT) framework, following the works of Refs. [5, 6]. The EFT approach allows one to study the baryon-baryon interactions in a more systematic and less model dependent way. The main idea is to separate a soft scale, which contains only the degrees of freedom that are relevant in the interaction (i.e. pions and kaons) from a hard scale (of the order of the mass of the nucleon). Then, one constructs the potential only with these degrees of freedom while respecting the symmetries of the underlying theory, as for example, chiral symmetry. The potential is built as an expansion on a parameter which is a ratio of the low scale over the hard scale. In baryon-baryon interactions, this expansion parameter is usually the transferred momentum between the baryons over the average mass of the baryons. To organize this expansion, we use the Weinberg power counting scheme [7] and the heavy baryon expansion [8].

The description for the $\Lambda N \to NN$ interaction is constrained by the value of the hypernuclear decay observables. When a hypernucleus decays non-mesonically, we can observe, mainly, three different quantities: the total non-mesonic decay rate, Γ_{nm}; the partial decay rates, $\Gamma_n = \Gamma(\Lambda n \to nn)$ and $\Gamma_p = \Gamma(\Lambda p \to np)$, with $\Gamma_{nm} = \Gamma_n + \Gamma_p$; and the asymmetry in the distribution of protons coming from the decay of polarized hypernuclei.

2. Leading order Effective field theory

The leading order (LO) EFT for the $\Lambda N \to NN$ interaction involves the long-ranged one-pion exchange (OPE), the medium-long ranged one-kaon exchange (OKE), and two contact operational structures. The Feynman diagrams for these contributions are depicted in Fig. 2. The kaon is included as an explicit degree of freedom since its mass (~ 495 MeV) is comparable to the momentum transferred to the final baryons (~ 400 MeV).

The strong vertex $NN\pi$ and the weak $\Lambda N\pi$ one are experimentally known from the corresponding decays. The kaon vertices, however, are not available experimentally, and one needs to use $SU(3)$ symmetry to relate them to the pionic ones. We use the values given by the Nijmegen Soft-Core 97 f [9] and the Jülich B [10] models.

The contact potential is made up of all the possible operational structures that connect initial and final spin, isospin and angular momentum states. Taking into account the antisymmetry of the final nucleons, and at zeroth order in the momentum expansion, we have two independent

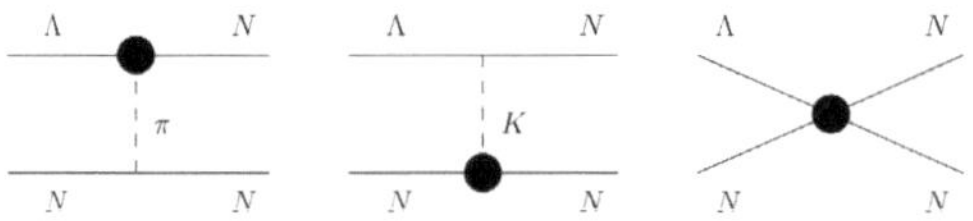

Figure 2. One-pion exchange, one-kaon exchange and contact contributions to the transition. The solid circle represents the weak interaction vertex.

structures:

$$V_{4P}(\vec{q}) = C_{00} + C_{01}\ \vec{\sigma}_1\vec{\sigma}_2\,. \tag{1}$$

The two low-energy constants (LEC's) appearing in the contact potential are not known and must be fitted to the experiment. Therefore, we need to relate the two-body transition to the hypernuclear decay rate, which involves the knowledge of the initial and final many-body states. For the initial state we use a shell model approach, and assume that the Λ is coupled to the lowest energy level. To uncouple a nucleon from the core we use the technique of the coefficients of fractional parentage. The interacting Λ and nucleon are described by harmonic oscillator wave functions, with the parameters b_N and b_Λ such that the core and hypernucleus binding energies are reproduced, respectively. Regarding the final wave function, we have not included the propagation of the two primary outgoing nucleons within the residual medium. In our approximation, the residual $(A-2)$ nucleus acts as a spectator, and only the strong interaction among these two primary nucleons is considered. More specifically, we solve a Lippmann-Schwinger scattering equation with the input of modern potential models (Nijmegen Soft-Core 97 f and Jülich B).

We have fitted the low-energy constants to reproduce the experimental data for three different hypernuclear observables and for three different hypernuclei — ${}^{5}_{\Lambda}He$, ${}^{11}_{\Lambda}B$ and ${}^{12}_{\Lambda}C$ —. In Fig. 3 we show the experimental data points together with the results obtained from this fit. We obtain two sets of data points, one for each of the two models used in the strong part of the calculation. The results obtained from the Jülich model fit a little better the data, but both models show that the data is fairly well reproduced by a LO EFT [11].

An alternative way to obtain information about the LEC's is to compare the EFT description with the OME one. In both formalisms the long and medium range is described by the one-pion and one-kaon exchanges, therefore we only need to match the short range terms in both formalisms. While in the EFT the operational structures are already organized in powers of momentum, the heavy meson exchange potentials in the OME approach are not. To obtain these different operational structures, we expand the OME potentials in powers of $\frac{q}{m}$ and $\frac{q}{\Lambda}$,

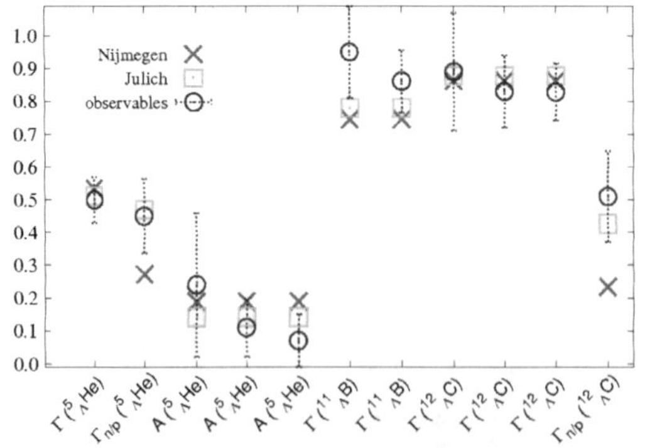

Figure 3. Hypernuclear decay observables (total and partial decay rates and asymmetry for ${}^{\Lambda}_{5}$He, ${}^{\Lambda}_{11}$B and ${}^{\Lambda}_{12}$C), including their error bars and their fitted values. The total decay rates are in units of the Λ decay rate in free space ($\Gamma_\Lambda = 3.8 \times 10^{9}\text{s}^{-1}$). All the quantities are adimensional.

Journal of Physics: Conference Series **503** (2014) 012033 doi:10.1088/1742-6596/503/1/012033

	Nijmegen		Jülich	
	OME expansion	LO PC calculation	OME expansion	LO PC calculation
C_0^0	1.07 ± 0.88	(4.01 ± 0.23)	-1.7 ± 2.6	(4.03 ± 0.50)
C_0^1	0.02 ± 0.36	(0.02 ± 0.33)	0.12 ± 0.37	(-0.30 ± 0.28)

Table 1. Values for the LEC's obtained from the two sources: OME expansion and LO (PC) EFT calculation, using the Nijmegen and Jülich strong interaction models. All the quantities are in units of $G_F = 1.166 \times 10^{-11}$ MeV^{-2}.

where m and Λ are the meson mass and cut-off used in the applicable potential. Thus, we can write the LEC's appearing in the LO EFT in terms of the heavy meson parameters. This way of assigning values to the LEC's is called resonance saturation and it has been previously applied in the strong NN case [12].

In Table 1 the values obtained for the LEC's according to each approach and for each model are shown. The larger discrepancy is found to affect the LEC C_{00}. This partly motivated us to include a sigma meson in the OME picture and to study how it affects the numerical values of LEC's. When the potential for the isoscalar sigma meson is expanded, and then included in the resonance saturation study, the LEC C_{01} stays the same while the LEC C_{00} has an extra term that depends linearly on the unknown parity-violating $\Lambda N \sigma$ coupling, A_σ. Thus, we can use this procedure to set bounds on the values of A_σ. For example, taking the values of $m_\sigma = 550$ MeV and $g_{NN\sigma} = 8.8$ from Ref. [13], and working within the Nijmegen strong potential model, we find $3.3 \leq A_\sigma \leq 7.3$ [11].

3. Next-to-leading order

The following order in the EFT includes all the possible operational structures at orders $\mathcal{O}(|\vec{q}|)$ and $\mathcal{O}(\vec{q}^{\,2})$ and the contributions coming from the two-pion exchange diagrams. In total we have 13 independent operational structures, and therefore 13 LEC's that, in principle, should be fitted to the experiment. However, this is not feasible with the current hypernuclear experimental data. One way to reduce this number of LEC's is to neglect the initial momentum (of the order of a few tens of MeV) in front of the final one (of the order of 400 MeV). Within this approximation, the contact potential at $\mathcal{O}(|\vec{q}|)$ and $\mathcal{O}(\vec{q}^{\,2})$ contains only 6 LEC's and reads (in units of G_F):

$$V_{4P}(\vec{q}) = C_{10}\,\frac{\vec{\sigma}_1\vec{q}}{2M_N} + C_{11}\,\frac{\vec{\sigma}_2\vec{q}}{2M_N} + \mathrm{i}\,C_{12}\,\frac{(\vec{\sigma}_1 \times \vec{\sigma}_2)\,\vec{q}}{2M_N} + C_{20}\,\frac{\vec{\sigma}_1\vec{q}\,\vec{\sigma}_2\vec{q}}{4M_N^2} + C_{21}\,\frac{\vec{\sigma}_1\vec{\sigma}_2\,\vec{q}^{\,2}}{4M_N^2} + C_{22}\,\frac{\vec{q}^{\,2}}{4M_N^2}\,.$$

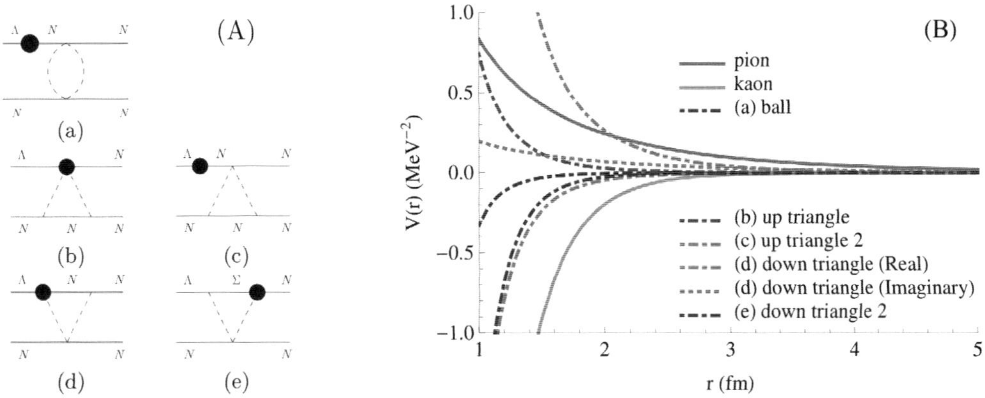

Figure 4. (Color online) (A) Feynman diagrams for the ball and triangle diagrams contributing to the $\Lambda N \to NN$ interaction. (B) Medium-long range part of the potentials for the one-pion-exchange, one-kaon-exchange, ball diagram and triangle diagrams

The two-pion exchange diagrams can be organized according to their topologies: balls, triangles and boxes. The momenta carried by the baryons in the loops of these diagrams are not constrained by any external momenta, and all their possible values must be integrated. These integrals have been calculated using dimensional regularization, with the minimal substraction scheme. Additionally, the two-pion exchange contributions involve the Σ hyperon as an explicit degree of freedom. The $\Sigma N\pi$ vertex is experimentally known from the weak decay of the Σ. The other vertices that appear in the loop diagrams and are not present in the one-meson exchanges, such as $\Lambda N\pi\pi$, are derived from the $SU(3)$ chiral Lagrangian. Full details of this calculation can be found in Ref. [14]. As a sample, the ball and triangle diagrams are depicted in Fig. 4 (A), and their potentials driving the ${}^3S_1 \rightarrow {}^3S_1$ transition are compared to the corresponding ones for the OPE and the OKE in Fig. 4 (B). At medium-long range, the two-pion exchange diagrams are comparable to the OPE and OKE mechanisms. However, they decay much faster than the OPE, which is the main contribution at long range.

4. Summary and perspectives

We have developed an EFT for the weak $\Lambda N \rightarrow NN$ transition up to $\mathcal{O}(\vec{q}^2)$. The numerical values for the LO LEC's have been obtained by two means. First, by a fit to the experimental data on the decay of hypernuclei, and second, by a comparison to the OME model describing the same interaction. Additionally, we have studied the possible inclusion of a σ meson in the OME picture, extracting thus a range for corresponding PV coupling. The next-to-leading order part of the EFT has been derived by systematically including all the possible two-pion exchange diagrams and operational structures up to second order in the momentum expansion.

In the future, we plan to reduce the model dependency of our approach by also using the EFT formalism in the strong sector. This is feasible for $A = 3, 4$ hypernuclei, where the few-body initial and final states can be solved using the Faddeev-Yakubovsky scheme [15]. Currently we are describing the non-mesonic decay of the hypertriton with our LO EFT weak potential and by using strong EFT potentials in both the initial hypernuclear wave function and the final state interactions. Once this calculation is completed, it can be easily extended to $A = 4$ hypernuclei.

Acknowledgments

This work is partly supported by grants FPA2010-21750-C02-02, FIS2008-00784 TOQATA and FIS2011-24154 from MICINN, 283286 from European Community-Research Infrastructure Integrating Activity 'Study of Strongly Interacting Matter', CSD2007-00042 from Spanish Ingenio-Consolider 2010 Program CPAN, and 2009SGR-1289 from Generalitat de Catalunya. A.P-O. acknowledges support by the APIF Ph.D. program of the University of Barcelona. B.J.D. is supported by the Ramon y Cajal program.

References

[1] T R Saito et al., Nucl. Phys. A **853**, 110 (2010).
[2] W Erni et al., Eur. Phys. J. A **49**, 25 (2013).
[3] O Hashimoto and H Tamura, Prog. Part. Nucl. Phys. **57**, 564 (2006).
[4] A Parreño, A Ramos, and C Bennhold, Phys. Rev. C **56**, 339 (1997).
[5] J-H Jun, Phys. Rev. C **63**, 044012 (2001).
[6] A Parreño, C Bennhold, and B R Holstein, Phys. Rev. C **70**, 051601 (2004); A Parreño, C Bennhold, and B R Holstein, Nucl. Phys. A **754**, 127c (2005).
[7] S Weinberg, Phys. Lett. B **251**, 288 (1990); S Weinberg, Nucl. Phys. B **363**, 3 (1991).
[8] E Jenkins and A Manohar, Phys. Lett. B **255**, 558-562 (1991).
[9] V G J Stoks and T A Rijken, Phys. Rev. C **59**, 3009 (1999); T A Rijken, V G J Stoks and Y Yamamoto, Phys. Rev. C **59**, 21 (1999).
[10] B Holzenkamp, K Holinde, and J Speth, Nucl. Phys. A **500**, 485 (1989).
[11] A Pérez-Obiol, A Parreño, and B Juliá-Díaz, Phys. Rev. C **84**, 024606 (2011).
[12] G Ecker, J Kambor, and D Wyler, Nucl. Phys. B **394**, 101 (1993).
[13] R Machleidt, Adv. Nucl. Phys. **19**, 189 (1989).
[14] A Pérez-Obiol, D R Entem, B Juliá-Díaz, and A Parreño, Phys. Rev. C **87**, 044614 (2013).
[15] H Kamada and W Glöckle, Nucl. Phys. A **548**, 205 (1992).

Polyakov loop susceptibilities in pure gauge system

P M Lo

GSI, Helmholzzentrum für Schwerionenforschung, Planckstrasse 1, D-64291 Darmstadt, Germany

E-mail: pmlo@gsi.de

Abstract. We present new lattice results on Polyakov loop susceptibilities in the SU(3) pure gauge system. Ratios of these susceptibilities are useful probes for deconfinement. Progress toward understanding these ratios in the presence of quarks are briefly discussed.

1. Introduction

Deconfinement can be described by the spontaneous breaking of $\mathcal{Z}_3$ center symmetry [1, 2, 3, 4, 5]. This symmetry, however, is explicitly broken by dynamical quarks. The result is the flattening of the static quark potential at long distances, even when the temperature is below criticality. If one aims at studying the confining part of the heavy quark potential, and probe the related symmetry breaking phase transition, it is useful to study QCD in the limit of exact $\mathcal{Z}_3$ symmetry —SU(3) pure gauge theory.

The relevant observables to study deconfinement are the Polyakov loop and its susceptibilities. The Polyakov loop, which measures the free energy of a static quark immersed in a hot gluonic medium [5, 6], defines an order parameter for the deconfinement transition. At low temperatures its thermal expectation value vanishes, signaling color confinement, while at high temperatures it is nonzero, resulting in a finite energy of a static quark and consequently the deconfinement of color. The Polyakov loop susceptibility, on the other hand, represents fluctuations of the order parameter. It features a peak at the transition temperature, and a width that signals the temperature window in which phase transition takes place.

While the basic thermodynamic functions of the SU(3) pure gauge theory, such as pressure and entropy, are well established within the lattice approach [7, 8], it is less clear for the temperature dependence of the renormalized Polyakov loop and its susceptibilities. Careful study of these quantities can enhance our understanding of the QCD phase structure.

2. Polyakov loop and its susceptibilities on the lattice

On an $N_\sigma^3 \times N_\tau$ lattice, the Polyakov loop is defined as the trace of the product over temporal gauge links,

$$L_{\vec{x}}^{\text{bare}} = \frac{1}{3} Tr \prod_{\tau=1}^{N_\tau} U_{(\vec{x},\tau),4}\,, \tag{1}$$

$$L^{\text{bare}} = \frac{1}{N_\sigma^3} \sum_{\vec{x}} L_{\vec{x}}^{\text{bare}}\,. \tag{2}$$

Journal of Physics: Conference Series **503** (2014) 012034 doi:10.1088/1742-6596/503/1/012034

The bare Polyakov loop needs to be renormalized to give a physical, N_τ-independent result. We perform the following multiplicative renormalization [9]:

$$L^{\rm ren} = \left(Z(g^2)\right)^{N_\tau} L^{\rm bare}. \tag{3}$$

and introduce the ensemble average of its modulus, $\langle|L^{\rm ren}|\rangle$. This quantity is well defined in the continuum and thermodynamic limits and is an order parameter for the spontaneous breaking of $\mathcal{Z}_3$ center symmetry.

We now define the corresponding susceptibilities. In the SU(3) gauge theory, the Polyakov loop operator is complex. One can therefore explore its fluctuations along the longitudinal and transverse directions, as well as that of its absolute value,

$$T^3\chi_L = \frac{N_\sigma^3}{N_\tau^3}\left[\langle(L_L^{\rm ren})^2\rangle - \langle L_L^{\rm ren}\rangle^2\right], \tag{4}$$

$$T^3\chi_T = \frac{N_\sigma^3}{N_\tau^3}\left[\langle(L_T^{\rm ren})^2\rangle - \langle L_T^{\rm ren}\rangle^2\right], \tag{5}$$

$$T^3\chi_A = \frac{N_\sigma^3}{N_\tau^3}\left(\langle|L^{\rm ren}|^2\rangle - \langle|L^{\rm ren}|\rangle^2\right), \tag{6}$$

where $L_L = \mathrm{Re}(\tilde{L})$ and $L_T = \mathrm{Im}(\tilde{L})$. Here we have introduced the $\mathcal{Z}_3$-transformed Polyakov loop, $\tilde{L} = L\, e^{2\pi n i/3}$, with $n = 0, \pm 1$. The phase of the transformation is chosen such that the transformed Polyakov loop is located in the main sector, defined by $-\pi/3 < \arg(\tilde{L}) < \pi/3$.

Our lattice results [10] for these quantities are collected in Fig. 1.

3. The ratios of susceptibilities

The renormalization of gluonic correlation functions in general, Polyakov loop susceptibility in particular, is still an unsolved problem. One way to get around this is by considering ratios of these susceptibilities.

Fig. 2 shows the temperature dependence of the ratios $R_A = \chi_A/\chi_L$ and $R_T = \chi_T/\chi_L$, obtained in the SU(3) pure gauge theory. Also shown in the figure are the analogous results extracted from simulations of (2+1)-flavor QCD by the HotQCD collaboration, using the highly improved staggered quark (HISQ) action with almost physical quark masses on a $32^3 \times 8$ lattice [11].

In the pure gauge limit ($N_f = 0$), ratios R_A and R_T exhibit a θ-like discontinuity at T_c, and change only weakly with temperature on either side of the transition. These features make them ideal for probing deconfinement.

We now interpret their limiting values, deep in the confining and deconfining phases, based on general symmetry arguments and the characteristics of Polyakov loop distribution function. Consider first the confining phase at low temperatures. Due to the $\mathcal{Z}_3$-symmetric vacuum, the expectation value of any symmetry breaking operator must vanish. In particular,

$$V(\langle\tilde{L}^2\rangle - \langle\tilde{L}\rangle^2) = \chi_L - \chi_T = 0. \tag{7}$$

It follows that $R_T = 1$ in the confining phase.

In the same temperature range, R_A closely follows the Gaussian distribution result of $(2-\pi/2) \simeq 0.429$ [12]. This suggests that the quadratic term in the effective action contributes dominantly to the susceptibilities deep in the confining phase. However, we expect non-Gaussian terms to be crucial in determining higher-order cumulants.

At high temperatures, $\mathcal{Z}_3$ symmetry is spontaneously broken. This has been shown to yield $R_A \simeq 1$ [12]. On the other hand, R_T is not restricted by symmetries. Our result indicates a

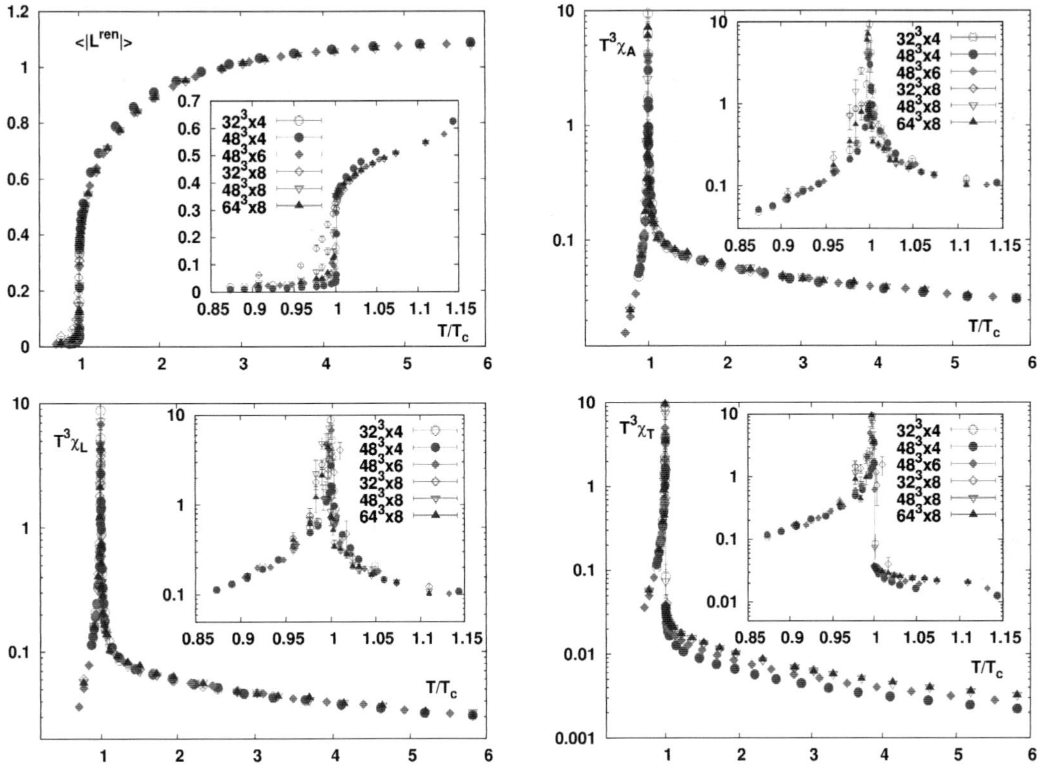

Figure 1. The temperature dependence of the modulus of renormalized Polyakov loop and its susceptibilities in the SU(3) pure gauge theory.

small value for this ratio. In the language of an effective Polyakov loop potential, this finding suggests that, around the global minimum associated with the symmetry-broken vacuum, the local curvature along the transverse (imaginary) direction is much steeper than that along the longitudinal (real) direction. We note, however, that a residual N_τ dependence remains in our results. Therefore we cannot yet draw any firm conclusions about the continuum extrapolation of this quantity.

In the presence of light quarks, the Polyakov loop is no longer an order parameter for deconfinement. Due to the explicit breaking of $\mathcal{Z}_3$ symmetry, the ratios are smoothened and vary continuously across the pseudocritical temperature. R_A interpolates between the two limits set by the pure gauge theory, joined by a crossover region. We expect the width of this crossover to depend on the number of flavors and the values of quark masses. We also observe that the value of R_T at high temperatures deviate substantially from the pure gauge limit. For this there is as yet no good theoretical understanding.

Further work is needed for the physical interpretations of these results, and to analyze in detail the systematic uncertainties involved in the extractions of these quantities.

4. Conclusions

Polyakov loop susceptibilities and their ratios provide excellent signals for deconfinement phase transition in the pure gauge system. Most of their features can be related to $\mathcal{Z}_3$ center symmetry and its spontaneous breaking. In this sense, they are the natural composite operators to study deconfinement.

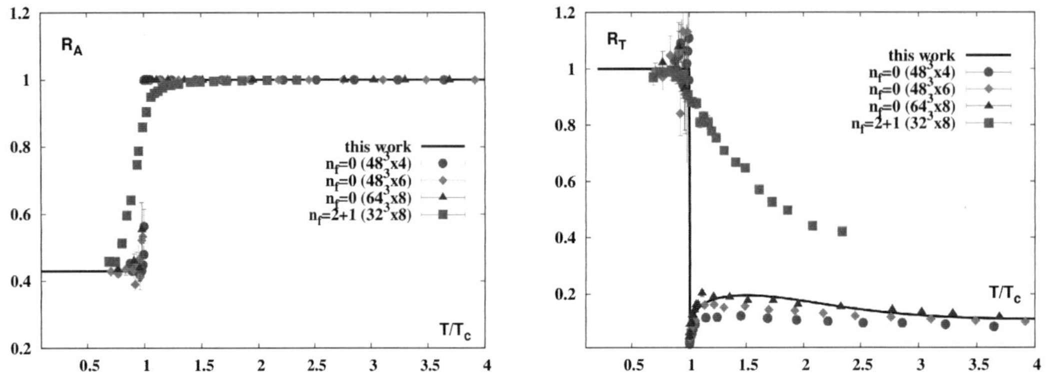

Figure 2. Lattice results on the ratios of Polyakov loop susceptibilities, $R_A = \chi_A/\chi_L$ and $R_T = \chi_T/\chi_L$, for the pure gauge system and (2+1)-flavor QCD. The temperature is normalized to its (pseudo)critical value for the respective lattice. The lines are results from the Polyakov loop model proposed in Ref. [10].

Many effective potentials have their model parameters tuned to match the lattice data of thermodynamic pressure and Polyakov loop [13, 14, 15, 16]. Fluctuation effects, on the other hand, have yet to be included. The new susceptibility results presented here can help to better constrain these seemingly arbitrary model parameters [10].

Our preliminary results for (2+1)-flavor QCD suggest that ratios of susceptibilities are considerably smoothened in the presence of light quarks. Also, the value of R_T in the high temperature phase deviates substantially from the pure gauge limit. To understand these issues, theoretical inputs are essential. In particular, it is important to understand the quark mass and flavor dependence of the Polyakov loop and the susceptibilities. An exploratory first step would be to include a small explicit breaking term in the pure glue potential and study how the picture changes from the pure gauge theory.

For the lattice calculations, more work is needed for the robust extractions of these gluonic correlation functions, as well as for the detailed understanding of their systematic uncertainties.

Acknowledgments
I am grateful to the organizers of FAIRNESS 2013 for arranging this excellent workshop. The support for accommodation and travel expenses from the GSI Helmholtz Center for Heavy Ion Research is greatly appreciated. I would like to thank Bengt Friman, Krzysztof Redlich, Olaf Kaczmarek and Chihiro Sasaki for the productive collaboration. I am also grateful to Frithjof Karsch for many illuminating discussions and to the HotQCD Collaboration for providing the data of the Polyakov loop. This work was supported by the Frankfurt Institute for Advanced Studies (FIAS).

References

[1] Greensite J 2003 *Prog. Part. Nucl. Phys.* **51** 1
[2] Swanson E S 2004 *AIP Conf. Proc.* vol 717, ed H Orth *et al.* (United States: American Institute of Physics) p 636
[3] 't Hooft G 1978 *Nucl. Phys.* B **138** 1
[4] Yaffe L G and Svetitsky B 1982 *Phys. Rev.* D **26** 963
[5] McLerran L D and Svetitsky B 1981 *Phys. Lett.* B **98** 195; 1981 *Phys. Rev.* D **24** 450
[6] Polyakov A M 1978 *Phys. Lett.* B **72** 477

[7] Boyd G, Engels J, Karsch F, Laermann E, Legeland C, Lutgemeier M and Petersson B 1996 *Nucl. Phys.* B **469** 419
[8] Borsanyi S, Endrodi G, Fodor Z, Katz S D and Szabo K K *J. High Energy Phys.* 07 (2012) 056
[9] Kaczmarek O, Karsch F, Petreczky P and Zantow F 2002 *Phys. Lett.* B **543** 41
[10] Lo P M, Friman B, Kaczmarek O, Redlich K and Sasaki C 2013 *Phys. Rev.* D **88** 074502
[11] Bazavov A *et al.* 2012 *Phys.Rev.* D **85** 054503
[12] Lo P M, Friman B, Kaczmarek O, Redlich K and Sasaki C 2013 *Phys. Rev.* D **88** 014506
[13] Ratti C, Thaler M A and Weise W 2006 *Phys. Rev.* D **73** 014019
[14] Rossner S, Ratti C and Weise W 2007 *Phys. Rev.* D **75** 034007
[15] Fukushima K 2004 *Phys. Lett.* B **591** 277
[16] Sasaki C, Friman B and Redlich K 2007 *Phys. Rev.* D **75** 074013

Journal of Physics: Conference Series **503** (2014) 012035 doi:10.1088/1742-6596/503/1/012035

Zero temperature properties of mesons and baryons from an extended linear sigma-model

P Kovács and Gy Wolf

Institute for Particle and Nuclear Physics, Wigner Research Centre for Physics, Hungarian Academy of Sciences, XII. Konkoly Thege Miklós út 29-33, 1121 Budapest, Hungary

E-mail: kovacs.peter@wigner.mta.hu

Abstract. An extended linear sigma model with mesons ($q\bar{q}$ states) and baryons (qqq states) is presented. The model contains a low energy multiplet for every hadronic particle type, namely a scalar, a pseudoscalar, a vector and an axialvector nonet, a baryon octet and a baryon decuplet. The model parameters are determined through a multiparametric minimalization with the help of well known physical quantities. It is found that the considered zero temperature quantities (masses and decay widths) can be described well at tree-level and are in good agreement with the experimental data.

1. Introduction

The vacuum properties of strong interaction are very hard to investigate within the framework of QCD – the fundamental theory of strong interaction – (see e.g.[1]), which is due to its subtlety at low energies. Consequently, instead of solving QCD one can set up an effective theory, which reflects some properties of the original theory. The underlying principle in the construction of such theories is that they share the same global symmetries as QCD.

For zero masses of the u, d and s quarks, the global symmetry of QCD is $U(3)_R \times U(3)_L$, the so-called chiral symmetry. In the vacuum, this chiral symmetry is spontaneously broken due to the existence of a quark-antiquark condensate. The chiral symmetry can be realized in different ways, nonlinearly [2] and linearly [3], which we choose here. Accordingly, in this paper we set up an extended linear sigma model, which contains mesonic and baryonic degrees of freedom. The previous versions of our model [4, 5] contained the scalar, pseudoscalar, vector- and axial-vector nonets. The vacuum phenomenology of mesons was described very well in that model. As an improvement, in this paper we include additionally the nucleon-octet and the Delta-decuplet to extend the vacuum phenomenology for baryons as well. Another approach to baryon phenomenology can be found in [6].

Our paper is organized as follows. In the Sec. 2 we briefly present the model, while in Sec. 3 we describe how to calculate various tree-level quantities. The Sec. 4 is dedicated to the results in the mesonic and baryonic sector and finally we conclude in Sec. 5.

2. The Model

The model can be defined through a Lagrangian consisting of a purely mesonic and a baryonic-mesonic part as $\mathcal{L} = \mathcal{L}_{\text{meson}} + \mathcal{L}_{\text{baryon}}$. The terms in the meson part are limited by chiral and dilaton symmetry (for details see [5]), while in case of the baryon part we included all the $SU(3)_V$

invariants which can produce baryon masses – with different masses for different particles in the given multiplet – and decuplet decays with the lowest possible dimension ($B-B-\Phi-\Phi$, $\Delta-\Delta-\Phi-\Phi$ and $\Delta-B-\Phi$ terms). The mesonic part has the following form

$$\begin{aligned}\mathcal{L}_{\rm meson} &= \mathrm{Tr}[(D_\mu\Phi)^\dagger(D_\mu\Phi)] - m_0^2\,\mathrm{Tr}(\Phi^\dagger\Phi) - \lambda_1[\mathrm{Tr}(\Phi^\dagger\Phi)]^2 - \lambda_2\,\mathrm{Tr}(\Phi^\dagger\Phi)^2\\ &-\frac{1}{4}\mathrm{Tr}(L_{\mu\nu}^2+R_{\mu\nu}^2) + \mathrm{Tr}\left[\left(\frac{m_1^2}{2}+\Delta\right)(L_\mu^2+R_\mu^2)\right] + \mathrm{Tr}[H(\Phi+\Phi^\dagger)]\\ &+c_1(\det\Phi-\det\Phi^\dagger)^2 + i\frac{g_2}{2}(\mathrm{Tr}\{L_{\mu\nu}[L^\mu,L^\nu]\}+\mathrm{Tr}\{R_{\mu\nu}[R^\mu,R^\nu]\})\\ &+\frac{h_1}{2}\mathrm{Tr}(\Phi^\dagger\Phi)\,\mathrm{Tr}(L_\mu^2+R_\mu^2) + h_2\,\mathrm{Tr}[(L_\mu\Phi)^2+(\Phi R_\mu)^2] + 2h_3\,\mathrm{Tr}(L_\mu\Phi R^\mu\Phi^\dagger).\\ &+g_3[\mathrm{Tr}(L_\mu L_\nu L^\mu L^\nu)+\mathrm{Tr}(R_\mu R_\nu R^\mu R^\nu)] + g_4[\mathrm{Tr}\,(L_\mu L^\mu L_\nu L^\nu)+\mathrm{Tr}\,(R_\mu R^\mu R_\nu R^\nu)]\\ &+g_5\,\mathrm{Tr}\,(L_\mu L^\mu)\,\mathrm{Tr}\,(R_\nu R^\nu) + g_6[\mathrm{Tr}(L_\mu L^\mu)\,\mathrm{Tr}(L_\nu L^\nu)+\mathrm{Tr}(R_\mu R^\mu)\,\mathrm{Tr}(R_\nu R^\nu)]\,,\end{aligned} \tag{1}$$

where

$$\begin{aligned} D^\mu\Phi &\equiv \partial^\mu\Phi - ig_1(L^\mu\Phi-\Phi R^\mu) - ieA^{e\,\mu}[T_3,\Phi]\,,\\ L^{\mu\nu} &\equiv \partial^\mu L^\nu - ieA^{e\,\mu}[T_3,L^\nu] - \{\partial^\nu L^\mu - ieA^{e\,\nu}[T_3,L^\mu]\}\,,\\ R^{\mu\nu} &\equiv \partial^\mu R^\nu - ieA^{e\,\mu}[T_3,R^\nu] - \{\partial^\nu R^\mu - ieA^{e\,\nu}[T_3,R^\mu]\}\,,\end{aligned}$$

The quantities $\Phi = \sum_{i=0}^8(S_i+iP_i)T_i$, $L^\mu/R^\mu = \sum_{i=0}^8(V_i^\mu \pm A_i^\mu)T_i$ represent the scalar-pseudoscalar nonets and the left-/right-handed vector nonets. $T_i\,(i=0,\ldots,8)$ denote the generators of $U(3)$, while S_i represents the scalar, P_i the pseudoscalar, V_i^μ the vector, and A_i^μ the axial-vector meson fields, and $A^{e\,\mu}$ is the electromagnetic field. H and Δ are some constant external fields. It should be noted that in the $(0-8)$ sector[1] of the scalars and pseudoscalars there is a mixing and it is more suitable to use the non strange – strange basis defined as $\varphi_N = 1/\sqrt{3}(\sqrt{2}\varphi_0+\varphi_8)$, $\varphi_S = 1/\sqrt{3}(\varphi_0-\sqrt{2}\varphi_8)$ for $\varphi_i \in (S_i, P_i, V_i^\mu, A_i^\mu)$.

Moving on to the baryonic-mesonic part, the Lagrangian is given by

$$\begin{aligned}\mathcal{L}_{\rm baryon} &= \mathrm{Tr}\left[\bar{B}\ i\not{D} - M_{(8)})\,B\right]\\ &-\mathrm{Tr}\left\{\bar{\Delta}_\mu\cdot\left[\left(i\not{D}-M_{(10)}\right)g^{\mu\nu} - i\left(\gamma^\mu D^\nu+\gamma^\nu D^\mu\right)+\gamma^\mu\left(i\not{D}+M_{(10)}\right)\gamma^\nu\right]\Delta_\nu\right.\\ &+C\,\mathrm{Tr}\left[\bar{\Delta}^\mu\cdot\left(-\frac{1}{f}(\partial_\mu - ieA_\mu^e[T_3,\Phi]) - \frac{1}{f}[\Phi,V_\mu]+A_\mu\right)B\right] + \mathrm{h.\ c.}\\ &-\xi_1\,\mathrm{Tr}\ \bar{B}B)\,\mathrm{Tr}\left(\Phi^\dagger\Phi\right) - \xi_2\,\mathrm{Tr}\left(\bar{B}\{\{\Phi,\Phi^\dagger\},B\}\right) - \xi_3\,\mathrm{Tr}\left(\bar{B}[\{\Phi,\Phi^\dagger\},B]\right)\\ &-\xi_4\left(\mathrm{Tr}\ \bar{B}\Phi)\,\mathrm{Tr}\left(\Phi^\dagger B\right)+\mathrm{Tr}\left(\bar{B}\Phi^\dagger\right)\mathrm{Tr}\left(\Phi B\right)\right) - \xi_5\,\mathrm{Tr}\left(\bar{B}\{[\Phi,\Phi^\dagger],B\}\right)\\ &-\xi_6\,\mathrm{Tr}\left(\bar{B}[[\Phi,\Phi^\dagger],B]\right) - \xi_7\left(\mathrm{Tr}\ \bar{B}\Phi)\,\mathrm{Tr}\left(\Phi^\dagger B\right)-\mathrm{Tr}\left(\bar{B}\Phi^\dagger\right)\mathrm{Tr}\left(\Phi B\right)\right)\\ &-\xi_8\left(\mathrm{Tr}\left(\bar{B}\Phi B\Phi^\dagger\right)-\mathrm{Tr}\left(\bar{B}\Phi^\dagger B\Phi\right)\right)+\chi_1\,\mathrm{Tr}\ \bar{\Delta}\cdot\Delta)\,\mathrm{Tr}\left(\Phi^\dagger\Phi\right)\\ &+\chi_2\,\mathrm{Tr}\left((\bar{\Delta}\cdot\Delta)\{\Phi,\Phi^\dagger\}\right)+\chi_3\,\mathrm{Tr}\left((\bar{\Delta}\cdot\Phi)(\Phi^\dagger\cdot\Delta)+(\bar{\Delta}\cdot\Phi^\dagger)(\Phi\cdot\Delta)\right)\\ &+\chi_4\,\mathrm{Tr}\left((\bar{\Delta}\cdot\Delta)[\Phi,\Phi^\dagger]\right),\end{aligned} \tag{2}$$

where $B = \sqrt{2}\sum_{i=1}^8 B_aT_a$ and Δ_μ stands for the baryon octet and decuplet. $M_{(8)}$ and $M_{(10)}$ are the bare masses of the baryon octet and decuplet. f is the pion decay constant, while [,]

[1] A 2 by 2 segment of the mass matrix consisting of the components: 00, 08, 80, 88

and { , } denote the commutator and the anticommutator. Here the meson-baryon interaction terms are all the possible $SU(3)_V$ invariants that can be written down with the given number of fields [7, 8]. The covariant derivatives are defined as

$$D_\mu B = \partial_\mu B + i[B, V_\mu] + \frac{1}{f}\{[A_\mu, \Phi], B\},$$

$$D_\mu \Delta_\nu^{ijk} = \partial \Delta_\nu^{ijk} + \left(\frac{1}{f}[A_\mu, \Phi]_l^i - i V_{\mu}{}_l^i\right)\Delta_\nu^{ljk} + \left(\frac{1}{f}[A_\mu, \Phi]_l^j - i V_{\mu}{}_l^j\right)\Delta_\nu^{ilk} + \left(\frac{1}{f}[A_\mu, \Phi]_l^k - i V_{\mu}{}_l^k\right)\Delta_\nu^{ijl},$$

and the following dot notation is used:

$$(\bar{\Delta}\cdot\Delta)_k^m \equiv \bar{\Delta}_{ijk}\Delta^{ijm}, \quad (\bar{\Delta}\cdot\Phi)_k^m \equiv \bar{\Delta}_{ijk}\Phi_l^i\epsilon^{jlm}, \quad (\Phi\cdot\Delta)_k^m \equiv \Delta^{ijm}\Phi_i^l\epsilon_{jlm}. \tag{3}$$

From the given Lagrangian various tree-level quantities such as masses and decay widths can be calculated and can be used to determine the unknown parameters of the model. During this parametrization process we can check how well the physical spectrum is reproduced. In the next section calculation of tree-level quantities and the parametrization are discussed.

3. Tree-level quantities and parametrization

As a standard procedure in a spontaneously broken theory we assume non-zero vacuum expectation values (vev) to certain fields, in our case to the $\sigma_N \equiv 1/\sqrt{3}(\sqrt{2}\sigma_0 + \sigma_8)$ and $\sigma_S \equiv 1/\sqrt{3}(\sigma_0 - \sqrt{2}\sigma_8)$ scalar fields, and denote their vev by ϕ_N and ϕ_S. After that the σ_N and σ_S fields are shifted with their non zero vev's ϕ_N and ϕ_S. Consequently, the quadratic and three-coupling terms of the Lagrangian can be determined from which the masses and the decay widths originate. However, it should be noted that as a technical difficulty – due to the σ_N and σ_S field shifts – different particle mixings emerge. In detail, there will be mixings in the $N - S$ (or $0 - 8$) sector of the scalar and pseudoscalar octets and between the vector-scalar and axialvector-pseudoscalar nonets. The $N - S$ mixings can be resolved by some orthogonal transformation, while the other mixings by redefinition of certain (axial-)vector fields. The details can be found in [5] together with explicit expressions for the meson masses and various decay widths.

Regarding the baryon sector the tree-level octet and decuplet masses – from the terms of the Lagrangian quadratic in the fields B and Δ_μ – are found to be

$$\begin{aligned}
m_p = m_n &= M_{(8)} + \frac{1}{2}\xi_2(\Phi_N^2 + 2\Phi_S^2) + \frac{1}{2}\xi_3(\Phi_N^2 - 2\Phi_S^2),\\
m_\Xi &= M_{(8)} + \frac{1}{2}\xi_2(\Phi_N^2 + 2\Phi_S^2) - \frac{1}{2}\xi_3(\Phi_N^2 - 2\Phi_S^2),\\
m_\Sigma &= M_{(8)} + \xi_2\Phi_N^2,\\
m_\Lambda &= M_{(8)} + \frac{1}{3}\xi_2(\Phi_N^2 + 4\Phi_S^2) + \frac{1}{3}\xi_4(\Phi_N - \sqrt{2}\Phi_S)^2, m_\Delta \quad = M_{(10)} + \frac{1}{2}\chi_2\Phi_N^2,\\
m_{\Sigma^\star} &= M_{(10)} + \frac{1}{3}\chi_2(\Phi_N^2 + \Phi_S^2) + \frac{1}{6}\chi_3(\Phi_N - \sqrt{2}\Phi_S)^2,\\
m_{\Xi^\star} &= M_{(10)} + \frac{1}{6}\chi_2(\Phi_N^2 + 4\Phi_S^2) + \frac{1}{6}\chi_3(\Phi_N - \sqrt{2}\Phi_S)^2,\\
m_\Omega &= M_{(10)} + \chi_2\Phi_N^2.
\end{aligned} \tag{4}$$

Beside the masses one can consider two-body decays of the decuplet baryons. According to PDG [9] there are four such physically allowed decays,

$$\Delta \to p\pi, \quad \Sigma^\star \to \Lambda\pi, \quad \Xi^\star \to \Xi\pi, \quad \Sigma^\star \to \Sigma\pi. \tag{5}$$

And the decay widths are given by

$$\Gamma_{\Delta\to\pi p} = \frac{k_\Delta^3}{24 m_\Delta}(m_p + E_p)G^2, \quad \Gamma_{\Sigma^\star\to\pi\Lambda} = \frac{k_{\Sigma^\star}^3}{48 m_{\Sigma^\star}}(m_\Lambda + E_\Lambda)G^2,$$
$$\Gamma_{\Xi^\star\to\pi\Xi} = \frac{k_{\Xi^\star}^3}{48 m_{\Xi^\star}}(m_\Xi + E_\Xi)G^2, \quad \Gamma_{\Sigma^\star\to\pi\Sigma} = \frac{k_{\Sigma^\star}^3}{72 m_{\Sigma^\star}}(m_\Sigma + E_\Sigma)G^2, \tag{6}$$

with

$$G^2 = C^2 Z_\pi^2 \left(w_{a_1}^2 + \frac{1}{f^2} \right).$$

As it can be seen from the Lagrangian there are 30 unknown parameters of the model, 14 in the meson sector and 16 in the baryon sector. However some of them can be set to zero without the loss of generality, some of them not even appear in the formulas of the physical quantities considered here, while some of them appear only in certain combinations. All in all there is 19 parameters which should be determined. These parameters are determined through the comparison of the calculated tree-level expressions – from which we have 23 – of the spectrum and decay widths with their experimental value taken from [9] with artificially increased errors (5% for the masses[2] and 10% for the decay widths, since we do not expect from a tree-level model to be more precise). Our strategy is that first we set the parameters of the meson sector [5] and then we fit the remaining parameters of the meson-baryon interaction terms. For this we used a χ^2-minimalization, which was realized with a multiparametric minimalization code (MINUIT [10]).

4. Results

Using the above mentioned χ^2-minimalization method in the meson sector we obtain results summarized in Table 1, which can also be found in [5].

In the baryon sector at first we only investigated the decuplet decays. The results are given in Table 2.

It can be seen that the meson observables are reproduced very well in general, while the decuplet decays show a fair correspondence. It is worth to note that in case of the the decuplet decays their tree-level expressions Eq. (6) only differ in their kinematic parts and we have only one parameter to fit for the four observables, which can cause the deviation from the experimental value.

5. Conclusion

We have presented an extended linear sigma model with meson and baryon degrees of freedom. This is a possible extension with baryon octet and decuplet of our previous meson model [5]. We included interaction terms, such as $\Delta - B - P$ and $B - B - \Phi - \Phi$ to describe Δ decays and baryon masses. We calculated the tree-level masses and physically relevant decuplet decay widths and we found that in general they are in good agreement with the experimental data taken from the PDG [9].

As a continuation we plan to include the baryon masses to the fit and to add other (higher dimension) interaction terms containing derivatives which are important in case of scattering processes [6]. Our further aim is to go to finite temperature and/or densities with all these fields included in our model.

Acknowledgments

P. Kovács and Gy. Wolf were partially supported by the Hungarian OTKA funds NK101438 and K109462.

[2] The isospin violation in some cases (e.g. for pion) has the order of 5%

Table 1. Calculated and experimental values of meson observables

Observable	Fit [MeV]	PDG [MeV]	Error [MeV]
m_π	141.0 ± 5.8	137.3	± 6.9
m_K	485.6 ± 3.0	495.6	± 24.8
m_η	509.4 ± 3.0	547.9	± 27.4
$m_{\eta'}$	962.5 ± 5.6	957.8	± 47.9
m_ρ	783.1 ± 7.0	775.5	± 38.8
$m_{K^\star}$	885.1 ± 6.3	893.8	± 44.7
m_ϕ	975.1 ± 6.4	1019.5	± 51.0
m_{a_1}	1186 ± 6	1230	± 62
$m_{f_1(1420)}$	1372.5 ± 5.3	1426.4	± 71.3
m_{a_0}	1363 ± 1	1474	± 74
$m_{K_0^\star}$	1450 ± 1	1425	± 71
$\Gamma_{\rho\to\pi\pi}$	160.9 ± 4.4	149.1	± 7.4
$\Gamma_{K^\star\to K\pi}$	44.6 ± 1.9	46.2	± 2.3
$\Gamma_{\phi\to\bar{K}K}$	3.34 ± 0.14	3.54	± 0.18
$\Gamma_{a_1\to\rho\pi}$	549 ± 43	425	± 175
$\Gamma_{a_1\to\pi\gamma}$	0.66 ± 0.01	0.64	± 0.25
$\Gamma_{f_1(1420)\to K^\star K}$	44.6 ± 39.9	43.9	± 2.2
Γ_{a_0}	266 ± 12	265	± 13
$\Gamma_{K_0^\star\to K\pi}$	285 ± 12	270	± 80

Table 2. Calculated and experimental values of baryon observables

Observable	Fit [MeV]	PDG [MeV]	Error [MeV]
$\Gamma_{\Delta\to p\pi}$	67.3	110.0	± 11.0
$\Gamma_{\Sigma^\star\to\Lambda\pi}$	27.0	32.0	± 3.2
$\Gamma_{\Sigma^\star\to\Sigma\pi}$	4.9	4.3	± 0.4
$\Gamma_{\Xi^\star\to\Xi\pi}$	11.2	9.5	± 1.0

References

[1] Peskin M E and Schroeder D V 1995*An Introduction To Quantum Field Theory* (Westview Press)
[2] Weinberg S 1968 *Phys. Rev.* **166** 1568
[3] Gasiorowicz S and Geffen D A 1969 *Rev. Mod. Phys.* **41** 531
[4] Parganlija D, Giacosa F and Rischke D H 2010 *Phys. Rev.* D **82** 054024
[5] Parganlija D, Kovacs P, Wolf G, Giacosa F and Rischke D H 2013 *Phys. Rev.* D **87** 014011
[6] Zhuang P 1999 *Commun. Theor. Phys.* **32** 395
[7] Dittner P 1971 *Commun. Math. Phys.* **22** 238
[8] Semke A 2010 *On the quark-mass dependence of baryon ground-state masses* (Ph.D. Thesis, TU Darmstadt) http://tuprints.ulb.tu-darmstadt.de/2360/
[9] Beringer J *et al* (Particle Data Group) 2012 *Phys. Rev.* D **86** 010001
[10] James F and Roos M 1975 *Comput. Phys. Commun.* **10** 343

Triplet based online track finding in the PANDA-STT

M C Mertens for the PANDA collaboration
II. Physikalisches Institut, Justus-Liebig-Universität Gießen, Germany
and IKP, Forschungszentrum Jülich GmbH, Germany
Present address: ZIM, Universität Duisburg-Essen, Germany

E-mail: marius.mertens@exp2.physik.uni-giessen.de

Abstract. The PANDA-Experiment at the future FAIR facility in Darmstadt will study antiproton-proton collisions in a fixed-target setup with a phase-space cooled antiproton beam with a momentum from 1.5 to 15 GeV/c at a nominal interaction rate of $2 \cdot 10^7$ s^{-1}. The data acquisition of the detectors has to run in a triggerless mode and the physics events of interest are identified by an online event filter. Tracking information is a key input for the event filter to distinguish signal events from background. A variety of tracking algorithms is foreseen to process the different track topologies. The so-called Triplet Finder, which is presented here, is a track finding algorithm based on the central straw tube tracker (STT) of PANDA. The algorithm focuses on mathematical simplicity and robustness to allow an online processing of the incoming detector hits. The algorithm and results of a proof-of-concept implementation are presented.

1. Introduction

The PANDA-Experiment will be an internal experiment in the High Energy Storage Ring (HESR) at the future Facility for Antiproton and Ion Research (FAIR) in Darmstadt. PANDA has a broad physics program in the charmonium energy region [1]. Of particular interest are channels with low production cross sections which require a high nominal interaction rate of $2 \cdot 10^7$ s^{-1} in order to be observed.

The PANDA data acquisition has to run in a continuous mode without a hardware trigger, i.e. the subdetectors continuously deliver data which is analyzed by an online event filter to identify the signal events to be stored. The main reasons for this type of operation are the high interaction rate and the fact that the topologies of signal events and hadronic background are very similar.

The online event filter uses high level information such as particle identification, momentum, invariant mass, etc. to distinguish signal events from background. Online track and event reconstruction is a key input to determine this high level information.

1.1. Charged particle tracking detectors in the target spectrometer

The PANDA apparatus consists of a target spectrometer with a 2 T solenoidal magnetic field which surrounds the interaction region and a forward spectrometer with a 2 Tm dipole field.

The main tracking detectors of PANDA's target spectrometer are the silicon sensor based Micro-Vertex-Detector (MVD, [2]) and the large-volume Straw Tube Tracker (STT, [3]) around the target interaction region together with a set of Gas Electron Multiplier (GEM) disks for the charged particle tracking within the 2 T solenoidal magnetic field.

The STT is a gas based detector which consists of 4636 cylindrical drift tubes (straws) of 1 cm diameter and 140 cm length, filling an almost cylindrical volume from 16 cm up to a radius of 42 cm around the MVD. The straw tubes are glued together to closed-packed multi-layers and operated at 1 bar overpressure which provides both the wire and tube tension without the need for a massive external frame (self supporting straws). This results in a radiation length of the STT of as low as 1.2% and a weight of about 15 kg. The distance of a track to the wire is determined by the drift time, i.e. the difference of the event start time and the recorded hit time, with a maximum value of 200 ns corresponding to a hit close to the tube wall.

1.2. Online tracking

Tracking information is required to determine the high level information used by the online event filter to identify signal events. Thus it is not known in advance whether a particular track belongs to a signal or a background event and we attempt to reconstruct all tracks in a continuous manner. The interplay of different tracking algorithms is required for an optimized reconstruction of the multitude of possible event topologies. As a consequence of the high interaction rate, the hits of tracks from different events may overlap in time and have to be associated to their respective events (event building) prior to the filter decision.

There is no distinct bunch structure in the quasi-continuous beam of the HESR, but the events can be organized in so-called bursts. One burst is equivalent to the events generated by one revolution (2000 ns) of the antiprotons stored in the HESR. At the nominal PANDA interaction rate of $2 \cdot 10^7$events per second, a burst comprises about 40 events. An estimated gap of 400 ns between each burst is sufficiently long (compared to the maximum drift time of 200 ns in the STT) to allow a separation into independent processing units without overlap.

PANDA's data acquisition will comprise FPGA (Field Programmable Gate Array)-based compute nodes for online data processing (Sec. 2.2.4 in [1]) and eventually conventional PC farms utilizing GPU (Graphics Processing Unit)-computing. The burst structure also allows for a natural way of parallel processing by parallelizing the analysis of the individual bursts. This strategy of data parallelism scales very well with added nodes and does not impose any special constraints on the tracking algorithms.

2. Triplet Finder

The Triplet Finder is a track reconstruction algorithm specifically designed for the PANDA-STT to reconstruct tracks of charged particles. However, it should also be applicable to other tracking detectors of similar geometry. The design of the Triplet Finder features the following main properties:

- Dependence only on information which is initially available from the detector: Absolute hit times and straw index. This enables the algorithm to be executed as hits arrive without any additional knowledge such as drift-times or event times (t_0). Based on the hit time distribution the algorithm can constrain the event start time to provide a seed value for subsequent algorithms.
- Mathematical simplicity: This facilitates the implementation of the algorithm not only in software but as well on FGPAs or GPUs as may be required to process the complete hit output of the STT in real-time.
- Robustness: It should be easy to add routines which deal with faulty (e.g. dead or noisy) channels or unfortunate hit patterns in the detector in order to optimize performance under imperfect operating conditions.

It consists of three stages Triplet finding, circle calculation and hit association.

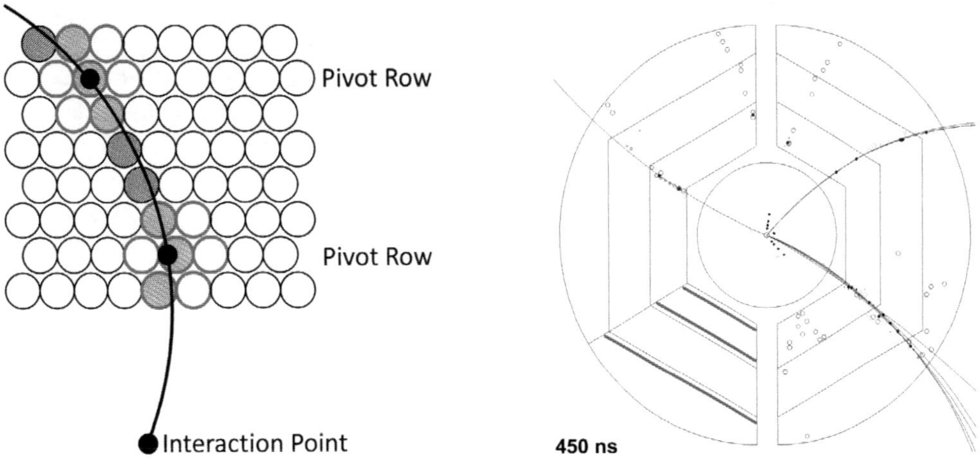

Figure 1. *Left:* Sketch (not to scale) of the Triplet finding and track calculation procedure in xy-projection with the second and second to last row being pivot rows. A hit tube is indicated by a shaded circle. Red outlines indicate straws which participate in the Triplet calculation and the small circles with black filling show the resulting Triplets. *Right:* Snapshot of an xy-view of the STT at 450 ns running time. Small circles indicated the detector hits as they are present at that particular timestamp. The thin blue lines are the found tracks as calculated by the Triplet Finder. The thick red lines indicate the location of the pivot rows for one sector. The pivot rows of the other sectors are distributed in the same way but not shown here.

2.1. Triplet finding
The Triplet finding stage analyzes the pattern of three up to seven adjacent hit straw tubes around selected pivot straws and calculates the center-of-mass coordinates of such a pattern which is called Triplet. The location of the pivot straws has been chosen as shown in Fig. 1 (right) in order to cover a large area with a large lever arm, but other patterns with more or less pivot straws are also possible. The Triplet Finder is designed to operate on a full burst of hits which results in hits from different events being present in the buffer at the same time. Therefore, only hits with timestamps which differ by at most one maximum drift-time window (200 ns and an additional safety margin of 50 ns which is subject to optimization) can be combined to one Triplet. Due to the finite number of possible patterns the calculation of the center-of-mass on an FPGA can be implemented as a lookup table.

2.2. Circle calculation
Then the center and radius of a circle through the coordinate origin and two Triplets are calculated. It is also possible to calculate the track circle through three arbitrary points. However one of the points being the origin serves as an additional simplification and allows for quick processing of the tracks originating from the primary vertex. A later processing of the remaining hits with three arbitrary points could be a useful extension of the Triplet Finder to identify also tracks from secondary vertices, e.g. Lambda-decays. In each case the purely analytic approach of a circle calculation though three points in a plane reduces the computational complexity of the algorithm.

2.3. Hit association
The final step is the hit association which accounts all hits whose distance from the circle origin differs by at most the tube radius (5 mm plus safety margin, subject to optimization) from

the circle radius. This step also considers only hits within a compatible drift-time window for association. The total number of hits (including the Triplets) then determines whether the track candidate is considered a valid track. The threshold for the minimum number of hits is another parameter to be optimized in order to obtain the best combination of efficiency and purity.

3. Performance

A proof of concept version of the Triplet Finder has been implemented and tested within the PandaROOT framework which is based on ROOT [4]. For testing and developing the Triplet Finder the main goal of benchmarking is to optimize the reconstruction efficiency while maintaining high purity.

In order to achieve realistic results, the Triplet Finder was tested with events generated according to the Dual Parton Model (DPM, [5]) with a beam momentum of 15 GeV/c at the nominal PANDA interaction rate of $2 \cdot 10^7$ events per second with a Poissonian event distribution taken into account. The DPM generator simulates elastic and inelastic, hadronic and Coulomb interactions and is used as a background generator for antiproton proton collisions. This setup ensures a realistic track distribution, event overlap and hit patterns similar to what will actually be seen by the detector.

The event overlap is also visible in Fig. 1 (right). The hit patterns which look like tracks with some missing hits are in fact early hits which actually belong to tracks from a later event. At such high rates, the overlap of events within a burst is unavoidable and therefore it must be verified that the track finding algorithm correctly processes such input.

The combination of having a continuous stream of hits and not a specific signal channel to be reconstructed but instead statistically distributed particles with very different momenta poses a challenge for the benchmark: In order to obtain realistic values for the reconstruction efficiency of the algorithm, the tracks which are reconstructable must be identified based on criteria such as momentum, number of hits, etc. so the reconstructed track ratio can be correctly normalized.

Obviously, tracks with too low momentum, missing the STT, simulated secondary or tertiary electrons, remnants of nuclear reactions, etc. are all valid tracks in the simulation (Fig. 2 upper left), but should not enter the benchmark because they are neither interesting nor reconstructable at all. As an initial selection, a reconstructable track is required to meet the following criteria:

- Charged tracks in the first and second generation.
- $p_t > 60$ MeV/c and $\beta\gamma > 0.4$.
- Track start at most 2 cm away from coordinate origin.
- At least one STT hit.

While this still allows tracks which cannot be reconstructed due to geometric issues (e.g. just hitting a single straw) to pass, these rather loose conditions ensure that the performance is not overestimated by removing tracks which are difficult to reconstruct but will occur in a real life setup. Fig. 2 shows the momentum distribution of all simulated tracks (top left) and those which match the above criteria (top right).

From the simulated sample, 9228 tracks were considered reconstructable, of which the Triplet Finder's raw output contained 6821 tracks. A comparison of the momentum distributions of the momentum distribution of reconstructable (Fig. 2 top right) and reconstructed (Fig. 2 lower left) tracks shows a similar shape in both cases, i.e. the efficiency distribution is flat.

Another aspect to be studied is the output quality of the Triplet Finder, i.e. how many hits have been correctly associated to the found tracks. Different combinations of hits can lead to correct tracks being found multiple times with different subsets of hits as well as fake tracks which do not correspond to an initially simulated track. With the current settings, the Triplet Finder outputs 38445 tracks of which 21137 contain more than 80 % correctly assigned hits. The deviation of the reconstructed ejection angles from the simulated ejection angles (Fig. 2,

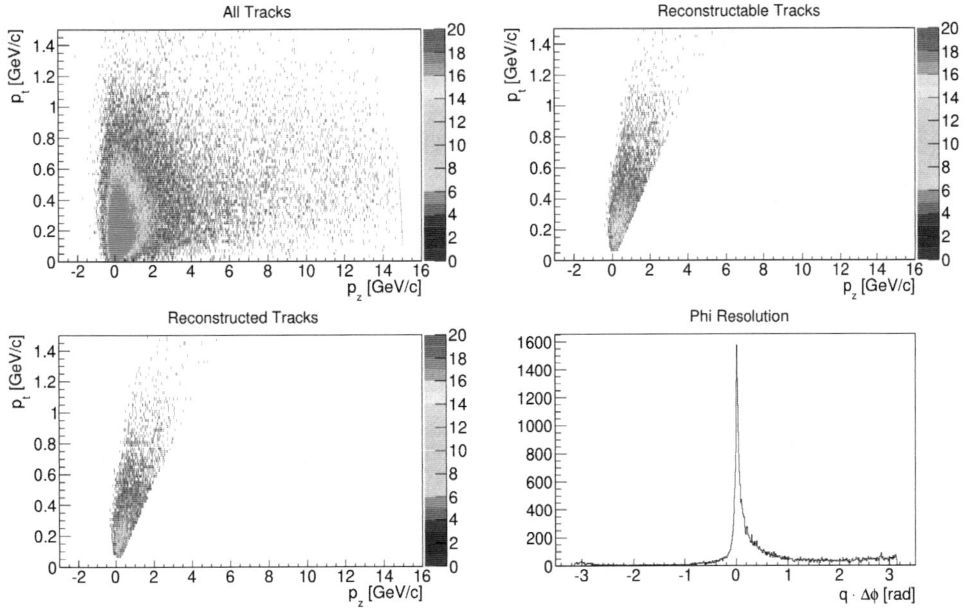

Figure 2. *Top left:* Momentum distribution of all simulated tracks. *Top right:* Momentum distribution of reconstructable tracks. *Bottom left:* Momentum distribution of reconstructed tracks. *Bottom right:* Deviation of the reconstructed track ejection angle from the MC-value at track start, separated by particle's charge.

lower right) is shown as an example for the track reconstruction properties. The plot is divided such that negatively charged particles are shown on the negative half of the x-axis and positively charged particles are shown on the positive half of the x-axis. The excess of positively charged particles is due to the large amount of elastically scattered recoil protons entering the detector.

4. Conclusion

The Triplet Finder has been designed as a track finding algorithm specifically for the PANDA-STT to identify tracks based on only the hit data with timestamps as it is generated by the detector without using further information which may require preprocessing of the data. The benchmark of its proof-of-concept implementation has shown that it is able to work under realistic operating conditions at the expected interaction rate with the resulting event overlap and delivers reasonable results. Since the development is just at its beginning, many parameters remain to be studied in order to optimize the performance of the Triplet Finder.

Acknowledgments

This work was supported by BMBF, HIC4FAIR and Forschungszentrum Jülich GmbH.

References

[1] PANDA Collaboration 2013 Physics performance report for PANDA: Strong interaction studies with antiprotons *Preprint* ArXiv:0903.3905v1[hep-ex]
[2] PANDA Collaboration 2013 Technical design report for the: PANDA micro vertex detector *Preprint* ArXiv:1207.6581[physics.ins-det]
[3] PANDA Collab. 2013 Technical design report for the PANDA straw tube tracker *Eur. Phys. J. A* **49** 25
[4] Antcheva I et al. 2009 *Computer Physics Communications* **180** (12) 2499–2512
[5] Capella A, Sukhatme U, Tan C I and Van J T T 1994 *Physics Reports* **236** (4-5) 225–329

Journal of Physics: Conference Series **503** (2014) 012037 doi:10.1088/1742-6596/503/1/012037

Status and future plan of the spectroscopy of pionic atoms

Y. N. Watanabe[1], G. P. A. Berg[2], M. Dozono[3], H. Fujioka[4], N. Fukuda[3], T. Furuno[4], H. Geissel[5], R. S. Hayano[1], N. Inabe[3], K. Itahashi[3], S. Itoh[1], D. Kameda[3], T. Kubo[3], H. Matsubara[6], S. Michimasa[7], K. Miki[8], H. Miya[7], Y. Murakami[1], M. Nakamura[3], N. Nakatsuka[4], T. Nishi[1], S. Noji[1], K. Okochi[1], S. Ota[7], H. Suzuki[3], K. Suzuki[9], M. Takaki[7], H. Takeda[3], Y. K. Tanaka[1], K. Todoroki[1], K. Tsukada[10], T. Uesaka[3], H. Weick[5], H. Yamada[1], H. Yamakami[4] and K. Yoshida[3]

[1] The University of Tokyo, 113-0033 Tokyo, Japan
[2] JINA and Department of Physics, University of Notre Dame, Indiana, 46556, USA
[3] Nishina Center for Accelerator-Based Science, RIKEN, Saitama, 351-0198, Japan
[4] Department of Physics, Kyoto University, Kyoto, 606-8502, Japan
[5] GSI Helmholtzzentrum fur Schwerionenforschung GmbH, D-64291 Darmstadt, Germany
[6] National Institute of Radiological Sciences (NIRS), Chiba 263-8555, Japan
[7] Center for Nuclear Study, University of Tokyo, Tokyo, 113-0033, Japan
[8] Research Center for Nuclear Physics, Osaka University, 567-0047 Osaka, Japan
[9] Stefan-Meyer-Institut fur subatomare Physik, 1090, Vienna, Austria
[10] Department of Physics, Tohoku University, 980-8578, Sendai, Japan

E-mail: `ywatanabe@nucl.phys.s.u-tokyo.ac.jp`

Abstract. The spectroscopy of deeply bound pionic atoms provides a way to understand the restoration of chiral symmetry breaking at finite density. We have been performing a series of experiments of missing-mass spectroscopy of the $(d,^3\mathrm{He})$ reaction at RIBF to investigate pionic atoms of several Sn isotopes. As a first step, we conducted a pilot experiment to measure deeply bound pionic states of ^{121}Sn and successfully observed the deeply bound pionic states. In addition to the experiment at RIBF, we are planning the spectroscopy of deeply bound pionic atoms in inverse kinematics and conducted a feasible study by simulations. We showed that by using a deuterium gaseous active target TPC and silicon detectors, the Q-value resolution is about 500 keV (FWHM) and the yield of the pionic 1s state is 20 counts/day, indicating the experiment is feasible.

1. Introduction

Chiral symmetry and its spontaneous breakdown plays an important role in the low energy quantum chromodynamics. The chiral symmetry is known to be restored in hot or dense medium and the order parameter of the chiral symmetry breaking $|\langle\bar{q}q\rangle|$ is reduced[1, 2]. The spectroscopy of deeply-bound pionic atoms is one of the methods that enables us to investigate experimentally the chiral symmetry restoration in the nuclear medium. The order parameter $|\langle\bar{q}q\rangle|$ is related to the isovector parameter b_1 of the s-wave part of the pion-nucleus optical potential[3, 4, 5, 6]

$$V_s(r) = -\frac{2\pi}{\mu}[\epsilon_1 b_0\{\rho(r) + b_1(\rho_n(r) - \rho_p(r))\} + \epsilon_2 B_0 \rho(r)^2].$$

In deeply-bound pionic states such as the 1s state, the pion probes a fraction of 0.6 of the normal nuclear density (ρ_0) due to the large overlap between the wavefunctions of the pion and the nucleus and the in-medium b_1 value is deduced from the binding energy and the width of the pionic states. Comparing it with the b_1 value in vacuum, the reduction of $|\langle\bar{q}q\rangle|$ is evaluated. In previous experiments at GSI with the missing-mass spectroscopy of $(d, {}^3\mathrm{He})$ pion transfer nuclear reaction, deeply-bound pionic ^{123,119,115}Sn atoms were measured and the results indicated symmetry restoration in the nuclear medium[7, 8, 9, 10, 11]. However, there were large statistical and systematic uncertainties in these experiments.

For further studies of pionic atom spectroscopy, there are two directions. The first is to determine the b_1 value more precisely. We are working for this purpose at the RI beam factory (RIBF) and will report preliminary results in Section 2.

The other is to evaluate b_1 value at density other than 0.6 ρ_0 and study the density dependence of $|\langle\bar{q}q\rangle|$. For this, we need to measure pionic atoms with unstable nuclei. This is because pionic atoms with stable nuclei probe only densities around 0.6 ρ_0[12], but with neutron rich unstable nuclei the pion is pushed outward and it is expected that they probe lower densities. We conducted a feasibility study for this experiment by simulations and will present the design of the experimental setup and the results of the study in Section 3.

2. Experiment at RIBF

We plan to perform a series of experiments at RIBF using the $(d, {}^3\mathrm{He})$ reaction near the pion emission threshold. Exploiting the intense deuteron beam, several targets around Sn nuclei will be studied to suppress the systematical errors mainly caused by the ambiguity of the nuclear properties. A first pilot experiment using the missing-mass spectroscopy of $^{122}\mathrm{Sn}(d, {}^3\mathrm{He})\pi^- \otimes {}^{121}\mathrm{Sn}$ reaction has been carried out.

2.1. Experimental Setup

Figure 1. shows the setup of the pilot experiment. The BigRIPS fragment separator is used as a high-resolution spectrometer with a newly developed optics. A 10^{12} particle per second, 250 MeV/u deuteron beam impinged on a ^{122}Sn target at the nominal target position and ^{3}He was produced by the $^{122}\mathrm{Sn}(d, {}^3\mathrm{He})\pi^- \otimes {}^{121}\mathrm{Sn}$ reaction. With this energy, the momentum transfer in the reaction is small and pionic s states are selectively populated. The horizontal position and the angle of the produced ^{3}He were measured at the F5 focal plane by two multi-wire drift chambers (MWDCs) which were aligned in the beam direction. To determine the reaction Q-value, the momentum of ^{3}He was deduced by the position and the angle at F5.

The main background was protons produced by nuclear-breakup reaction and it was about 10^3 times more abundant than the ^{3}He events. The background was clearly identified against the ^{3}He events by measuring the time of flight between F5 and F7 and the energy loss in the scintillation counters.

2.2. Results and discussion

Figure 2. shows the acceptance-corrected Q-value spectrum of the $^{121}\mathrm{Sn}(d, {}^3\mathrm{He})$ reaction with ^{3}He particles emitted at angles smaller than 1°. The red line indicates the pion emission threshold. The peaks of interest ride on a constant background of ^{3}He particles that are produced in nuclear reactions without pion production. On the right side of the threshold, peak structures corresponding to the pionic bound states were observed. The spectrum is in

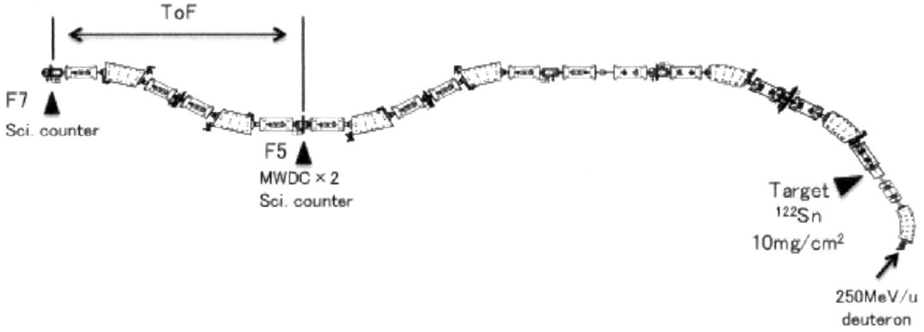

Figure 1. The setup of the experiment. A deuteron beam is injected into the BigRIPS spectrometer and the position and the angle of the produced ^{3}He particles are measured at the F5 dispersive focal plane using MWDCs.

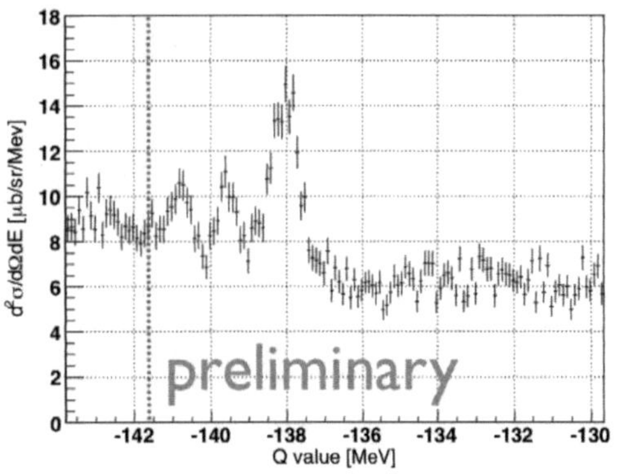

Figure 2. Preliminary Q-value spectrum of the^{121}Sn(d,^{3}He) reaction at the forward angles. The red dotted line corresponds to the pion emission threshold.

qualitatively agreement with the theoretical calculation including the angular dependence of the emitted ^{3}He[13].

3. Feasibility study of the spectroscopy using inverse kinematics

We are now planning to conduct the spectroscopy of pionic atoms with unstable nuclei. In order to use unstable nuclei, the inverse kinematics method is adopted where heavy ions are used as a beam and deuterons are used as a target. A dedicated setup is needed for the inverse kinematics measurement because the momentum of the emitted ^{3}He is only about 60 MeV with the recoilless condition in the center of momentum frame. Therefore, the signal ^{3}He particles are sensitive to the energy loss in the target so that the information of the reaction is easily lost. To suppress the influence of the energy loss, we plan to employ an active-target time projection chamber (AT-TPC) with silicon strip detectors.

3.1. Design of the experiment

Figure 3. shows the conceptual design for the inverse kinematics measurement with AT-TPC and silicon detectors. A 250MeV/u heavy ion beam is bombarded onto a deuterium-filled AT-

TPC that is placed in a uniform magnetic field of about 1T to separate the signal ^{3}He from the heavy ion beam. Inside the AT-TPC, 1500μm thick silicon strip detectors are installed to measure the full energy of the ^{3}He.

To determine the reaction Q-value, the momentum of ^{3}He at the reaction point should be measured in a similar way in experiments with normal kinematics. In this experiment, the momentum is deduced from three observables listed below:

- the energy loss along the path from the reaction point to the silicon detector
- the full energy at the silicon detector
- the angle of the emitted ^{3}He at the reaction point

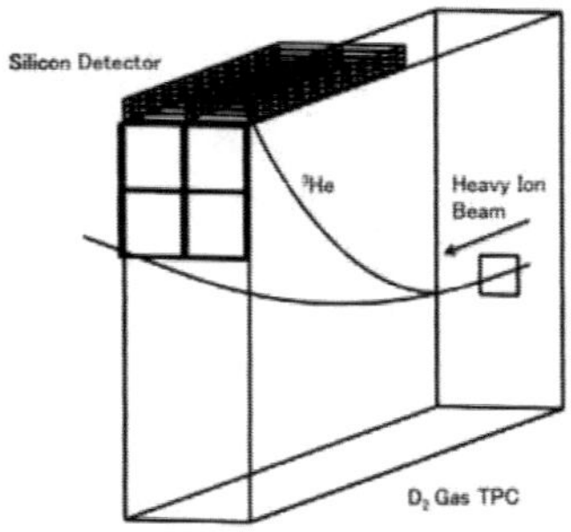

Figure 3. Conceptual design of the TPC and the silicon detectors. A heavy ion beam is injected into the TPC and the energy of the emitted ^{3}He is measured by the silicon detectors. Magnetic field is applied in the perpendicular direction to the beam.

3.2. Yield and resolution estimation

In such a setup, the yield of the pionic 1s state and the Q-value resolution were estimated by simulations with the condition shown in Table 1. The total cross section is estimated from past experiments.There are three main components contributing to the Q-value resolution:

- resolution of silicon detectors
- vertex position resolution
- energy straggling along the path from reaction point to the silicon detector

Besides these quantities, the energy spread of the incident beam may affect the resolution, but this contribution is relatively small.

Using these conditions, the yield of the pionic 1s state is calculated to be 20 counts/day. The estimated Q-value resolution from each of the components is shown in Table 2. The total resolution is 500 keV (FWHM).

Beam intensity	10^6 /s
D_2 gas thickness	1 m
Cross section	$5\times10^{-2}\mu$b
AT-TPC position resolution	500 μm
Energy resolution of Si detector at 60 MeV^{3}He	0.1 %

Table 1. Simulation conditions.

Resolution contribution	Contribution to Q-value resolution (FWHM) [keV]
Energy straggling	350
Vertex position resolution	150
Energy resolution of Si detectors	350
Total	500

Table 2. Estimate of Q-value resolution from each components.

3.3. Discussion

The resolution of 500 keV (FWHM) is sufficient to separate the peaks as the energy difference between typical pionic states is about 1MeV. The yield of 20 counts/day for pionic 1s state is not so large, but we expect the peak structures observed for about ten days measurements. Therefore, this result indicates that the spectroscopy of pionic atom with inverse kinematics is feasible using the presented conceptual design.

4. Summary and future prospect

A pilot experiment of the spectroscopy of deeply bound pionic ^{121}Sn was conducted at RIBF and the pionic states were successfully observed. We are now finalizing the analysis of the pilot experiment to determine the isovector parameter b_1 and preparing for the main experiment with several targets.

We conducted a feasibility study of the spectroscopy of pionic atom in inverse kinematics using AT-TPC. The results indicate that the experiment is feasible in terms of resolution and yield. Preparations for this experiment are presently under way. We have started to develop the AT-TPC and are now working on the silicon detectors.

References

[1] U. Vogl and W. Weise, Prog. Part. Nucl. Phys. **27**, 195 (1991) and references therein.
[2] T. Hatsuda and T. Kunihiro, Phys. Rept. **247**, 221 (1994) and references therein.
[3] M. Gell-Mann, R.J. Oakes and B. Renner, Phys. Rev.**175** (1968) 2195.
[4] E. E. Kolomeitsev, N. Kaiser, and W. Weise, Phys. Rev. Lett.**90**, 092501 (2003).
[5] D. Jido, T. Hatsuda and T. Kunihiro, Phys. Lett. **B670**, 109 (2008).
[6] N. Ikeno et al., Prog. Theor. Phys., **126**, 483 (2011).
[7] H. Gilg et al., Phys. Rev. **C62**, 025201 (2000).
[8] K. Itahashi et al., Phys. Rev. **C62**, 025202 (2000).
[9] H. Geissel et al., Phys. Rev. Lett. **88**, 122301 (2002).
[10] H. Geissel et al., Phys. Lett. B **549**, 64 (2002).
[11] K. Suzuki et al., Phys. Rev. Lett. **92**, 072302 (2004).
[12] N. Ikeno et al., Progress of Theoretical Physics, Vol. 126, No. 3, September 2011
[13] N. Ikeno, H. Nagahiro and S. Hirenzaki, Eur. Phys. J. **A47**, 161 (2011).

Theoretical nuclear structure and astrophysics at FAIR

Tomás R. Rodríguez
Institut für Kernphysik, Technische Universität Darmstadt, Schlossgartenstr. 2, D-64289, Darmstadt, Germany
E-mail: t.rodriguez@gsi.de

Abstract. Next generation of radioactive ion beam facilities like FAIR will open a bright future for nuclear structure and nuclear astrophysics research. In particular, very exotic nuclei (mainly neutron rich) isotopes will be produced and a lot of new exciting experimental data will help to test and improve the current nuclear models. In addition, these data (masses, reaction cross sections, beta decay half-lives, etc.) combined with the development of better theoretical approaches will be used as the nuclear physics input for astrophysical simulations. In this presentation I will review some of the state-of-the-art nuclear structure methods and their applications.

1. Introduction

The main goal of the nuclear astrophysics is the study of the origin of the elements that are found in Nature and their abundance. Hence, both the different nucleosynthesis processes, as well as the astrophysical scenarios where the proper conditions to obtain such processes are found, are investigated.

Concerning the nucleosynthesis processes, the existence of several of them are well established [1, 2], namely (see Fig. 1): a) Big Bang nucleosynthesis that produced basically hydrogen and helium; b) nuclear burning of light nuclei (hydrogen, helium, carbon, etc.) that produces elements up to the iron group where the nuclear fusion is not energetically favorable any longer; c) s-process, whose main mechanism is the slow neutron capture -slow compared to the beta decay half-lives- and produces the peaks observed in the elementary abundance curve around mass numbers $A = 90$, 138 and 208. This mechanism occurs through nuclei in the valley of stability; d) r-process or rapid neutron capture [3], which produces the peaks around $A = 80$, 130 and 196. This process flows through very neutron rich nuclei away from the stability; e) rp-process or rapid proton capture [4] and νp-process, where the antineutrino absorption plays a key role [5]. In both cases, the nuclei involved are neutron deficient.

On the other hand, each process requires compositional and thermodynamical conditions to make possible the nuclear and electroweak reactions which characterize them. Hence, while the nuclear burning and s-process occur mainly in the interior of stars, the other processes are found in X-Ray bursts in binary systems, supernova explosions or neutron star collisions, for instance.

To understand all of these processes and establish the astrophysical sites, the nucleosynthesis simulations require input parameters coming from nuclear physics, such as masses and half-lives, reaction rates, fission products, neutrino reactions, etc. These quantities are taken from

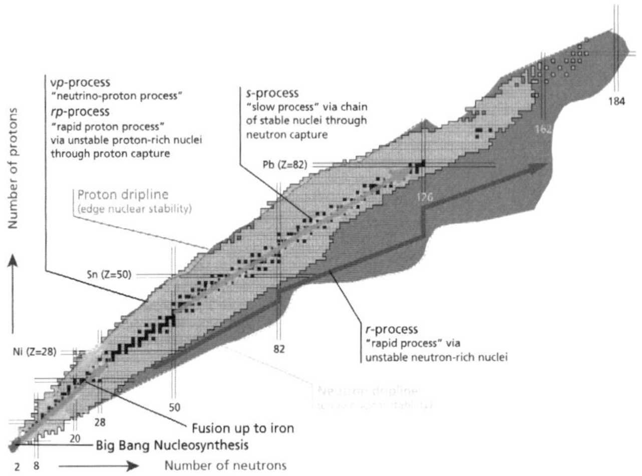

Figure 1. Chart of nuclei and the paths followed by the different nucleosynthesis processes. Black, orange and green colors represent stable, synthesized unstable and non-synthesized in earth nuclei, respectively. Figure adapted by G. Martínez-Pinedo from material available at https://www.gsi.de/en/start/fair/forschung_an_fair/nuclear_structure_physics.htm.

the experimental data where available. To do so, stable and radioactive ion beams facilities play a key role. In particular, one of the main goals of the NUSTAR collaboration at FAIR is the study of the properties of the most relevant nuclei in astrophysical processes, especially in the r-process [6]. Nevertheless, although in a near future both FAIR and other similar facilities such as FRIB (MSU-USA), SPIRAL-2 (GANIL-France) or RIBF (RIKEN-Japan) will reach a wide region of nuclei relevant for the r-process, they will not be able to cover the full range of nuclei involved in such a nucleosynthesis mechanism. Hence, the properties of these experimentally out of reach nuclei will have to be studied only theoretically. Therefore, it is important to continue developing nuclear models and theories to provide a reliable nuclear physics input for astrophysical simulations. In this contribution I will introduce briefly the most relevant theoretical approaches to study currently the nuclear structure and their range of applicability.

2. The road from QCD to the stars

The starting point of the microscopic models in nuclear physics is the definition of the interaction between the constituent nucleons (protons and neutrons). These interactions can be defined from bare interactions between the nucleons (sometimes called *realistic* interactions) or can take into account nuclear medium effects (*phenomenological* interactions). In any case, all of them have parameters or coupling constants that must be fitted to reproduce some sets of experimental data.

Hence, belonging to the first category, the bare interactions are fitted to reproduce with high precision the scattering phase shifts in nucleon-nucleon (NN) collisions and deuteron properties. If three body (3N) are included, this channel is normally fitted to reproduce properties of triton and ^{4}He. Therefore, the experimental information contained in this kind of interactions stops at four body systems at most since no additional many body forces beyond 3N have been taken into account so far. Among these interactions, those extracted from a chiral effective field theory (χEFT) are widely used and studied recently for their close connection to QCD [7, 8].

Here, the interactions are expanded systematically in powers of the pion mass with respect to the momentum scale of the effective theory, obtaining a hierarchy in the NN, 3N, etc. forces. Hence, at the lowest two orders of the expansion, only contact and one- and two- pion exchange NN forces are obtained, while 3N forces appear at next-to-next-to-leading order (N^2LO).

Concerning the phenomenological interactions, these are adjusted in different ways depending on the nuclear many body method used to solve the problem. For example, in the interactions used in large scale shell model (LSSM) calculations [9], the single particle energies are taken from the excitation energies of nuclei next to magic numbers -with one nucleon more or less. In the interactions used in energy density functional methods (EDF) [10], the parameters of the functional are fitted to reproduce masses and radii of nuclei in different regions of the nuclear chart.

In the following subsections I will give a brief description of the nuclear many body methods currently used applying either bare or in medium interactions as starting points, according to the previous classification.

2.1. Nuclear structure with bare nucleon interactions

The main obstacle for using bare interactions to calculate nuclei with mass numbers $A > 12$ is the fact that the nucleons with low momentum can be scattered to high momenta and viceversa due to the strong short-range repulsion and tensor terms present in those forces. This means that the number of states that have to be taken into account in the A-body Hilbert space explodes. Nevertheless, there are some methods like the Similarity Renormalization Group (SRG) -see Ref. [11] and references therein- or the Unitary Correlator Operator Method (UCOM) [12] where the original hamiltonian is unitary transformed to decouple high and low momentum states. The main drawback of these renormalization methods is that they induce many body forces even when the original hamiltonian only contains NN interactions. Therefore, truncations in those induced many body forces have to be done, although the contribution of these missing pieces can be estimated through the dependence of the observables on the renormalization parameters.

These soft interactions have been widely used recently with different many body methods and some of these calculations have been called as *ab-initio* nuclear structure. Hence, using the no-core shell model method (NCSM), where the hamiltonian is diagonalized in a large valence space without a core, nuclei up to ^{16}O including genuine and induced NN+3N forces have been calculated (see Ref. [13] and references therein). With other methods, which do not perform direct diagonalization, such as Couple-Cluster (CC) [14], in-medium SRG [15] or Self-consistent Green's function methods [16] heavier magic and semi magic nuclei up to $A \sim 70$ have been calculated. Recently, these interactions have been also used in combination with many body perturbation theory and standard large scale shell model techniques to compute neutron deficient semi magic nuclei $A < 50$ [17] or neutron rich calcium isotopes [18].

On the other hand, the neutron matter equation of state and its astrophysical implications in neutron stars have been studied using these soft interactions transformed from the bare one [19].

2.2. Nuclear structure with phenomenological interactions

Despite the range of applicability of the methods shown in the previous section has grown in the last years, most of the nuclei with astrophysical interest are out of reach for those methods. In addition, the precision obtained in the description of binding energies is unacceptable for nucleosynthesis simulations because the absence of forces beyond 3N interactions produces largely over-bound nuclei. The inclusion of these many body forces within the chiral effective field theory framework -and their corresponding coupling constants fitted to few body experimental data- is a work in progress but the results are expected in the long-range plan.

Therefore, to attain more accurate predictions of nuclear masses, spectra or reaction rates in a general way, many body methods based on phenomenological interactions are employed.

Among them, large scale shell model (LSSM) [9] and energy density functionals (EDF) [10] are the most common work horses.

The first one is based on the diagonalization of an effective interaction defined within a restricted valence space. These interactions are adapted specifically to the different valence spaces and fitted to reproduce the evolution of the single particle energies with the number of nucleons included in the space. This method is the best one in reproducing the nuclear spectra. It has been also used in nuclear astrophysics applications such as the calculation of beta decay rates including allowed and first forbidden transitions [20, 21]. Although in the last years the increase of computational power and the improvement in the algorithms have allowed the calculation of nuclei quite far away from shell closures, the main limitation of this method is the combinatorial grow of the number of configurations within the valence space.

On the other hand, the EDF methods do not have this restriction and can be applied all over the nuclear chart, for example, in neutron rich nuclei belonging to the r-process path. They are based on the variational principle assuming very simple many body trial states (Hartree-Fock-Bogoliubov -HFB-) and interactions valid in the whole nuclear table, being Skyrme [10], Gogny [22] or Relativistic [23] the most commonly used. Bulk properties such as ground state radii and masses are very well described within this framework [24, 25]. The latter can be used to deduce neutron capture rates to be implemented in nucleosynthesis simulations. In the last years there have been several advances to extend these methods beyond the HFB approach, allowing the study of nuclear spectra (see Ref. [26] and references therein) and nuclear response to electromagnetic probes [27], for example. The most important improvements of the EDF methods which should be addressed to obtain the same degree of precision as the LSSM -in the regions where it can be applied- are the redefinition of the effective interactions including beyond mean field effects and the inclusion of single particle degrees of freedom.

2.3. Nuclear physics impact in r-process nucleosynthesis

Finally, I will mention a particular example where the relevance of the nuclear physics input in astrophysical simulations, the nuclear masses in this case [28], is shown. More than the half of elements heavier than iron are produced in the r-process through very neutron rich nuclei. Nuclear masses are needed both for computing neutron capture rates and beta decay Q-values that enter in the nucleosynthesis simulations. In Fig. 2(a) the two-neutron separation energy S_{2n} for the Cadmium isotopic chain using two mass models, FRDM [29] and Duflo-Zuker [30], is represented. Both mass models reproduce the experimental data where available. However, they differ significantly in the region around neutron number $N = 90$, where FRDM has a minimum

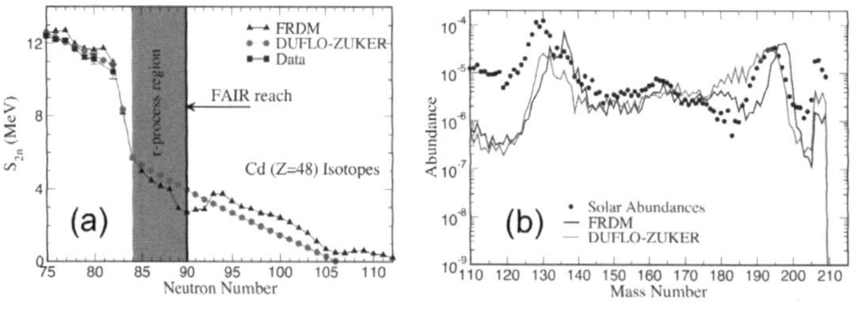

Figure 2. (a) Two neutron separation energies for the cadmium isotopic chain calculated for FRDM and Duflo-Zuker mass models. (b) Final elementary abundances obtained by using FRDM or Duflo-Zuker models as part of the nuclear physics input in r-process nucleosynthesis simulations (r-process solar abundances are plotted to guide the eye). This figure is courtesy of Joel Mendoza-Temis.

and Duflo-Zuker continuously decreases. In Fig.2(b) the elementary abundances obtained after an r-process nucleosynthesis simulation with the same astrophysical conditions, using these two mass models as inputs, are represented. Noticeable differences around the peaks $A = 130$ and 196 are observed. In the FRDM case, an overproduction around $A = 138$ and an underproduction around $A = 190$ are obtained. This structure is ultimately linked to the minimum observed in the S_{2n} for the FRDM model because in this point the neutron capture is inhibited and modify the dynamics of the r-process. This 'waiting point' observed in the FRDM produces after beta decays the overabundance at $A = 138$ and, correspondingly, a 'hole' in the second peak abundances because the matter cannot continue capturing neutrons producing heavier nuclei. More interestingly, the origin of this minimum is a change in the intrinsic deformation in those specific isotopes.

This figure summarizes the necessity of having, on the one hand, experimental results in neutron rich regions to constrain the uncertainties in the astrophysical simulations. This is one of the main goals of NUSTAR at FAIR. On the other hand, we observe that the future facilities will not cover the whole range of nuclei involved in the r-process and further research will be necessary to improve the nuclear theory and make the models and predictions more reliable.

This work was supported from BMBF-Verbundforschungsprojekt number 06DA7047I.

References

[1] E. M. Burbidge, G. R. Burbidge, W. A. Fowler, and F. Hoyle, Rev. Mod. Phys. 29, 547 (1957).
[2] A. G. W. Cameron, Report CRL-41, Chalk River (1957).
[3] F.-K. Thielemann, K.-L. Kratz, B. Pfeiffer, T. Rauscher, L. van Wormer, M. C. Wiescher, Nucl. Phys. A 570, 329 (1994).
[4] H. Schatz, A. Aprahamian, J. Görres, M. Wiescher, T. Rauscher, J.F. Rembges, F.-K. Thielemann, B. Pfeiffer, P. Möller, K.-L. Kratz, H. Herndl, B.A. Brown, H. Rebel, Phys. Rep. 294, 167 (1998).
[5] C. Fröhlich, G. Martínez-Pinedo, M. Liebendörfer, F.-K. Thielemann, E. Bravo, W. R. Hix, K. Langanke, and N. T. Zinner, Phys. Rev. Lett. 96, 142502 (2006).
[6] J. Gerl, A. Herlert and T. Nilsson, NUSTAR Progress Report, http://www.fair-center.eu/fileadmin/fair/experiments/NUSTAR/Pdf/NUSTAR_ProgressReport_SC_Nov2012_final.pdf
[7] D. R. Entem and R. Machleidt, Phys. Rev. C 68, 041001(R) (2003).
[8] E. Epelbaum, H.-W. Hammer, and Ulf-G. Meißner, Rev. Mod. Phys. 81, 1773 (2009).
[9] E. Caurier, G. Martínez-Pinedo, F. Nowacki, A. Poves, and A. P. Zuker Rev. Mod. Phys. 77, 427 (2005).
[10] M. Bender, P.-H. Heenen, and P.-G. Reinhard Rev. Mod. Phys. 75, 121 (2003).
[11] S.K. Bogner, R.J. Furnstahl, A. Schwenk, Prog. Part. Nucl. Phys. 65, 94 (2010).
[12] H. Feldmeier, T. Neff, R. Roth, J. Schnack, Nucl. Phys. A 632, 61 (1998).
[13] R. Roth, J. Langhammer, A. Calci, S. Binder, and P. Navrátil, Phys. Rev. Lett. 107, 072501 (2011).
[14] D. J. Dean and M. Hjorth-Jensen, Phys. Rev. C 69, 054320 (2004).
[15] H. Hergert, S. Binder, A. Calci, J. Langhammer, and R. Roth, Phys. Rev. Lett. 110, 242501 (2013).
[16] V. Somà, C. Barbieri, and T. Duguet, Phys. Rev. C 87, 011303 (2013).
[17] J. D. Holt, J. Menéndez, and A. Schwenk, Phys. Rev. Lett. 110, 022502 (2013).
[18] A. T. Gallant *et al.*, Phys. Rev. Lett. 109, 032506 (2012).
[19] K. Hebeler, J. M. Lattimer, C. J. Pethick, and A. Schwenk, Phys. Rev. Lett. 105, 161102 (2010).
[20] K. Langanke, and G. Martínez-Pinedo, At. Data. Nucl. Data Tables 79, 1 (2001).
[21] Q. Zhi, E. Caurier, J. J. Cuenca-Garca, K. Langanke, G. Martínez-Pinedo, and K. Sieja, Phys. Rev. C 87, 025803 (2013).
[22] J. -P. Delaroche, M. Girod, J. Libert, H. Goutte, S. Hilaire, S. Péru, N. Pillet, and G. F. Bertsch, Phys. Rev. C 81, 014303 (2010).
[23] P. Ring, Prog. Part. Nucl. Phys. 37, 193 (1996).
[24] S. Goriely, N. Chamel, and J. M. Pearson, Phys. Rev. Lett. 102, 152503 (2009).
[25] S. Goriely, S. Hilaire, M. Girod, and S. Péru. Phys. Rev. Lett. 102, 242501 (2009).
[26] T. R. Rodríguez and J. L. Egido, Phys. Rev. C 81, 064323 (2010).
[27] E. Litvinova, P. Ring, and V. Tselyaev, Phys. Rev. C 75, 064308 (2007).
[28] A. Arcones and G. Martínez-Pinedo, Phys. Rev. C 83, 045809 (2011).
[29] P. Möller, J. R. Nix, W. D. Myers, and W. J. Swiatecki, Atomic Data Nucl. Data Tables 59, 185 (1995).
[30] J. Duflo and A.P. Zuker, Phys. Rev. C 52, R23 (1995).

Journal of Physics: Conference Series **503** (2014) 012039 doi:10.1088/1742-6596/503/1/012039

Dilepton Production in Transport Calculations and Coarse-Grained Dynamics

Stephan Endres[1,2], Hendrik van Hees[1,2], Janus Weil[2] and Marcus Bleicher[1,2]

[1]Institut für Theoretische Physik, Universität Frankfurt, Max-von-Laue-Straße 1, 60438 Frankfurt, Germany

[2]Frankfurt Institute for Advanced Studies, Ruth-Moufang-Straße 1, 60438 Frankfurt, Germany

E-mail: endres@th.physik.uni-frankfurt.de

Abstract. We present transport calculations with the Ultra-relativistic Quantum Molecular Dynamics approach (UrQMD) for dilepton spectra at SIS energies and simulations for NA60 within a coarse-graining approach. While the transport model obtains a good agreement with experiment for elementary reactions, in heavy-ion collisions an excess in the invariant mass spectra is observed which cannot be described by the model. As the pure transport calculations do not include explicit in-medium effects and are limited to hadronic degrees of freedom, we present an alternative approach that uses coarse-grained output from transport calculations to determine thermal dilepton emission rates. For this we apply the medium-modified ρ spectral function by Eletsky et al. In a first exemplary comparison to data from the NA60 experiment we find that the coarse-graining approach gives reasonable results.

1. Introduction

Dileptons are a unique tool to study the properties of hot and dense matter created in nuclear collisions. They might serve as probes for the in-medium properties of vector mesons and the predicted restoration of chiral symmetry [1–3]. Unlike hadrons, the lepton pairs produced in the nuclear fireball do not participate in the strong interaction and therefore penetrate the strongly interacting medium with negligible final-state reactions. Thus we gain insight into all the different stages of a nuclear collision, from the first nucleon-nucleon interactions to the final freeze-out. But this also means that in experimental measurements we obtain time-integrated spectra only, stemming from a broad variety of sources. In consequence we need good models that help to understand the production mechanisms and their contribution to the total spectra. Especially at lower bombarding energies, where the dominant dilepton contribution originates from hadronic decays, transport models have been successful in describing the experimentally measured dilepton spectra [4–7]. In these proceedings we present recent results from our calculations with the UrQMD model at SIS energies. However, for a hot and dense evironment as created in ultra-relativistic heavy-ion collisions it is supposed that medium effects play a crucial role for dilepton production, but it is highly difficult to implement them in a transport model. We argue here that a coarse-graining approach is a good way to apply in-medium spectal functions within an underlying microscopic description of the reaction dynamics.

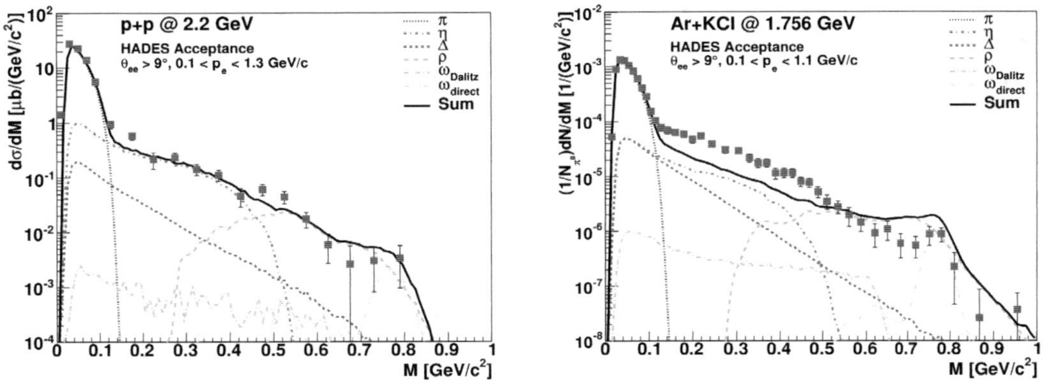

Figure 1. Invariant mass spectra for p+p at 2.2 GeV (left) and Ar+KCl at 1.756 AGeV (right) obtained with the UrQMD transport model. The results are corrected for the HADES acceptance, detector resolution and compared with data from the HADES Collaboration [17, 18].

2. Transport Approach

For our calculations of dilepton spectra we use the Ultra-relativistic Quantum Molecular Dynamics model (UrQMD), which is a hadronic non-equilibrium transport approach and includes all baryons and mesons with masses up to 2.2 GeV [8–10]. In this model the production of resonances takes place either via inelastic collisions or the decay of higher resonances. As dilepton-production channels the Dalitz decays of the π^0, η and ω meson and of the $\Delta(1232)$ resonance are included, as well as the direct decay channels of the ω, ρ and ϕ vector mesons. The Dalitz decays are treated in a two-step process, first the decay into a virtual photon and as second step the electromagnetic conversion into a lepton pair. In contrast to previous calculations we now apply the parametrization for the radiative decay width of the $\Delta(1232)$ by Krivoruchenko [11]. However, the difference in the yield is small (only few percent) if compared to the previously used old parametrization by Wolf [12]. Dalitz decays of other baryonic states than the $\Delta(1232)$ are not treated explicitly, but these resonances contribute to the dilepton spectra via their decays into light mesons. For further details we refer the reader to [6, 13].

There are several parameters that strongly influence the dilepton production within our model: Especially the number and types of baryonic resonances which are included, their production cross-sections and their branching ratios into light mesons. These parameters have to be adapted to experimental constraints. However, in case of the resonance production cross-sections the data are very limited, and almost no information is available in the HADES energy regime up to $\sqrt{s} = 3$ GeV [14]. The situation is not better for many decay-branching ratios: When comparing the results of different partial-wave analyses one finds big differences as well [15, 16]. (Furthermore, the issue of a Δ form factor is still unclear, which we do not consider in this work.) We reevaluated all the available data and adjusted some of the branching ratios and N^* and Δ^* production cross-sections within the model to get the best possible consistency with the available experimental data, e.g. the inclusive and exclusive resonance procuction cross-sections. The results presented in Figure 1 show the invariant mass spectra of dileptons produced in proton-proton reactions at 2.2 GeV (left plot) and in Ar+KCl at 1.756 AGeV (right plot) compared to HADES data [17, 18]. We see that the description of the elementary p+p reactions is in a good agreement with experiment, while in the larger system Ar+KCl we find on the one side that the ρ contribution in our calculation overshoots the data in the pole mass region and on the other hand that the experimental results show an excess between 150 and 500 MeV mass which is not reproduced with our model. There may be several possible explanations for this

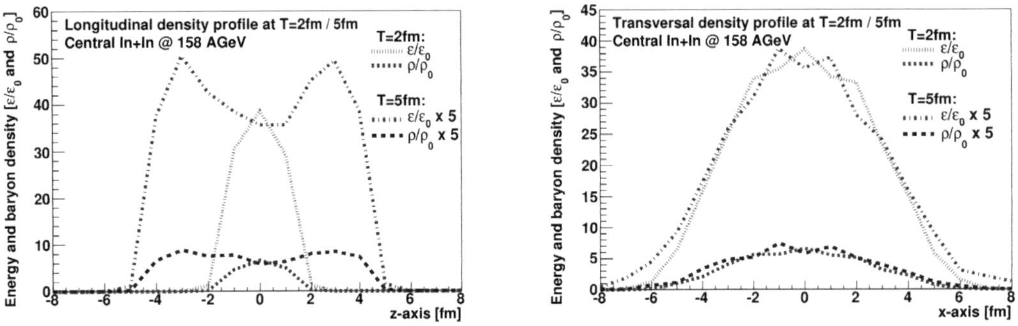

Figure 2. (Left) Longitudinal profile (i.e. along the beam axis) for the energy and baryon desities in a central In+In collisions at 158 AGeV, as obtained from the coarse-graining of UrQMD events. The values here are given in units of the nuclear-matter ground state energy and baryon densities. (Right) The same plot but for the transverse energy and baryon densities.

excess, e.g. bremsstrahlung effects or medium modifications of the vector mesons. However, a different study with the GiBUU model [19], which includes the bremsstrahlung contribution in soft-photon approximation, still underestimates the yield in this low-mass region [20].

3. Coarse-Graining Approach

It is important to note that our present transport calculations do not include explicit in-medium modifications, except for reabsorption and rescattering, which are dynamically implemented. For an entirely proper calculation it would be necessary to include off-shell effects and multi-particle interactions, which is a highly non-trivial challenge. Another problem arises at higher bombarding energies, where dilepton emission from thermal radiation comes into play, as the UrQMD model only includes hadronic degrees of freedom. One method to avoid these problems is the integration of dilepton radiation from an intermediate hydrodynamic phase within a hybrid model, using in-medium dilepton rates [21]. In the following we sketch a different approach that uses coarse-grained transport dynamics instead of hydrodynamics to calculate in-medium emission rates. A similar procedure was first proposed in [22].

As first step we put the UrQMD output on a 3+1 dimensional space-time grid and determine baryon and energy density for each cell as an average over several hundred events. Assuming thermal equilibrium we transform into the local rest frame of the cell, according to Eckart's definition that requires the net baryon flow to be zero. Then we apply an equation of state to extract temperature T and baryon chemical potential μ_B. For SPS and higher energies we use a chiral equation of state which includes chiral symmetry restoration and a deconfinement phase transition. It shows a crossover between the hadronic and QGP phase with a coexistence region [23].

Assuming vector meson dominance we can calculate the equlibrium l^+l^- emission rates per volume and momentum in each cell according to [2] as

$$\frac{\mathrm{d}^8 N_{V\to ll}}{\mathrm{d}^4x\mathrm{d}^4q} = -\frac{\alpha^2 m_V^4}{\pi^3 g_V^2}\frac{L(M^2)}{M^2} f_B(q_0;T)\,\mathrm{Im}\,D_V^{(\mathrm{ret})}(M,q;T,\mu_B), \tag{1}$$

in terms of the imaginary part of the meson's retarded propagator $\mathrm{Im}\,D_V^{(\mathrm{ret})}$. Here $L(M^2)$ accounts for the lepton phase space and the function f_B is the Bose-Einstein distribution. As the largest in-medium contribution is expected to stem from the ρ meson we concentrate on this contribution in our considerations. For a complete study it will be necessary to include the

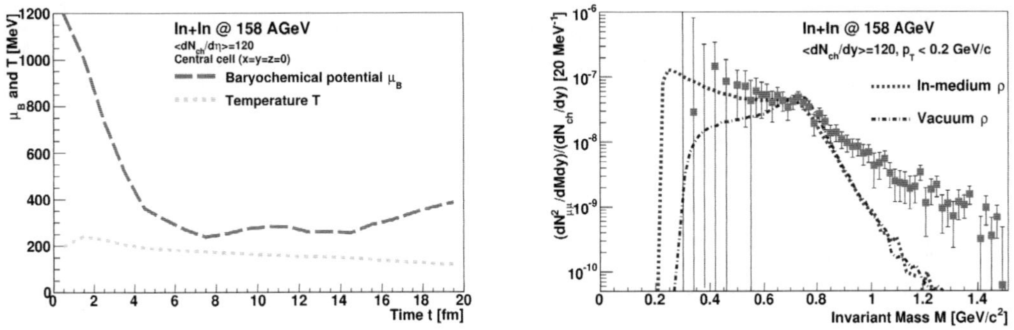

Figure 3. (Left) Time evolution of the baryochemical potential μ_B and temperature T for a central cell in our grid of In+In collisons at 158 AGeV as obtained with a chiral equation of state. (Right) Dimuon excess spectrum from coarse-grained dynamics for In+In at $E_{lab} = 158$ AGeV with $\mathrm{d}N_{ch}/\mathrm{d}y = 120$, here in the $p_t < 0.2$ GeV region. The in-medium ρ and, for comparison, the pure vacuum contribution are shown. The results are compared to NA60 data [27].

in-medium modifications of the ω and the ϕ meson, too.
Regarding the ρ in-medium spectral function several approaches exist, relying either on calculations from hadronic many-body theory or on empirical scattering amplitudes obtained from experiment. Here we apply the description by Eletsky and Kapusta [24], which uses experimental data in linear density approximation to include the scattering from pions and nucleons in the vector meson's retarded self-energy, i.e. $\Sigma_\rho = \Sigma^{\mathrm{vac}} + \Sigma^{\rho N} + \Sigma^{\rho\pi}$. These additional effects on the self energy result in a broadening of the spectral width in a hot and dense medium. At high collision energies, it is not sufficient to look only at the modifications of the vector mesons, but also additional thermal sources must be taken into account. If a deconfined phase is created during the heavy-ion collision we will also find a contribution from $q\bar{q}$ annihilation in the Quark-Gluon Plasma. The rates can be calculated using the electromagnetic cross-section for the annihilation of a quark-antiquark pair into dileptons [25]. Similarly we expect multi-particle interactions in the medium. While two-pion interactions are included in the ρ spectral function, it has been shown that also four-pion reactions play a crucial role for the production of lepton pairs [26]. However, as these contributions will become dominant only above 1 GeV in the invariant mass spectra we postpone a detailed study to a future work.
The advantage of the coarse-graining approach described above is that we can simulate a realistic evolution of temperature and baryochemical potential during the nuclear collision related to the underlying microscopic dynamics. This is an important aspect as only a reliable description of the different stages enables to understand the role of the different dilepton-production channels. To check our approach we make a first comparison with NA60 data on dimuon production in In+In collisions at 158 AGeV [27]. In Figure 2 the longitudinal and the transverse profile of energy and baryon density (ϵ respectively ρ_B) are shown, resulting from the coarse graining of central UrQMD events at $t = 2$ fm and 5 fm after the beginning of the collision. At the later time time the nuclei have already passed through each other as indicated by the two peak structures in the profile along the beam axis. In the center of the collision we find a hot and dense region with values of ϵ and ρ_B reaching up to several times the ground-state densities. The resulting time evolution of temperature and baryon chemical potential is shown in Figure 3 (left), here exemplarily for a cell at the center of the collision. The baryon chemical potential decreases very fast after the beginning of the collision and then remains at a level of 300-400 MeV, while the temperature shows an initial peak around 250 MeV followed by a cooling down to around 100 MeV at $t = 20$ fm. These results are in line with similar previous studies [22,28].

Looking at the resulting dilepton spectrum (here for $p_t < 0.2$ GeV) in the right plot of Figure 3, we see that the low-mass tail of the excess spectrum (with all hadronic cocktail contributions already subtracted) is nicely filled up by a broadened in-medium ρ. Applying only a vacuum spectral function, in comparison, underestimates the contribution in the region below the ρ pole mass significantly. Above a mass of 1 GeV the ρ no longer fills up the spectrum, here one expects other thermal sources to be dominant (mainly QGP and four-pion emission) that are not included in the present calculations.

4. Summary & Outlook

In summary we showed that our UrQMD transport calculations give a good description of experimental dilepton spectra, but the lack of a correct implementation of medium modifications remains a problem. To account for this we proposed that a coarse-graining approach might serve to obtain a realistic description of the reaction dynamics and the resulting dilepton rates. Our exemplary comparison to the data from the NA60 experiment shows a good agreement. Further studies have to be performed, e.g. by investigating different equations of state and spectral functions. Of special interest would be a check of the many-body calculation for the ρ spectral function by Rapp and Wambach [29], which has proven successful in describing the dilepton results from NA60 and RHIC in a fireball model [30], within our coarse-graining approach.

Acknowledgments

We thank the HADES Collaboration, especially T. Galatyuk and C. Sturm, for providing their acceptance filters and data. This work was supported by BMBF, HIC for FAIR and H-QM.

References

[1] Xia L H, Ko C M, Xiong L and Wu J Q 1988 *Nucl. Phys.* A **485** 721
[2] Rapp R and Wambach J 2000 *Adv. Nucl. Phys.* **25** 1 (*Preprint* hep-ph/9909229)
[3] Leupold S, Metag V and Mosel U 2010 *Int. J. Mod. Phys.* E **19** 147 (*Preprint* arXiv:0907.2388 [nucl-th])
[4] Weil J *et al.* 2012 *Eur. Phys. J.* A **48** 111; Erratum-ibid. 2012 A **48** 150 (*Preprint* arXiv:1203.3557 [nucl-th])
[5] Bratkovskaya E L and Cassing W 2008 *Nucl. Phys.* A **807**, 214 (*Preprint* arXiv:0712.0635 [nucl-th])
[6] Schmidt K, Santini E *et al.* 2009 *Phys. Rev.* C **79** 064908 (*Preprint* arXiv:0811.4073 [nucl-th])
[7] Vogel S, Petersen H, *et al.* 2008 *Phys. Rev.* C **78** 044909 (*Preprint* arXiv:0710.4463 [hep-ph])
[8] Bass S A *et al.* 1998 *Prog. Part. Nucl. Phys.* **41** 255 (*Preprint* nucl-th/9803035)
[9] Bleicher M *et al.* 1999 *J. Phys.* G **25** 1859 (*Preprint* hep-ph/9909407)
[10] Petersen H, Bleicher M, Bass S A and Stoecker H 2008 *Preprint* arXiv:0805.0567 [hep-ph]
[11] Krivoruchenko M I and Faessler A 2002 *Phys. Rev.* D **65** 017502 (*Preprint* nucl-th/0104045)
[12] Wolf G *et al.* 1990 *Nucl. Phys.* A **517** 615
[13] Endres S and Bleicher M 2013 *J. Phys. Conf. Ser.* **426** 012033
[14] Flaminio V, Moorhead W G, Morrison D R O and Rivoire N 1984 *CERN-HERA-84-01*
[15] Manley D M and Saleski E M 1992 *Phys. Rev.* D **45** 4002
[16] Anisovich A V *et al.* 2012 *Eur. Phys. J.* A **48** 15 (*Preprint* arXiv:1112.4937 [hep-ph])
[17] Agakishiev G *et al.* (HADES Collab.) 2011 *Phys. Rev.* C **84** 014902 (*Preprint* arXiv:1103.0876 [nucl-ex])
[18] Agakishiev G *et al.* (HADES Collab.) 2012 *Phys. Rev.* C **85** 054005 (*Preprint* arXiv:1203.2549 [nucl-ex])
[19] Buss O, Gaitanos T, Gallmeister K *et al.* 2012 *Phys. Rept.* **512** 1 (*Preprint* arXiv:1106.1344 [hep-ph])
[20] Weil J and Mosel U 2013 *J. Phys. Conf. Ser.* **426** 012035 (*Preprint* arXiv:1211.3761 [nucl-th])
[21] Santini E, Steinheimer J *et al.* 2011 *Phys. Rev.* C **84** 014901 (*Preprint* arXiv:1102.4574 [nucl-th])
[22] Huovinen P, Belkacem P *et al.* 2002 *Phys. Rev.* C **66** 014903 (*Preprint* nucl-th/0203023)
[23] Steinheimer J, Schramm S and Stocker H 2011 *J. Phys.* G **38** 035001 (*Preprint* arXiv:1009.5239 [hep-ph])
[24] Eletsky V L *et al.* 2001 *Phys. Rev.* C **64** 035202 (*Preprint* nucl-th/0104029)
[25] Cleymans J, Fingberg J and Redlich K 1987 *Phys. Rev.* D **35** 2153
[26] Huang Z 1995 *Phys. Lett.* B **361** 131 (*Preprint* hep-ph/9506399).
[27] Arnaldi R *et al.* (NA60 Collab.) 2009 *Eur. Phys. J.* C **61** 711 (*Preprint* arXiv:0812.3053 [nucl-ex])
[28] Bass S A, Weber H, Ernst C, Bleicher M *et al.* 1999 *Prog. Part. Nucl. Phys.* **42** 313 (*Preprint* nucl-th/9810077)
[29] Rapp R, Chanfray G and Wambach J 1997 *Nucl. Phys.* A **617** 472 (*Preprint* hep-ph/9702210)
[30] van Hees H and Rapp R 2008 *Nucl. Phys.* A **806** 339 (*Preprint* arXiv:0711.3444 [hep-ph])

Journal of Physics: Conference Series **503** (2014) 012040 doi:10.1088/1742-6596/503/1/012040

3+1 dimensional viscous hydrodynamics at high baryon densities

Iu.A. Karpenko[1,2] **M. Bleicher**[1,3] **P. uovinen**[1] **and . Petersen**[1]

[1] Frankfurt Institute for Advanced Studies, Ruth-Moufang-Straße 1, 60438 Frankfurt am Main, Germany

[2] Bogolyubov Institute for Theoretical Physics, 14-b, Metrolohichna str., 03680 Kiev, Ukraine

[3] Institute for Theoretical Physics, Johann Wolfgang Goethe Universität, Max-von-Laue-Str. 1, 60438 Frankfurt am Main, Germany

E-mail: `karpenko@fias.uni-frankfurt.de`

Abstract. We apply a 3+1D viscous hydrodynamic + cascade model to heavy ion collision reactions with $\sqrt{s_{NN}} = 6.3 \ldots 39$ GeV. To accommodate the model for a given collision energy range, the initial conditions for the hydrodynamic phase are taken from Ur MD, and the equation of state at finite baryon density is based on a Chiral model coupled to the Polyakov loop.

We study the collision energy dependence of pion and kaon rapidity distributions and m_T-spectra, as well as charged hadron elliptic flow and how shear viscosity affects them. The model calculations are compared to the data for Pb-Pb collisions at CERN SPS, as well as for Au-Au collisions in the Beam Energy Scan (BES) program energies at BNL RHIC. The data favours the value of shear viscosity $\eta/s \gtrsim 0.2$ for this collision energy range.

1. Introduction

Hydrodynamic approaches are well established for the description of hot and dense matter created in ultrarelativistic heavy ion collisions at BNL Relativistic Heavy Ion Collider (RHIC) and CERN Large Hadron Collider (LHC). Its success is based on the reproduction of bulk observables such as radial flow, elliptic and higher order flow harmonics [1], femtoscopy [2] etc. in so-called hybrid models coupling a viscous (or inviscid) hydrodynamic phase to a hadron cascade.

In this paper we report on the application of the viscous+cascade model tuned to the maximum RHIC energy of $\sqrt{s_{NN}} = 200$ GeV to heavy ion collisions at lower collision energies. To accommodate to the different collision energy regimes, we change the equation of state (EoS) and the initial conditions, and keep all the other parameters fixed.

2. Viscous hydro+cascade model

The approximation of boost-invariant scaling flow is well-justified for the hydrodynamic evolution of matter around midrapidity at top RHIC energies and above. However one should not pursue this assumption at lower energies, since at $\sqrt{s_{NN}} \approx 20$ GeV and below it becomes invalid because of the absence of a plateau in the rapidity distribution of produced particles.

We avoid this issue by using rapidity-dependent initial conditions together with a 3D hydrodynamic evolution. The model employed in the present studies consists of components,

Journal of Physics: Conference Series **503** (2014) 012040 doi:10.1088/1742-6596/503/1/012040

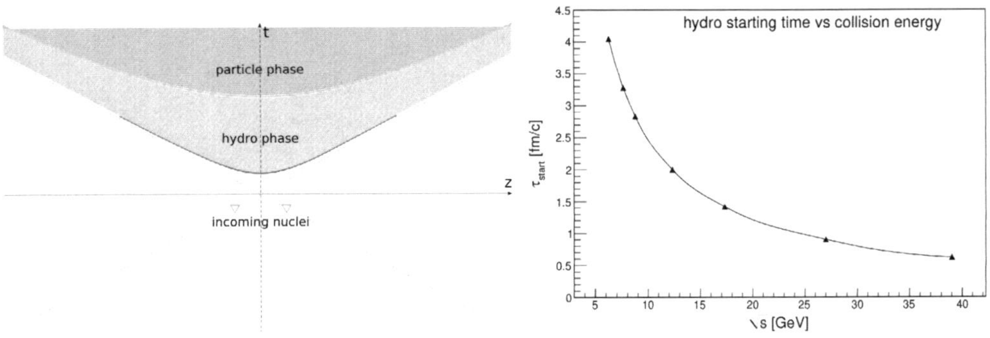

Figure 1. Left: Sketch of space-time picture of evolution, the different phases are marked by different colour. Right: starting time for the hydrodynamic phase as a function of $\sqrt{s_{NN}}$ according to Eq. 1.

each corresponding to a certain stage of the nucleus-nucleus collision. The space-time picture is sketched on Fig. 1.

Pre-thermal phase. The Ultrarelativistic Quantum Molecular Dynamics (UrQMD) model [3, 4] is used for the description of initial stage dynamics. The two nuclei are initialized according to Woods-Saxon distributions and the binary interactions are taken into account until a hypersurface at constant $\tau = \sqrt{t^2 - z^2}$ is reached. We generate a large enough set of UrQMD events to get smooth averaged 3-dimensional distribution of particles which cross this hypersurface. The energy and momentum of particles is then converted to energy and momentum densities of the fluid. In addition to energy/momentum densities, initial baryon and charge densities are non-zero and obtained from UrQMD, evolved in the hydro stage and accounted for in the particlization procedure. On the other hand, net strangeness density is set to zero.

ydrodynamic phase. We start the hydrodynamic phase at

$$\tau_0 = 2R/\sqrt{(\sqrt{s}/2m_N)^2 - 1} \tag{1}$$

where R is a radius of nucleus and m_N is a nucleon mass. This corresponds to the time when the two nuclei have passed through each other. The values of the starting time for different collision energies are shown in Fig. 1, right. The hydrodynamic equations are solved in Milne $(\tau - \eta - r_x - r_y)$ coordinates.

Another important ingredient of the model relevant to lower collision energies is the equation of state (EoS) for finite baryon density. We employ the chiral model based EoS [8], which features correct asymptotic degrees of freedom, i.e. quarks and gluons at high temperature and hadrons in the low-temperature limits, crossover-type transition between confined and deconfined matter for all values of μ_B and qualitatively agrees with lattice QCD data at $\mu_B = 0$.

We employ the Israel-Stewart framework for relativistic viscous hydrodynamics [5]. Several variants of Israel-Stewart equations exist in literature, the differences between them coming from slightly different assumptions in the derivation [7]. In particular we stick to the following choice for equations for the shear stress tensor:

$$\langle u^\gamma \partial_{;\gamma} \pi^{\mu\nu} \rangle = -\frac{\pi^{\mu\nu} - \pi^{\mu\nu}_{\text{NS}}}{\tau_\pi} - \frac{4}{3}\pi^{\mu\nu}\partial_{;\gamma}u^\gamma \tag{2}$$

where $\partial_{;\gamma}$ denotes covariant derivative and the brackets $\langle A^{\mu\nu}\rangle$ denote the symmetric, traceless and orthogonal to u^μ part of $A^{\mu\nu}$. For the current study we consider only the effects of shear viscosity, fixing bulk viscosity to zero, $\zeta/s = 0$. We do not include the baryon/electric charge diffusion either. For viscous hydro simulations, we initialize the shear stress tensor to zero. The relaxation time for shear, τ_π, is taken as $\tau_\pi = 3\eta/(sT)$, as in [6].

Particli ation and hadron corona. Motivated by previous studies at $\sqrt{s_{NN}} = 200$ A GeV RHIC and LHC energies [2], we fix the transition from fluid to particle description (so-called particlization) to take place at a constant energy density $\epsilon_{\rm sw} = 0.5$ GeV/fm^3, when the relevant degrees of freedom are hadrons. The Cornelius subroutine [9] is used to calculate the 3-volume elements $d\sigma_\mu$ of the transition hypersurface. It is important to note that while we use fixed energy density as transition criterion for all collision energies, the average net baryon density increases with decreasing collision energy. This corresponds to the increase of average temperature and decrease of baryon chemical potential with increasing collision energy, resembling the results for the collision energy dependence of chemical freezeout parameters from thermal model studies [10]. Since the EoS used has a mean field included, which somewhat shift the values of thermodynamic quantities from those in a free hadron-resonance gas (HRG) approximation, we switch to the free HRG EoS to sample the hadrons according to Cooper-Frye prescription. We recalculate the energy density, pressure, flow velocity and corresponding thermodynamical quantities from the energy-momentum tensor using a free HRG EoS, and employ them when sampling. To account for the effect of shear viscosity, we use the same corrections to the local equilibrium distribution functions for all hadron species:

$$f_i(x,p) = f_{i,\rm eq}(x,p)\left[1 + (1 \mp f_{i,\rm eq})\frac{p_\mu p_\nu \pi^{\mu\nu}}{2T^2(\epsilon+p)}\right] \quad (3)$$

Finally, UrQMD is employed to calculate the further evolution of the hadron corona.

3. Results

We fix most of the parameters of the model as described above. The only free parameter left is the shear viscosity to entropy density ratio η/s in the hydro stage. Finally we perform simulations corresponding to various collision energy and η/s values. Since a bulk of data from SPS exists for different observables (dN/dy, p_T-distributions, pion femtoscopy) one can first check how well the overall collective matter dynamics is described in the model. Thus we simulate Pb-Pb collisions at energies $E_{\rm lab} = 158, 80, 40, 30$ and 20 A GeV (i.e. $\sqrt{s_{NN}} = 17.3, \ldots, 6.3$ GeV), and compare m_T and dN/dy distributions with the data from the NA49 collaboration [11, 12].

With decreasing beam energy, the passage time calculated according to (1) increases, whereas the average initial energy density for hydrodynamic expansion becomes smaller. The latter leads to a shorter duration of the hydro phase, provided that the criterion for fluid to particle transition is kept the same. Consequently, at the lowest SPS energy ($\sqrt{s_{NN}} = 6.3$ GeV) the lifetimes of pre-hydro (4.1 fm/c) and hydro stages are comparable, and the pre-hydrodynamic stage gives a significant contribution to the development of transverse flow.

In the case of zero shear viscosity in the hydrodynamic phase we underestimate the particle yields and radial flow at midrapidity. However, the inclusion of shear viscosity in the hydro phase increases the yields at midrapidity due to viscous entropy production, see Fig. 2. One observes that shear viscosity makes the dN/dy profile narrower. The longitudinal expansion is weaker, and the expansion tends to be more spherical, as seen in the comparison of the p_T-spectra, Fig. 3: p_T-spectra become flatter, which is due to stronger transverse expansion and larger radial flow.

In Fig. 4 we show p_T-differential charged hadron elliptic flow at collision energies $\sqrt{s_{NN}} =$ 7.7, 27 and 39 A GeV compared with the results from the RHIC BES. An inviscid hydrodynamic phase leads to an overestimate of the data, while choosing $\eta/s = 0.2$ greatly improves agreement

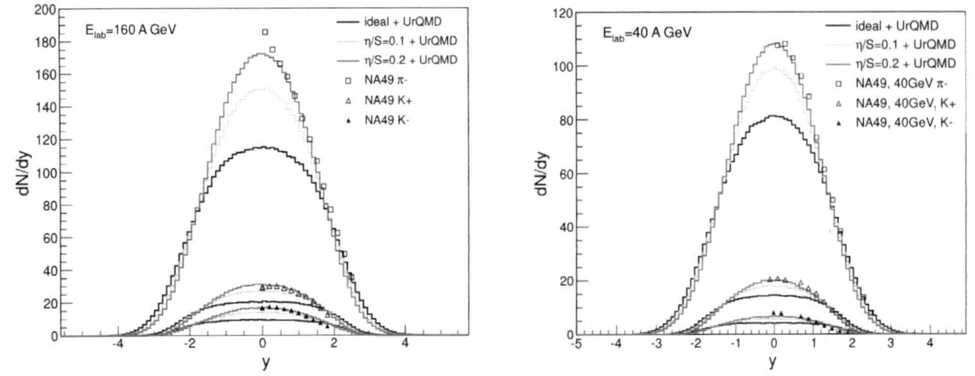

Figure 2. Rapidity distributions for π^-,K^+ and K^- for Pb-Pb collisions at $E_{\rm lab} = 158$ and 40 A GeV (corresponding to $\sqrt{s_{NN}} = 17.3$ and 6.3 GeV). Model calculations are compared to NA49 data [12]. The figures are taken from [13].

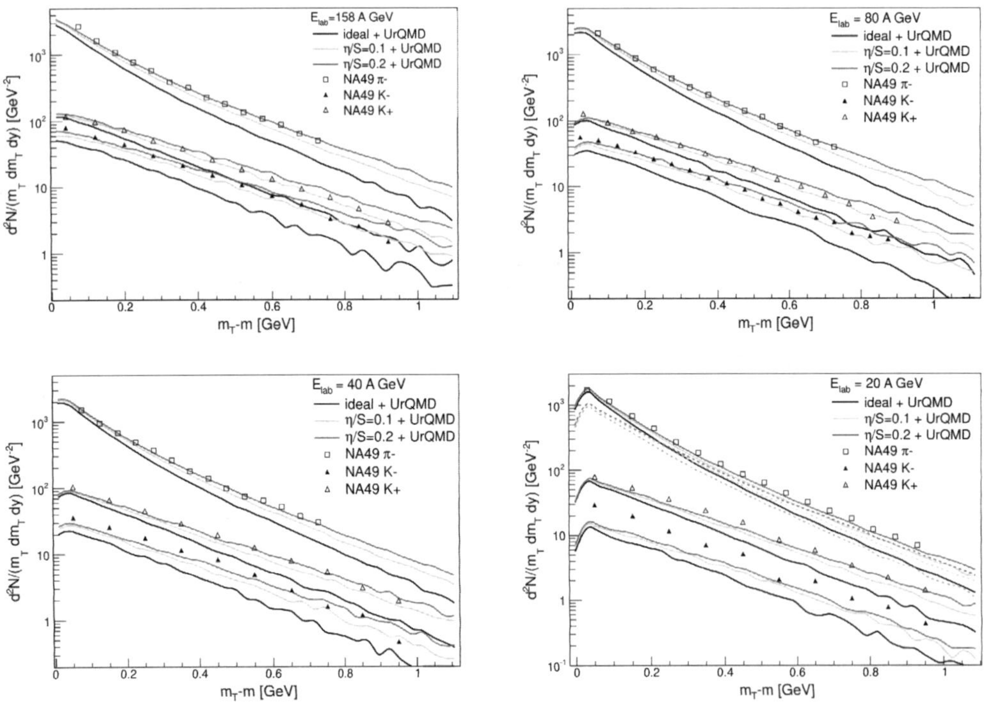

Figure 3. m_T-spectra for π^-,K^+ and K^- for Pb-Pb collisions at $E_{\rm lab} = 158, 80, 40, 20$ A GeV ($\sqrt{s_{NN}} = 17.3, 12.3, 8.8, 6.3$ GeV, respectively). Model calculations are compared to NA49 data [11, 12]. Dashed lines on bottom right plot: model calculations for π^+. The figures are taken from [13].

with it. Here no attempt to fit the data is made; we only demonstrate the model results for

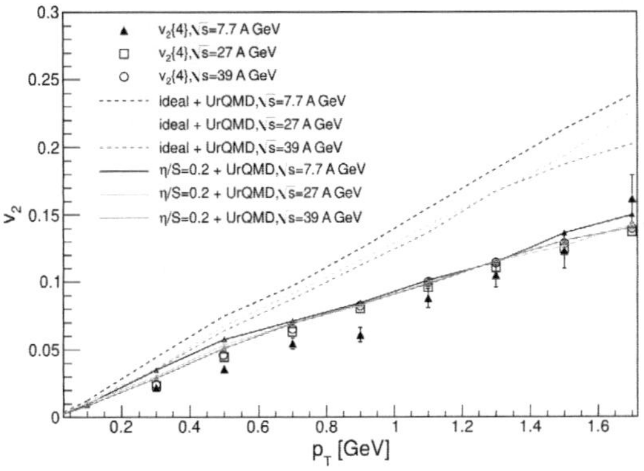

Figure 4. p_T-differential elliptic flow of all charged hadrons for 20-30% central Au-Au collisions at $\sqrt{s_{NN}} = 7.7, 27$ and 39 GeV. The data are from STAR collaboration [14] The figures are taken from [13].

$\eta/s = 0, 0.1$ and 0.2. So one can conclude that a consistent description of v_2, dN/dy, and p_T-spectra requires a value of η/s which is somewhat larger than 0.2, especially for the lowest collision energy points. $\eta/s = 0.2$ also makes $v_2(p_T)$ almost independent of the collision energy.

We conclude that the introduction of shear viscosity in the hybrid model consistently improves the description of the data in the low collision energy region. With the averaged initial state from UrQMD used, the suggested value of the effective shear viscosity is $\eta/s \geq 0.2$ for both Pb-Pb collisions with $E_{\text{lab}} = 20 \ldots 158$ GeV at SPS and Au-Au collisions with $\sqrt{s} = 7.7 \ldots 39$ GeV in the BES program at RHIC. For comparison, the typical η/s value obtained from the fit to elliptic flow at $\sqrt{s_{NN}} = 200$ GeV RHIC energy is 0.08 for the case of a Monte Carlo Glauber initial state [15].

Acknowledgements

IK and HP acknowledge the financial support by the ExtreMe Matter Institute EMMI and Hessian LOEWE initiative. HP acknowledges funding by the Helmholtz Young Investigator Group VH-NG-822. Computational resources have been provided by the Center for Scientific Computing (CSC) at the Goethe-University of Frankfurt.

References

[1] Hirano T, Huovinen P, Murase K and Nara Y 2013, *Prog. Part. Nucl. Phys.* **70** 108-158
[2] Karpenko Iu A, Sinyukov Yu M, Werner K 2013, *Phys. Rev. C* **87** 24914
[3] Bass S A et al. 1998, *Prog. Part. Nucl. Phys.* **41** 255-369
[4] Bleicher M et al. 1999, *J. Phys. G* **25** 1859
[5] Israel W and Stewart J M 1979, *Ann. Phys.* **118** 341
[6] Song H and Heinz U 2008, *Phys. Rev. C* **77** 064901
[7] Denicol G S, Molnár E, Niemi H, Rischke D H, *Eur. Phys. J. A* 48 11 (2012) 170
[8] Steinheimer J, Schramm S and Stocker H 2011, *J. Phys. G* **38** 35001
[9] Huovinen P and Petersen H 2012, *Eur. Phys. J. A* **48** 171
[10] Andronic A, Braun-Munzinger P, Stachel J 2006, *Nucl. Phys. A* **772** 167-199
[11] Alt C et al. (NA49 Collaboration) 2008, *Phys. Rev. C* **77** 24093
[12] Afanasiev S V et al. (NA49 Collaboration) 2002, *Phys. Rev. C* **66** 54902
[13] Karpenko Iu A, Bleicher M, Huovinen P, Petersen H 2013, *Preprint* arXiv:1310.0702 [nucl-th]
[14] Adamczyk L et al. (STAR Collaboration) 2012, *Phys. Rev. C* **86** 54908
[15] Song H, Bass S A, Heinz U, Hirano T, Shen C 2011 *Phys. Rev. Lett.* **106** 192301

MPD Detector at NICA

L Yordanova and V Vasendina
MPD Collaboration, Joint Institute for Nuclear Research, Dubna, Russia

Abstract: The goal of this article is to give information about the new accelerator complex NICA at JINR, Dubna and especially, to provide overview of the MultiPurpose Detector (MPD) and its subdetectors. The current results of the MPD performance for dileptons, hyperons, hypernuclei and φ-meson are presented.

The Nuclotron-based Ion Collider fAcility (NICA), shown in figure 1, is a new accelerator complex being constructed at JINR, Dubna, Russia. The global scientific goal of the NICA/MPD project is to explore the phase diagram of strongly interacting matter in the region of highly compressed baryonic matter. The study of hot and dense baryonic matter provides relevant information on the in-medium properties of hadrons and nuclear matter equation of state; allows a search for deconfinement and/or chiral symmetry restoration, phase transition, mixed phase and critical end-point, possible strong P- and CP violation; gives information about the evolution of the Early Universe and the formation of neutron stars. [1]

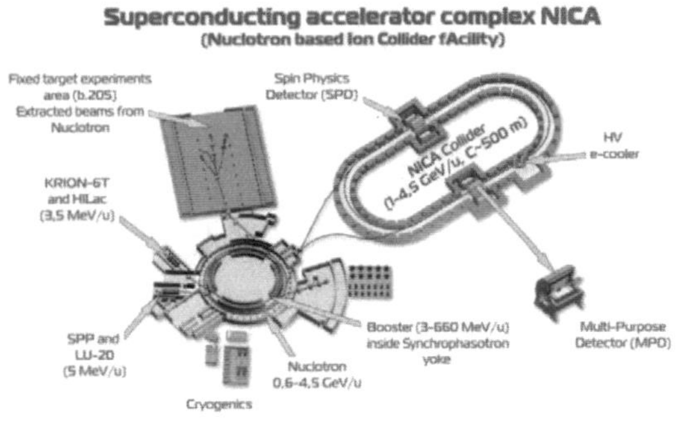

Figure 1. NICA accelerator complex at JINR

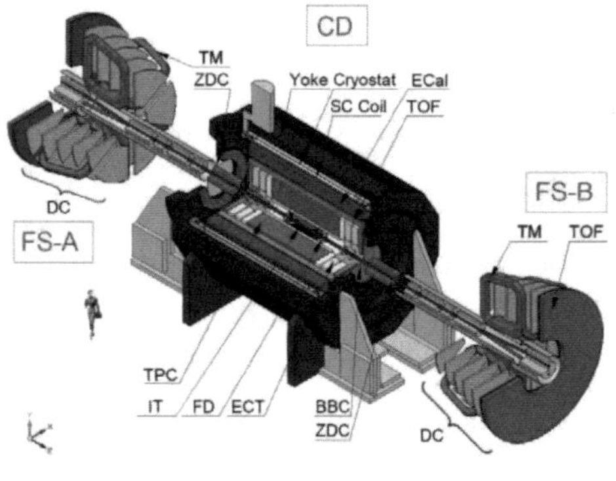

Figure 2. General view of the MPD detector

NICA's aim is to provide collisions of heavy ions over a wide range of atomic masses, from Au+Au collisions at $\sqrt{s_{NN}}$ = 11 GeV (for Au^{79+}) and an average luminosity of $L = 10^{27}$ cm^{-2} s^{-1} to proton-proton collisions with $\sqrt{s_{pp}}$ = 20 GeV and $L \approx 10^{32}$ cm^{-2} s^{-1}. Two interaction points are foreseen at NICA which provide a possibility for two detectors to operate simultaneously - MultiPurpose Detector (MPD) and Spin Physics Detector (SPD). This overview is focused on the MPD detector.

The MPD experimental program includes simultaneous measurements of observables that are presumably sensitive to high nuclear density effects and phase transitions. In the first stage of the project (starting in 2017) are considered - multiplicity and spectral characteristics of the identified hadrons including strange particles, multi-strange baryons and antibaryons; event-by-event fluctuations in multiplicity, charges and transverse momentum; collective flows (directed, elliptic and higher ones) for observed hadrons. In the second stage (starting in 2020) the electromagnetic probes (photons and dileptons) will be measured. It is proposed that along with heavy ions NICA will provide proton and light ion beams including the possibility to use polarized beams.

The software framework for the MPD experiment (MpdRoot) is based on FairRoot and provides a powerful tool for detector performance studies, development of algorithms for reconstruction and physics analysis of the data. [2] Extended set of event generators for heavy ion collisions is used (UrQMD, LAQGSM, HSD). [3]

The detector for exploring phase diagram of strongly interacting matter in a high track multiplicity environment has to cover a large phase space, be functional at high interaction rates and comprise high efficiency and excellent particle identification capabilities. The MPD detector, shown in figure 2, matches all these requirements. It consists of central detector (CD) and two optional forward spectrometers (FS-A and FS-B). CD consists of a barrel part and two end caps. The barrel part is a set of various subdetectors. The main tracker is the time projection chamber (TPC) supplemented by the inner tracker (IT). IT and TPC have to provide precise tracking, momentum determination and vertex reconstruction. The time of flight (TOF) system must be able to identify charged hadrons and nuclear clusters in a broad pseudorapidity range. The electromagnetic calorimeter (ECAL) should identify electrons, photons and measure their energy with high precision. The zero degree calorimeter (ZDC) should provide event centrality and event plane determination, and also measurement of the energy deposited by spectators. There are also a straw-tube tracker (ECT) and a fast forward detector (FFD).

The magnet of MPD is a solenoid with a thin superconducting NbTi winding and a flux return iron yoke. The magnet should provide a homogeneous magnetic field of 0.5 T. The field inhomogeneity in the tracker area of the detector is about 0.1%.

The MPD time projection chamber (TPC) is the main tracking detector of the central barrel and has to provide charged particles momentum measurement with sufficient resolution (about 2% at p_t=300 MeV/c), particle identification and vertex determination, two track separation (with a resolution <1 cm) and dE/dx measurement (dE/dx resolution better than 8%) for hadronic and leptonic observables at pseudorapidities $|\eta|<2.0$ and p_t >100 MeV/c. TPC readout system is based on Multi-Wire Proportional Chambers (MWPC) with cathode readout pads.

The inner tracker system (ITS) should enhance track reconstruction. ITS is able to restore tracks of particles with momentum less than 100 MeV/c and provides precise primary and secondary vertexes reconstruction with an accuracy of ~ 40 μm. ITS should also identify relatively rare events with production of hyperons. ITS is based on the Silicon Strip Detector (SSD) technology. It consists of a silicon barrel and discs which will register particles with large pseudorapidity $|\eta| < 2.5$.

The identification of charged hadrons (PID) at intermediate momentum (0.1–3 GeV/c) is achieved by the time-of-flight (TOF) measurements which are complemented by the energy loss (dE/dx) information from the TPC and IT detector systems. TOF system should provide a large phase space coverage $|\eta| < 3.0$, high combined geometrical and detection efficiency (better than 80%), identification of pions and kaons with $0.1< p_t < 2$ GeV/c and (anti)protons with $0.3 < p_t < 3$ GeV/c. The choice for the TOF system is Multigap Resistive Plate Counters (MRPC) which have good time resolution of $\sigma < 70$ ps. The barrel covers the pseudorapidity region $|\eta| < 1.5$ and the geometry efficiency in it is above 90%. The end cap system covers the pseudorapidity region $1.5< |\eta| <3.0$.

The fast forward detector (FFD) should provide fast determination of a nucleus-nucleus interaction in the center of MPD, generation of a start pulse for TOF and production of L0-trigger signal. The proposed FFD design is a granulated Cherenkov detector which has a high efficiency for the high energy photons and for ultra-relativistic charged particles as well. Its acceptance in pseudorapidity is $2.0 \leq |\eta| \leq 4.0$. FFD has excellent time resolution of 38 ps.

The end cap tracker (ECT) has to provide charged particle identification and momentum measurement in the pseudorapidities $1 < |\eta| < 2.2$. ECT has high track reconstruction efficiency and momentum resolution about 10%. ECT consists of two end cap parts which are made with modules, containing layers of straw tubes.

The primary role of the electromagnetic calorimeter (ECAL) is to measure the spatial position and energy of electrons and photons. ECAL should have a high segmentation, should provide good space resolution, energy resolution about 3% and should allow a separation of overlapping showers. The Pb-scintillator ECAL of the “shashlyk” type will be used.

The zero degree calorimeter (ZDC) should provide a classification of events by centrality, event plane determination and measurement of the energy deposited by spectators. ZDC consists of modules of 60 lead-scintillator tile "sandwiches" with lead and scintillator plates. The light readout is provided by the wave-length shifting fibers (WLS-fibers) and the micropixel avalanche photodiodes (MAPDs).

More detailed description of the MPD detector can be found in 'MPD Conceptual Design Report'. [1]

Measurements of the production of strange particles such as the φ-meson can provide important information on the properties of the medium and particle production mechanisms in ultra-relativistic Au-Au collisions. Measurements of the φ-meson p_T spectra and their dependence in terms of shape and normalization on centrality may shed light on the constituents of the medium at the time of φ formation as well as the mechanism through which the φ-mesons are formed. Further insight into mechanisms of particle production for strange particles compared to non-strange particles can be gained through measurement of the particle ratios of multistrange hadrons. The medium produced at NICA/MPD can be also probed by measuring the elliptic flow of the φ-meson. Since multistrange hadrons and particles with hidden strangeness are assumed to freeze out early and undergo fewer interactions in the hadronic stage, their v_2 signals should provide a clean signal from the early stage of the system's evolution. [4] The φ vector meson is the lightest bound state of hidden strangeness, consisting of a quark-antiquark pair. Although it is a meson, it is heavy in comparison with mesons consisting of u and d quarks, having a mass ($m_\varphi = 1019.456 \pm 0.020$ MeV/c^2) comparable to the proton and Λ baryons. The φ-meson is expected to have a very small cross-section for interactions with non-strange hadrons. [4] [5] It is unstable and its existence can only be detected by its decay products such as K+K−, $K^0_L K^0_S$, μ+μ−, etc.

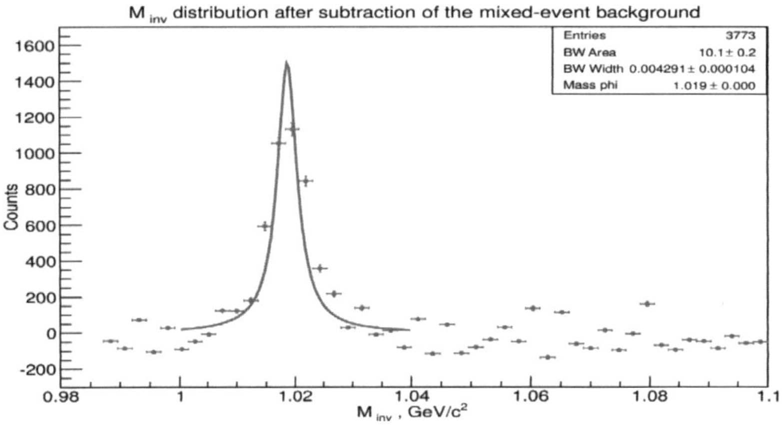

Figure 3. Invariant mass distribution of φ- meson

In our study we use the channel decay Φ→K+K− to detect the formation of the φ-meson. This channel is chosen because it has a high branching ratio (49.1%) and kaons are easy to detect. UrQMD event generator is used and central events at $\sqrt{s}$ = 11 GeV are analyzed. In the analysis is used selection of kaon pairs by track quality cuts and particle identification (PID). The invariant mass of the kaon pairs is calculated and then the combinatorial background (mixed-event technique) is subtracted. The obtained peak from the invariant mass distribution is then fitted by a Breit-Wigner function and the characteristics of the φ-meson such as its mass and its width are found. The recent results of the φ-meson study are shown in figure 3. The values of the parameters obtained by the fit (Width = 4.291 ± 0.104 MeV/c^2 and M_{inv} = 1019.995 ± 0.022 MeV/c^2) are consistent with the values given in literature. This study shows that the measurement of φ-mesons is feasible and we can expect detection of them when NICA/MPD will be put in operation.

The correlated e+e− or μ+μ− pairs (dileptons), especially those from decays of vector mesons (ρ, ω, φ), are the best candidates to relate medium modifications of hadronic spectral function to the

Journal of Physics: Conference Series **503** (2014) 012041 doi:10.1088/1742-6596/503/1/012041

restoration of the chiral symmetry in A+A collisions. The invariant mass distribution of electron-positron pair reflects the mass distribution of the vector meson at the moment of the decay and since the decay products interact only electromagnetically, they escape the interaction region unaffected by subsequent strong interactions in dense hadronic matter and carry to the detectors information about the conditions and properties of the medium at the time of their creation.

The experimental study of dileptons in heavy-ion collisions is a challenging task. The main difficulty is a huge combinatorial background of uncorrelated lepton pairs which mainly come from π^0 Dalitz decays and photon conversion in the detector material. A special attention should be paid to reduce this background as much as possible. The configuration used in the study includes TPC, TOF and ECAL covering the pseudorapidity range $|\eta| < 1.2$.

The study of dielectron production in central (0-3 fm) Au-Au collisions at $\sqrt{s}$ = 7A GeV was performed using the Pluto code [6] generating a cocktail of hadrons decaying into the electron-positron or Dalitz electron-positron pairs. The multiplicities of dilepton sources ($\pi^0,\eta,\rho^0,\omega,\varphi$) were predicted by the statistical thermal model [7]. The background was calculated by the UrQMD generator [8] which produced hadrons and photons, and mixed with the Pluto output. The event samples have been transported through the detector using the Geant 3.21 transport package (containing photon conversions, etc.). Then the events were reconstructed and analysed [9].

Figure 4 shows the signal-to-background ratio (S/B) in invariant mass bins of 20 MeV/c^2 obtained for $2x10^7$ central events, corresponding to about 20 hours of running time at the NICA collision rate of 6 kHz. The signal-to-background ratio is defined as the ratio of the number of e+e− combinations from meson decays to that of all the other electron-positron combinations. The obtained results for the signal-to-background ratio in the invariant mass interval 0.2-1.5 GeV/c^2 are shown in figure 5 along with the published data from other experiments.

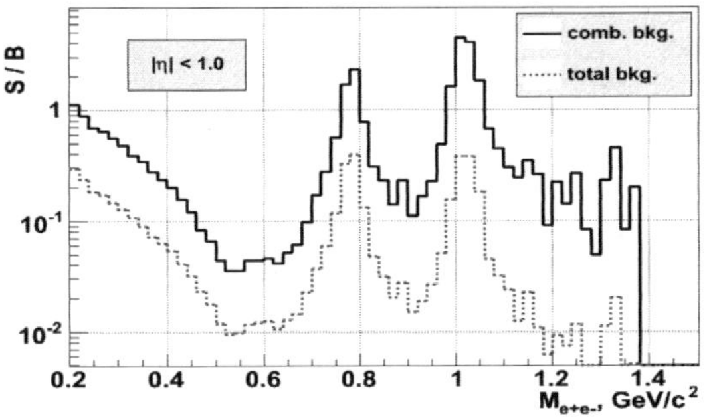

Figure 4. Signal-to-background ratio in invariant mass bins of 20 MeV/c^2 for pure combinatorial (Pluto originated part) and total background contributions.

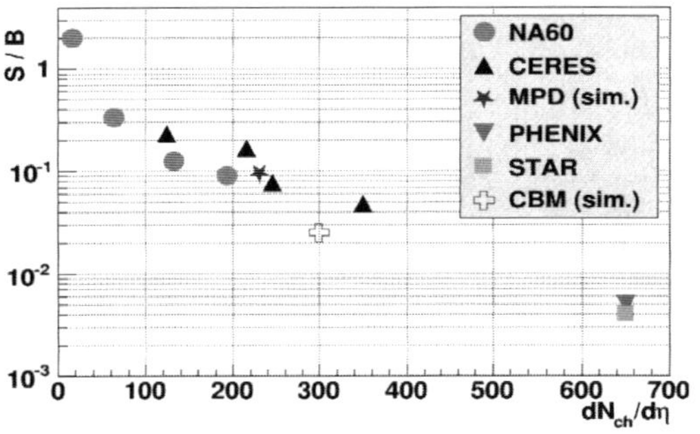

Figure 5. Signal-to-background ratios obtained in different experiments versus the number of charged particles per unit pseudorapidity. (Not that simulated results do not include in-medium modification effects.)

One of the main tasks of the NICA/MPD physics program is a study of the strangeness production (hyperons and hypernuclei) in heavy ion collisions. Since hyperons are created during the early stages of the fireball evolution, they could provide essential signatures of the excited and compressed baryonic matter. The heaviest species, multistrange Ξ− and Ω−, are especially interesting for physics analyses. Their identification and reconstruction is based on the reconstruction of the lightest hyperon Λ0 . In addition, since Λ0 hyperons are produced in relatively large quantities and have very attractive experimental features (resonance structure and simple decay mode), they can be used as a tool to study the detector performance. Therefore, the development of the Λ0 reconstruction techniques is a very important task.

The production of hypernuclei in relativistic heavy ion collisions is also interesting for understanding the strangeness degrees of freedom in hadronic systems. [10] The study of the production of the light hypernuclei ${}^{3}_{\Lambda}H$ is essential for understanding the production mechanism of exotic objects such as multi-hypernuclei and the strangelets they might decay into if the latter are more stable. The coalescence process for the formation of hypernuclei requires that nucleons and hyperons be in proximity in phase space (i.e., in coordinate and momentum space). Equilibration among the strange quark flavors and light quark flavors is one of the proposed signatures of QGP formation, which would result in high hypernucleus yields.

The detector configuration used in the study includes TPC and TOF for track reconstruction and particle identification in the pseudorapidity range $|\eta| < 1.3$. The gold-gold collisions were generated using the LAQGSM code [11]. Λ0 hyperons and ${}^{3}_{\Lambda}H$ hypernuclei were reconstructed using their decay mode into a negative pion and a proton (for hyperon) or He^3 (for hypernuclei). The signal event topology (decay of a relatively long-lived particle into a pair of charged tracks) defines the selection criteria: relatively large distance of the closest approach (DCA) to the primary vertex of decay products, small two-track separation in the decay vertex, relatively large decay length of the mother particle. The exact values of selection cuts were found by performing a multidimensional scan over the whole set of selection criteria with a requirement to maximize the invariant mass peak significance, defined as $S/\sqrt{(S+B)}$ where S and B are total numbers of signal (described by the gaussian) and background (polynomial function) combinations inside +- 2σ interval around the peak position. The obtained results are presented in figure 6 and figure 7 where invariant mass distributions are plotted. One can see that MPD can do a good job in reconstructing strange objects.

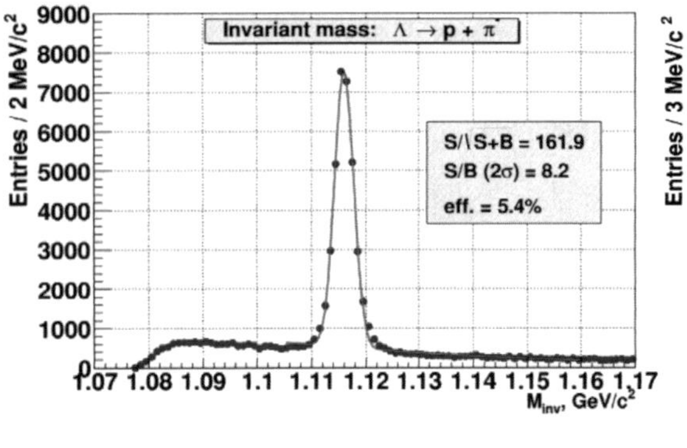

Figure 6. Reconstructed invariant mass of proton-pi- pairs in central Au+Au collisions at 5A GeV obtained for 10^4 events (30 seconds of running time).

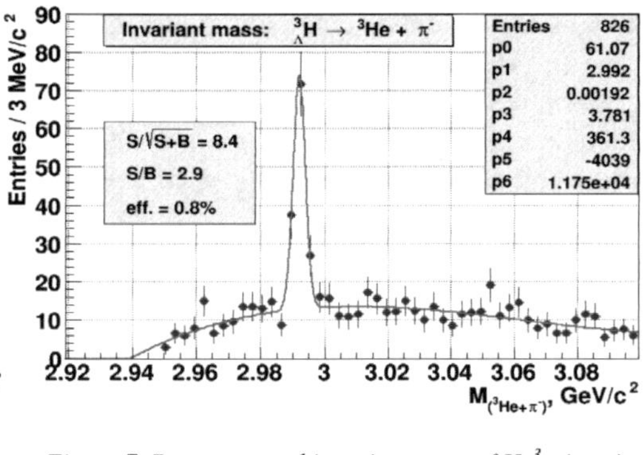

Figure 7. Reconstructed invariant mass of He^3-pi- pairs in central Au+Au collisions at 5A GeV obtained for $5x10^5$ events (30 minutes of running time).

In conclusion, it should be said that the MPD detector has many advantages and meets all the ambitious physics requirements for exploring phase diagram of strongly interacting matter in a high track multiplicity environment. The MPD detector's advantages comprise coverage of a large phase space, functionality at high interaction rates, high efficiency and excellent particle identification capabilities. The study of hyperons, hypernuclei, dileptons and vector mesons such as the φ-meson, based on the current analyses, is feasible and gives quite good results. Therefore, NICA facilities provide unique capabilities for studying fundamental properties of the theory of strong interactions.

[1] MPD Collaboration, MPD Conceptual Design Report, version 1.4, 2011
[2] http://fairroot.gsi.de/
[3] http://nica.jinr.ru/ ; http://mpd.jinr.ru/dr/
[4] Sarah-Louise Blyth, Using the φ-meson to probe the medium created in Au-Au collisions at RHIC, 2007
[5] D. Randrianasoloharisoa, F. Lahatra Razafindramisa, Detection of the φ-meson by the decay Φ→K+K−
[6] I. Froehlich et al. // arXiv:0708.2382v2 [nucl-ex]
[7] P. Braun-Munzinger, K. Redlich, and J. Stachel // Invited review for Quark Gluon Plasma 3, eds. R. C. Hwa and Xin-Nian Wang, World Scientific Publishing, 2004. P.491; arXiv:nucl-th/0304013
[8] http://urqmd.org/
[9] V. Vasendina et al. // Phys. of Part. and Nucl. Lett. 2013. V.10. No.7. P. 769.
[10] A. Andronic et al. // Phys. Lett. B. 2011. V.697. P. 203.; Yu. Zhu // Nucl. Phys. A. 2013. V.904-905. P.551c.
[11] L.V. Bravina et al. // Nucl. Phys. A. 1994. V.566. P.461c

AUTHOR INDEX

Institute of Physics
Dirac House, Temple Back
Bristol BS1 6BE UK

CURRAN ASSOCIATES INC
proceedings
.com